CELL BIOLOGY

细胞生物学

（第三版）

主　编　李　瑶

副主编　黄　燕

编　者（按姓氏字母排序）

戴亚蕾　同济大学
黄　燕　复旦大学
李　瑶　复旦大学
曲志才　曲阜师范大学
任　华　华东师范大学
孙　媛　大连医科大学
张　儒　同济大学
赵基源　宁波大学

復旦大學出版社

内容提要

本教材由复旦大学生命科学学院细胞教学组牵头，组织国内多年从事细胞和分子生物学科研及教学的教授和专家改编完成。在系统阐述细胞各部分的结构和功能的基础上，重点介绍细胞内外的物质运输、信息传递、能量转换、周期调控、分化发育、癌变、免疫、衰老与凋亡等细胞的重大生命活动。全书用英文撰写，语言简练、通俗，知识体系精确、先进，既介绍细胞生物学的基本概念和基本原理，又反映各领域的发展前沿。各章结尾附有思考题，附录为细胞生物学常用词汇的中文解释，便于读者自学，同时掌握中英文专业知识。

本教材为英文教材，可供综合性大学、师范院校、医学院校、农林院校的研究生使用，也可供教师、科研人员与科教管理人员参考。

本书配有PPT课件，欢迎来函免费获取：xdxtzfudan@163.com。

三版前言

细胞生物学是一门生命科学的重要基础学科,有关细胞的基础知识和相关研究是现代生命科学的基石。该课程的目的是使大学本科学生初步掌握细胞生物学的基本内容、基本原理、基本知识,为深入学习生命科学各科类奠定基础。我们编写英文版细胞生物学的初衷是为了让同学们在学习知识的同时,能够尽快跟国际接轨,提高专业英语水平,适应国际化的需求。

很多高校都开设了细胞生物学的双语课程。国内的中文版细胞教材有很多版本,但我们发现还缺乏英文版的教材。虽然国外有相关的原版英语教材,但篇幅太长,很多内容是分子生物学的知识点,跟其他专业课程重复太多,不适合中国的细胞生物学的教学特点。为了我国开展"细胞生物学"双语教学,我们于2006年正式出版英语教材*Cell Biology*,并配合教材编写了配套PPT课件。出版以来深受读者的欢迎,为了满足广大读者的需要,在第一版的基础上,我们于2013年更新了第二版。这两版教材共印刷9次,发行量两万册左右,是多所高校多年选择的正式教材。

由于细胞生物学发展迅猛,知识更新速度快,另外,第二版教材在使用过程中仍发现了一些不足。为此,我们着手修订第三版教材。大部分编写人员都参与了第一、二版的编写工作,都是活跃在科研、教学第一线的老师;为了更好地完成任务,我们又邀请了几位有丰富教学经验的老师加入了改编工作,为本小组增添了活力。在内容上,为了进一步精炼教材,在保持原书的框架下,删减了部分原来版本中不重要的内容,甚至有些章节的内容作了很大的调整;同时,参考了最新的国外教科书和科学杂志、重要科学论文,适当增加了最新的研究成果,在学习知识的同时也提高了同学们的专业英语水平,使学生增强专业论文的阅读能力;对专业词汇注明了中文解释,有助于学生的理解。考虑到不同院校的本科教学要求和学时数的限制,我们在内容上尽可能拓宽知识面,尽量避免与其他学科如生化、分子生物学等学科的重复,强调基础及面,尽量控制其深度和难度,使之更适合本科的教学要求。因此,该教材普遍合适农、林、医、师范及综合性大专院校的本科双语教学的教材,并可以作为中文教学的参考书。

由于工作等多方面原因,参与第一、二版编写的部分编委未能加入第三版修订工作,我们感谢所有参与过编书的老师的付出。我尤其要感谢我的两位恩师、第一版和第二版的主编沈大棱和吴超群两位教授,由于年龄的关系,他们不再参与第三版的编写工作,而且强调不再做第三版的主编。但正是有了前面两版的基础,才使得第三版的改编工作得以顺利完成。他们的这种精神也给晚辈们树立了学习的榜样,希望我们的这本教材能够按照这种方式承传下去,后继有人。另外,在编写过程中,一直得到了复旦大学出版社的宫建平老师的组织、建议和帮助,在此一并表示衷心地感谢!

本教材由多位作者共同参与编写,虽经轮流传阅,相互修改,最后由主编统稿而成,但由于水平有限,时间仓促,难免有疏漏和错误之处,欢迎热心的同行及读者批评指正,以便我们在今后的修订工作中不断改进。

编　者
2021 年 10 月

Contents

Chapter 1

Introduction to cell biology

1.1 What is cell biology?

Cell biology is the application of molecular biological approaches to an understanding of life at the cellular level. Knowledge of the molecular basis of cell structure, cell function and cell interactions is fundamental to an understanding of whole organisms, since the properties of organisms are dependent upon the properties of their constituent cells.

1.1.1 Cell biology is the basis of modern biology

Cell biology is modern biology, an academic discipline which studies the structure and physiological properties of cells, as well as their behaviors, interactions, and environment on a microscopic and molecular level. But two main features should be stressed in the modern cell biology.

(1) Study the molecules within cells Cell biology is a modern science, which is rooted in an understanding of the molecules within cells, and of the interactions between cells that allow construction of **multicellular organism**s. The more we learn about the structure, function, and development of different organisms, the more we recognize that all life processes exhibit remarkable similarities.

Cell biology concentrates on: ① Macromolecules and reactions, investigated by biochemists; ②The processes described by cell biologists; ③The gene control pathways identified by molecular biologists and geneticists.

(2) Study the molecular similarities and differences between cell types Understanding the composition of cells and how cells works is fundamental to all of the biological sciences. Appreciating the similarities and differences between cell types is particularly important to the fields of molecular cell biology. These fundamental similarities and differences provide a unifying theme, allowing the principles learned from studying one cell type to be extrapolated and generalized to other cell types. Research in cell biology is closely related to genetics, biochemistry, molecular biology and developmental biology.

1.1.2 Cell biology is in progress

All the concepts of cell biology continue to be derived from computational experiments and laboratory experiments. The powerful experimental tools that allow the study of living cells and organisms at higher and higher levels are being developed constantly. In this chapter, we address the current state of cell biology and look forward to what further exploration will uncover in the twenty-first century.

In this millennium, two gathering forces will reshape cell biology: ①The **genomics**, the complete DNA sequence of many organisms; ②The **proteomics**, the knowledge of all the possible shapes and functions that proteins employ.

1.2 The cell theory

The cell theory, or cell doctrine, states that all organisms are composed of similar units of organization, called cells. The concept was formally articulated in 1839 by Schleiden and Schwann, and has remained as the foundation of modern biology. The idea predates other great paradigms of biology including Darwin's theory of evolution (1859), Mendel's laws of inheritance (1865), and the establishment of comparative biochemistry (1940).

Ultrastructural research and modern molecular biology have added many tenets to the cell theory, but it remains as the preeminent theory of biology. The cell theory is to biology what atomic theory is to physics.

Just as an atom is the smallest particle of a chemical element, which can exist either alone or in combination and still possess the chemical and physical properties of that element, so then, a cell is the smallest entity, which can exhibit the characteristic of life.

1.2.1 Formulation of the cell theory

In 1663, an English scientist, Robert Hooke, discovered cells in a piece of cork, which he examined under his primitive microscope(Figure 1-1). Actually, Hooke only observed cell walls because cork cells are dead and without cytoplasmic contents. Hooke drew the cells he saw and also coined the word "cell". The word cell is derived from the latin word "cellula" which means small compartment. Hooke published his findings in his famous work, "Micrographia: Physiological Descriptions of Minute Bodies made by Magnifying Glasses (1665)."

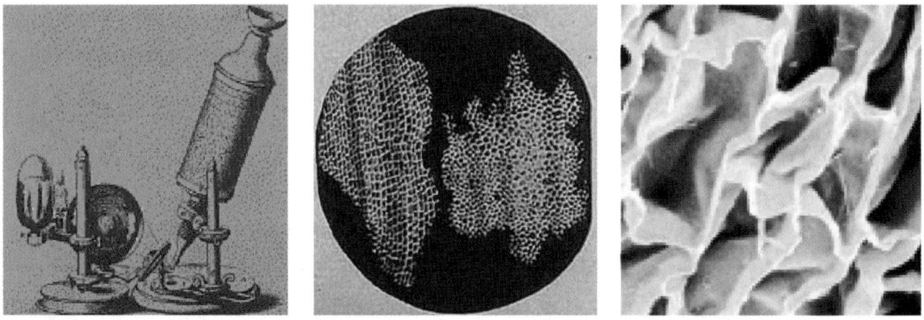

Figure 1-1 Hooke and his microscope. In 1663 Hooke examined under his primitive microscope (left). The small cell structures did not show up well or remained invisible (middle). The electron microscope not only showed more detail of previously known parts of the cell but also revealed new parts. Cells and cell structures can now be examined at magnifications of up to 500,000 times and more (right).

Ten years later, Anton van Leeuwenhoek (1632~1723), a Dutch businessman and a contemporary of Hooke used his own (single lens) monocular microscopes and was the first person to observe bacteria and protozoa. He looked at everything from rain water to tears. He saw moving objects that he termed "animalcules." The tiny creatures appeared to be swimming.

Between 1680 and the early 1800's, it appears that not much was accomplished in the study of cell structure. This may be due to the lack of quality lens for microscopes and the dedication to spend long hours of detailed observation over what microscopes existed at that time. Leeuwenhoek did not record his methodology for grinding quality lenses and thus microscopy suffered for over 100 years.

It is upon the works of Hooke, Leeuwenhoek, Oken, and Brown that Schleiden and Schwann built their cell theory. It was the German professor of botany at the university of Jena, Dr. Schleiden, who brought the nucleus to popular attention, and to asserted its all importance in the function of a cell.

The location of these nuclei at comparatively regular intervals suggested that they are found in definite compartments of the tissue, as Schleiden had shown to be the case with vegetables; indeed, the walls that separated such cell-like compartments one from another were in some cases visible. Soon Schwann was convinced that his original premise was right, and that all animal tissues are composed of cells not unlike the cells of vegetables. Adopting the same designation, Schwann propounded what soon became famous as the cell theory. So expeditious was his observations that he published a book early in 1839, only a few months after the appearance of Schleiden's paper. A most important era in cell biology dates from the publication of his book in 1839.

Schwann summarized his observations into three conclusions about cells: ①The cell is the unit of structure, physiology, and organization in living things; ②The cell retains a dual existence as a distinct entity and a building block in the construction of organisms; ③Cells form by free-cell formation, similar to the formation of crystals (**spontaneous generation**).

For a long time, people believed in spontaneous generation. They believed flies came from rotting meat and frogs from mud. It took a hundred years and many experiments to disprove those ideas and confirm that every cell comes from a pre-existing cell.

1.2.2 Modern tenets of cell theory

The cell doctrine reached its present-day eminence in 1896 with the publication of Wilson's "*The Cell in Development and Heredity*", which was an accumulation of what was known about the roles of cells in embryology and chromosomal behavior.

For the first 150 years, the cell theory was primarily a structural idea. This structural view, which is found in most textbooks, describes the components of a cell and their fate in cell reproduction. Since the 1950's, however, cell biology has focused on DNA and its informational features. Today we look at the cell as a unit of self-control. The description of a cell must include ideas about how genetic information is converted to structure and function.

The modern tenets of the cell theory include: ①All known living things are made up of cells; ②The cell is structural and functional unit of all living things; ③All cells come from pre-existing cells by division (not by spontaneous generation); ④Cells contain hereditary information which is passed from cell to cell during cell division; ⑤All cells are basically the same in chemical composition; ⑥All energy flow (metabolism & biochemistry) of life occurs within cells.

1.3 Cell is the basic unit of life

According to the cell theory, all living things are composed of one or more cells. Cells fall into **prokaryotic** and **eukaryotic** types. Prokaryotic cells are smaller (as a general rule) and lack much of the internal compartmentalization and complexity than eukaryotic cells are. No matter which type of cells we are considering, all cells have certain features in common: cell membrane, DNA, cytoplasm, and ribosome.

1.3.1 The structure of cell

The cell is one of the most basic units of life. There are millions of different types of cells. There are cells that are organisms onto themselves, such as microscopic amoeba and bacteria cells. And there are cells that only function when part of a larger organism, such as the cells that make up your body.

The cell is the smallest unit of life in human bodies. In the body, there are brain cells, skin cells, liver cells, stomach cells, and the list goes on. All of these cells have unique functions and features. All have some recognizable similarities (Figure 1-2).

(1) Plasma membrane All cells have a "skin", called the **plasma membrane**, protecting it from the outside environment. The cell membrane regulates the movement of water, nutrients and wastes into and out of the cell. Inside of the cell membrane are the working parts of the cell.

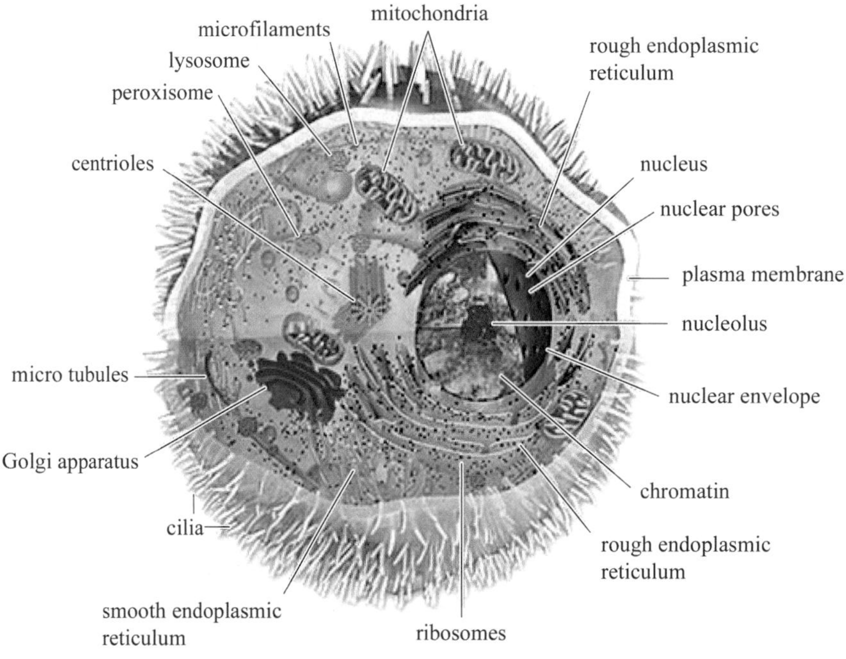

Figure 1-2 The structure of an animal cell (From: Michael W. Davidson and The Florida State University Research Foundation)

(2) Nucleus At the center of the cell is the cell **nucleus**. The cell nucleus contains the cell's DNA, the genetic code that coordinates protein synthesis.

(3) Organelles There are many organelles inside of the cell — small structures that help carry out the normal operations of the cell. One important cellular organelle is the ribosome. **Ribosomes** participate in protein synthesis. The transcription phase of protein synthesis takes places in the cell nucleus. After this step is complete, the mRNA leaves the

nucleus and travels to the cell's ribosomes, where translation occurs. Another important cellular organelle is the **mitochondrion**. Mitochondria (many mitochondrion) are often referred to as the power of the cell because many of the reactions that produce energy take place in mitochondria. Also important in the life of a cell are the **lysosomes**. Lysosomes are organelles that contain enzymes that aid in the digestion of nutrient molecules and other materials.

1.3.2 Three things make cell different from non-cell system

Life requires a structural compartment separate from the external environment in which macromolecules can perform unique functions in a relatively constant internal environment. These "living compartments" are cells. The cells differ from non-cell systems through three things: ①The capacity for replication from one generation to another. Most organisms today use DNA as the hereditary material, although recent evidence suggests that **RNA** (ribozyme) may have been the first nucleic acid system to have formed. Nobel laureate Walter Gilbert refers to this as the RNA world. ②The presence of enzymes and other complex molecules essential to the processes needed by living systems. Miller's experiment showed how these could possibly form. ③ A membrane that separates the internal chemicals from the external chemical environment. This also delimits the cell from not-cell areas.

1.3.3 Microscope is needed to visualize cells

The small size of cells makes the use of microscopes necessary to view them (Figure 1-3). If two objects are too close together, they start to look like one object. With normal human vision the smallest objects that can be resolved (i. e., distinguished from one another) are about 200 μm (0.2 mm) in size.

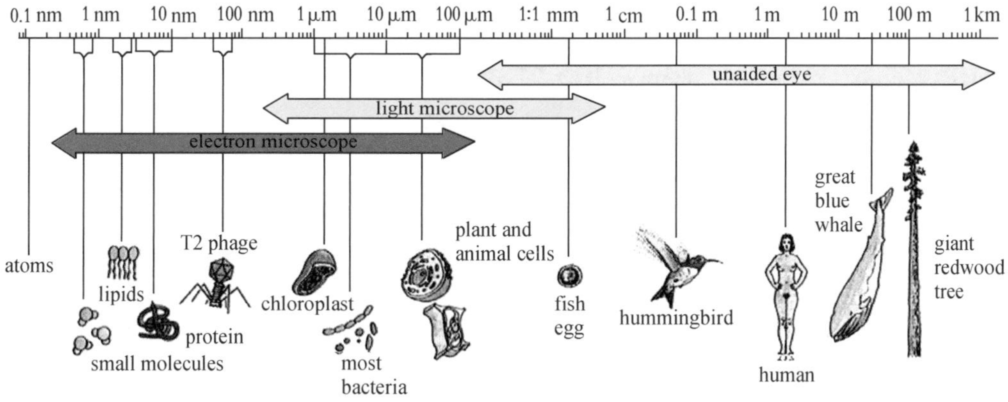

Figure 1-3 The relative sizes of biological objects ranging from atoms to tree (From: Farabee MJ. 2002)

Light microscopes use glass lenses and visible light and typically have a resolving power of 0.2 μm (0.2×10^{-6} m). Resolution depends on the wavelength of the illuminating light, but in general, resolution is about 1,000 times better than that of an unaided human eye. Living or killed and fixed cells may be viewed with light microscopes.

Electron microscopes have magnets, rather than glass lenses, to focus an electron beam. The wavelength of the electron beam is far shorter than that of light, and the resulting image resolution is far greater. This image is not visible without the use of either film or a fluorescent screen. Resolution is about 0.5 nm or 400,000 times finer than that of the human eye. Subcellular features can be seen only if the cells are killed and fixed with special fixatives and stains.

1.3.4 Cells show two organizational patterns

Every cell has a plasma membrane, a continuous membrane that surrounds the fluids and other structures of a cell. The membrane is composed of a lipid bilayer with proteins floating within it and protruding from it. Living organisms can be classified into one of two major categories based on the location within the cell where the most genetic material is stored.

Prokaryotes have no nucleus or other membrane-bounded compartments. They lack distinct organelles, although some do have invaginated membrane structures.

Eukaryotes have a membrane-bounded nucleus and usually have other membrane-bounded compartments or organelles as well.

1.4 Diversity of cells

Cells fall into prokaryotic and eukaryotic types. Prokaryotic cells are smaller (as a general rule) and lack much of the internal compartmentalization and complexity of eukaryotic cells. No matter which type of cell we are considering, all cells have certain features in common: cell membrane, DNA, cytoplasm, and ribosomes. Both prokaryotic and eukaryotic cells include tremendous diversities in shape, structure, metabolism, and biological activity.

1.4.1 Diversity of prokaryotic cell

The cyanobacteria, formerly known as the blue-green algae, are the largest of the prokaryotes (organisms having prokaryotic cells). They contain chlorophyll and other pigments for photosynthesis. The pigments are not in membrane-bound chloroplasts, those organelles found in many plant cells. The pigments permeate throughout the entire cytoplasm.

(1) Bacteria are the most abundant of all organisms They are ubiquitous in soil, water, and as symbionts of other organisms. Many pathogens are **bacteria**. Most are minute, usually only 0.5~5.0 μm in size (though one type, Thiomargarita namibiensis, reaches 0.5 mm in diameter, and has a volume up to a million times that of the typical bacterium). They generally have cell walls, like plant and fungal cells, but with a very different composition (peptidoglycans). Many move around using flagella, which are different in structure from the flagella of other groups.

(2) Diversity of bacteria Bacteria are classfied depending on morphology, but biologically bacteria are divided to subtypes, including *Actinobacteria*, *Aquificae*, *Bacteroidetes/Chlorobi*, *Chlamydiae/Verrucomicrobia*, *Chloroflexi*, *Chrysiogenetes*, *Cyanobacteria*, *Deferribacteres*, *Deinococcus-Thermus*, *Dictyoglomi*, *Fibrobacteres/Acidobacteria*, *Firmicutes*, *Fusobacteria*, *Gemmatimonadetes*, *Nitrospirae*, *Omnibacteria*, *Planctomycetes*, *Proteobacteria*, *Spirochaetes*, *Thermodesulfobacteria*, *Thermomicrobia*, *Thermotogae* and so on (Figure 1-4).

1.4.2 Eukaryotic diversity

The **diversity** derives from both variety in gene expression and the patterns of cellular

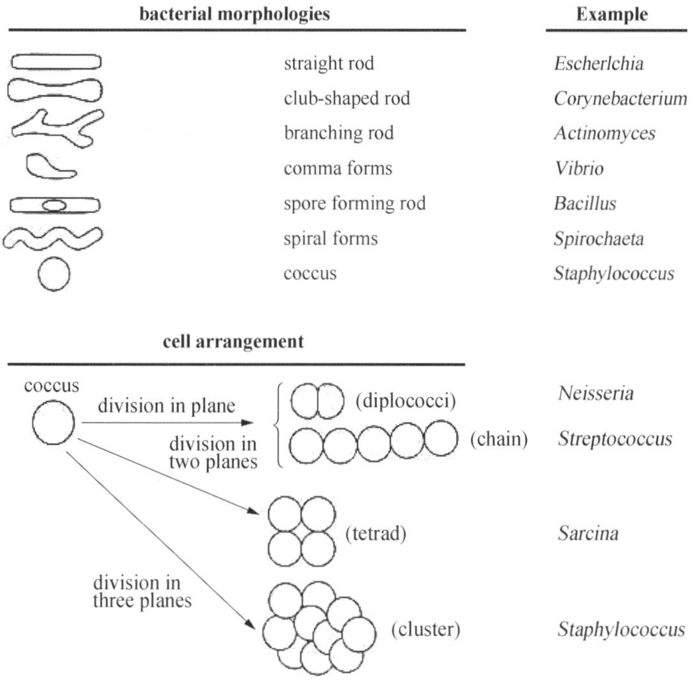

Figure 1-4 Shapes and grouping forms of various bacteria (From: http://129.109.136.65/microbook/ch002.htm)

control for gene product behavior. One can understand complex developmental processes only by considering them in the context of the cells that comprise the objects being formed, be their limbs, guts, or brains. Thus, developing a keen understanding of cells and how they work are essential for almost all cell biologists.

(1) Algae A variety of thallus formats characterize algae. it can consist of a single cell, or many cells in varying arrangement. Four types of algae are recognized, based on the following body structures: unicellular, colonial, filamentous, and multicellular. Algae are classified into seven phyla, based on their color, type of chlorophyll, form of food storage substance and cell wall composition.

Multicelular algae are largest marine algae, including brown, red, and green algae, are known collectively as seaweeds. Unicellular algae consists of a single cell; most are aquatic. They form the phytoplankton, a population of photosynthetic organisms that forms the foundation of the food chain, and produce half of the worlds carbohydrates and are among the major producers of oxygen in the atmosphere.

(2) Protists The first protist fossils occur in rocks approximately 1.2~1.4 billion years old from the Bitter Springs Formation in Australia. Multicellular protists appeared in the fossil record more than 600 million years ago. The protists include heterotrophs, autotrophs, and some organisms that can vary their nutritional mode depending on environmental conditions. Protists occur in freshwater, saltwater, soil, and as symbionts within other organisms. Due to this tremendous diversity, classification of the protist is difficult (Figure 1-5).

(3) Fungi Fungi are almost entirely multicellular (with yeast, *Saccharomyces cervisiae*, being a prominent unicellular fungus) heterotrophic (deriving their energy from another organism, whether alive or dead) and usually having some cells with two nuclei (multinucleate, as opposed to the more common one, or uninucleate) per cell.

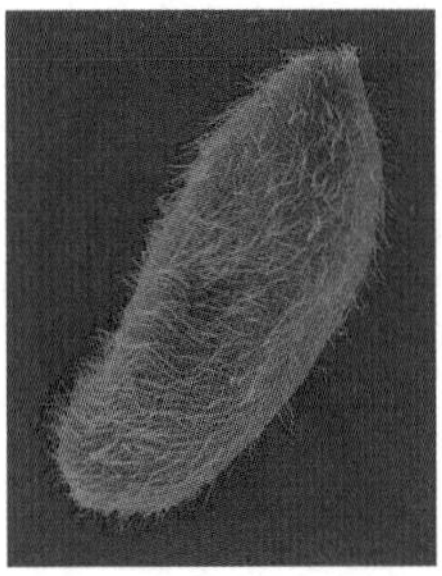

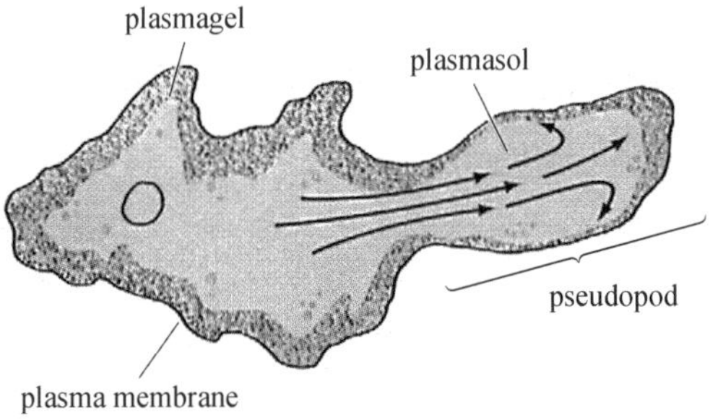

Figure 1-5 Ciliated Protozoan (SEM ×1,600) (left). Formation of pseudopodium by an amoeba (right) (From: http://www.denni-skunkel.com)

Evolution of multicellular eukaryotes increased the size and complexity of organisms, allowing them to exploit the terrestrial habitat. Fungi first evolved in water but made the transition to land through the development of specialized structures that prevented their drying out. First classified as plants, fungi are now considered different enough from plants to be placed in a separate kingdom.

(4) Multicellular organisms consist of differentiated cell types The living world is all about cells working alone and cells working together. All life is cells. Even the biggest animal started off as just a single cell. Multicellular organisms consist of differentiated cell types, with rather complex biochemical dynamics exhibited by complex metabolic and genetic networks. Through the course of development, cells differentiate into several types and often form complex patterns. The molecular mechanisms existing of each stage of development have been elucidated experimentally. Now, researchers explore how an animal's cells are adapted to its active, mobile lifestyle, and examine the different groups of cells that make up animals. These cells in multicellular organisms are specialized to perform different tasks from fighting infection to transporting nutrients. All the living things around you are made up of millions and millions of cells working together. It's important to understand what these cells are doing and how they do it.

Many cells, for example, most animal, plant and bacterial cells, have a fixed shape. However, multicellular and large organisms contain many cells. Examples, the human body contains billions of cells of many different types. In plants this is exemplified within a leaf where different cell types contain chloroplasts of different sizes and structures, enabling photosynthesis to be efficient. Indeed, about 200 different types of cells — many highly specialized — make up the tissues and organs of the human body.

1.4.3 Genetic control of cell diversity

The role of the genome in development is one of the great challenges of contemporary molecular biology. More and more researches provide the evidence that (with few exceptions) the genome remains intact during development of specialized cell types, without losing the information necessary for development of alternative cell types. The integrity of the genome implies that diversity is achieved not by loss of alternative genetic information but by differential use of the genetic material during development.

There are two primary means for regulating differential gene expression to achieve cellular diversity during development: ①The influence of distinct cytoplasms on nuclei in different regions of the early embryo; ②Extrinsic factors, such as inductive signals from neighboring cells.

1.5 The eukaryotic cell

Animals, plants, fungi, and protists have a membrane-bounded nucleus in each of their cells and are classified as eukaryotes.

The autogenous and endosymbiotic hypotheses relate to possible ways in which eukaryote cells may have evolved from prokaryotic ancestors. The latter hypothesis is favored by most biologists. Briefly, this hypothesis suggests that the first eukaryotic cells resulted from symbiotic consortia between various types of prokaryotic cells. For example, the chloroplast resulted when a photosynthetic prokaryote entered another cell and survived. The mitochondrion is presumed to have evolved in a similar manner, but between non photosynthetic cells.

1.5.1 Eukaryotic cells share common features

Eukaryotic cells vary from animals, plants, to fungi and protists, but they have some common features which make them different from the prokaryotic cells.

(1) Common features The different types of eukaryotic cells show some similar features, they are: ①Eukaryotic cells tend to be larger than prokaryotic cells; ②Each of eukaryotic cells has a membrane-bounded nucleus; ③Eukaryotic cells have a variety of membrane-bounded compartments called **organelles**; ④Eukaryotes have a protein scaffolding called the cytoskeleton, which provides shape and structure to cells, among other functions.

(2) Compartmentalization is the key to eukaryotic cell function The subunits, or compartments, within eukaryotic cells are called organelles, which are responsible for specialized functions of the cell. The central organelle nucleus contains most of the cell's genetic material (DNA). The **mitochondrion** is a power plant and industrial park for the storage and conversion of energy. The endoplasmic reticulum and **Golgi apparatus** make up a compartment where proteins are packaged and sent to appropriate locations in the cell. The **lysosome** and vacuole are cellular digestive systems where large molecules are hydrolyzed into usable monomers. The **chloroplast** performs photosynthesis in plant cells. Membranes surrounding these organelles keep away inappropriate molecules that might disturb organelle function. They also act as traffic regulators for raw materials into and out of the organelle.

(3) Organelles can be studied by microscopy or chemical analysis ①Cell organelles were first detected by light and electron microscopy; ②The target specific macromolecules can be used to determine the chemical composition of organelle; ③The process of cell fractionation, another means by which cells can be examined; ④Microscopy and cell fractionation can be used as complements to each other, giving a complete picture of the structure and function of each organelle.

1.5.2 Plant cell structure

A plant has two organ systems: the shoot system, the root system. The shoot system is above ground and includes the organs such as leaves, buds, stems, flowers (if the plant has any), and fruits (if the plant has any). The root system includes those parts of the plant below ground, such as the roots, tubers, and rhizomes.

Plant cells are formed at **meristems**, and then develop into cell types which are grouped into tissues. Plants have only three tissue types: **dermal**, **ground**, and **vascular**. Dermal tissue covers the outer surface of herbaceous plants. Dermal tissue is composed of epidermal cells, closely packed cells that secrete a waxy cuticle that aids in the prevention of water loss. The ground tissue comprises the bulk of the primary plant body. Parenchyma, collenchyma, and sclerenchyma cells are common in the ground tissue. Vascular tissue transports food, water, hormones and minerals within the plant. Vascular tissue includes xylem, phloem, parenchyma, and cambium cells.

Like other eukaryotes, the plant cell is enclosed by a plasma membrane, which forms a selective barrier allowing nutrients to enter and waste products to leave. Unlike other eukaryotes, however, plant cells have retained a significant feature of their prokaryote ancestry, a rigid **cell wall** surrounding the plasma membrane. The cytoplasm contains specialized organelles, each of which is surrounded by a membrane. Plant cells differ from animal cells in that they lack centrioles and organelles for locomotion (cilia and flagella), but they do have additional specialized organelles. Chloroplasts convert light to chemical energy, a single large vacuole acts as a water reservoir, and plasmodesmata allow cytoplasmic substances to pass directly from one cell to another. There is only one nucleus and it contains all the genetic information necessary for cell growth and reproduction. The other organelles occur in multiple copies and carry out the various functions of the cell, allowing it to survive and participate in the functioning of the larger organism(Figure 1-6).

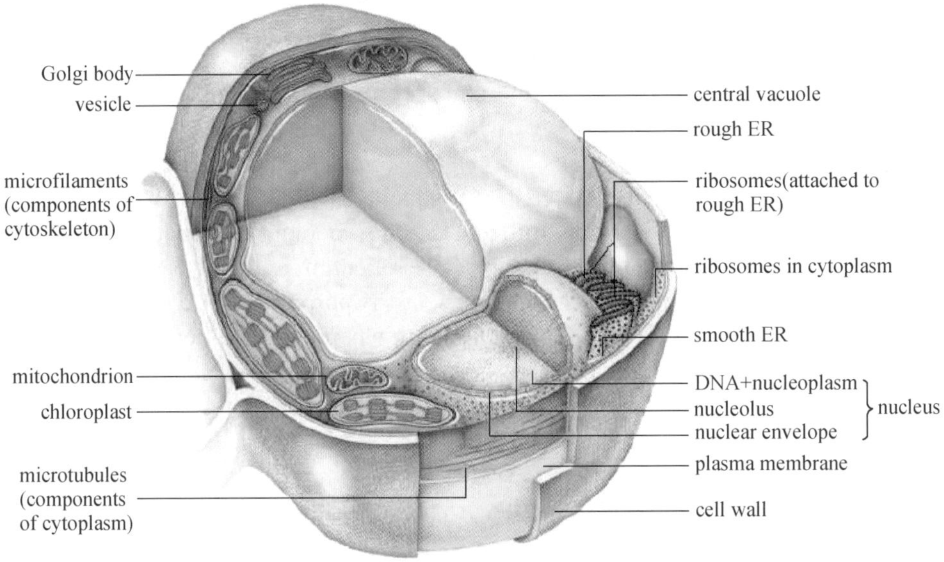

Figure 1-6 A typical structure of plant cell (From: Rosemary Richardson)

1.5.3 Animal cell structure

Animals are a large and incredibly diverse group of organisms. Making up about three-quarters of the species on Earth, they run the gamut from corals and jellyfish to ants, whales, elephants, and, of course, humans. Being mobile has given animals, which are capable of sensing and responding to their environment, the flexibility to adopt many different modes of feeding, defense, and reproduction. Unlike plants, however, animals are unable to manufacture their own food, and therefore, are always directly or indirectly dependent on plant life.

Animal cells are typical of the eukaryotic cell, enclosed by a plasma membrane and containing a membrane-bound nucleus and organelles. Unlike the eukaryotic cells of plants and fungi, animal cells do not have a cell wall. This feature was lost in the distant past by the single-celled organisms that gave rise to the kingdom Animalia. Most cells, both animal and plant, range in size between 1 and 100 micrometers and are thus visible only with the aid of a microscope(Figure 1-7).

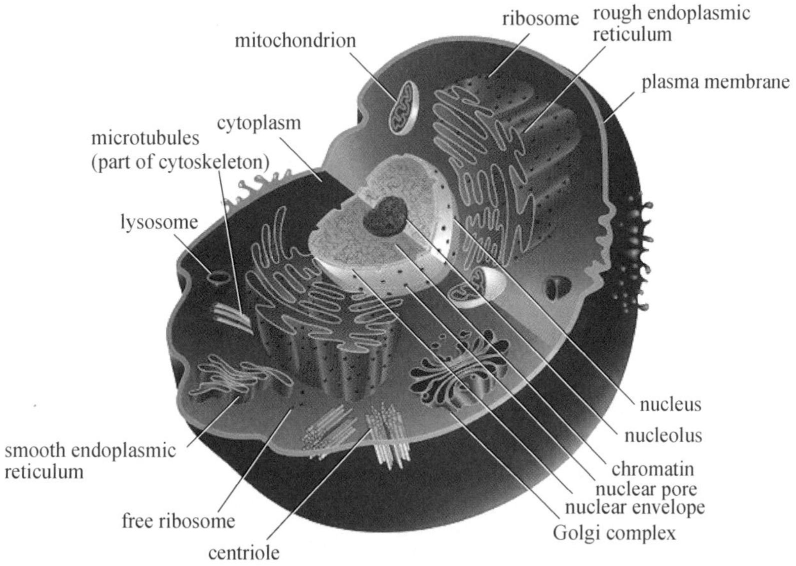

Figure 1-7 The typical structure of an animal cell (From: Audesirk T, Audesirk G, Byers BE. 2002)

Most animal cells are **diploid**, meaning that their chromosomes exist in homologous pairs. Different chromosomal ploidies are also, however, known to occasionally occur. The proliferation of animal cells occurs in a variety of ways. In instances of sexual reproduction, the cellular process of meiosis is first necessary so that haploid daughter cells, or gametes, can be produced. Two haploid cells then fuse to form a diploid zygote, which develops into a new organism as its cells divide and multiply.

1.6 Modern cell biology

Cell biology is an integrated view of cells at work, involving essential cell biology, molecular biology of cell and cell systems biology.

1.6.1 Essential cell biology

Essential cell biology is the scientific discipline that studies the cell as an individual unit and as a contributing part of a larger organism, and a science that studies the properties of cells: the functional structures within them, including organelles as mitochondria, chloroplasts, the Golgi apparatus, microsomes, and many others, the interactions and communication among cells, and the biochemical phenomena, which are shared by all cells.

(1) Study the cells as a unit of life A single cell is often considered a complete organism in itself, such as a bacterium or yeast. An individual cell is capable of digesting its own nutrients and transforming simple nutritional substances into cellular protoplasm, producing some of its own useable energy, and eventually replicating itself. A cell may be viewed as an enclosed vessel composed of smaller sub-cell parts, organelles, each of which has specific metabolic functions. Within the cell, countless chemical reactions take place simultaneously, all of them are controlled, so they contribute to the processes we call life and the eventual procreation of the cell.

(2) Study the cells in multicellular organism In a multi-cellular organism, individual cells, by differentiating in order to acquire particular functions, can become the building blocks of the larger multi-cellular organisms. In order to do this, each cell keeps in continuous communication with its neighboring cells. Cooperative assemblies of like cells make up tissues, and a collaboration between tissues helps forms organs, the functional units of a whole organism as complex as a human being.

The smallest known cells are a group of tiny bacteria called mycoplasmas. Some of these single-celled organisms are spheres about 0.3 μm in diameter with a total mass of 10^{-14} g. But human cells are about 400 times larger, they are about 20 μm across. It was estimated that a human being may be composed of more than 7.5×10^{13} cells, divided to more than 200 types with special structure and function. Cells in a multi-cellular organism are organized in strict roles to be fully harmonized the whole body, through differentiation, communication and interaction.

1.6.2 Molecular biology of cell

Although biochemistry and molecular biology, as disciplines, might have made substantial development without the advent of the cell theory, each science has strongly influenced the other almost from the start. Contemporary cell biology, often referred to as molecular cell biology, or cell and molecular biology, is then a compilation of four separation disciplines: ①Cell physiology, which takes a comparative approach looking at how cells answer universal problems, from water conservation to cell communication; ② Systemic physiology, which is the science of organ systems and whole organism physiology, such as insect, plant, fish or human physiology; ③Biochemistry, which looks at the chemical and physical commonality of the mechanisms of cellular reactions; ④Molecular biology, which studies the properties of organisms through the constituent molecules.

(1) Genetic control in cells The molecular basis of genetic control in cells, particularly in eukaryotic cells is one of the most basic active areas of molecular cell biology. Of particular interest is the understanding of the regulation mechanisms involved in the development of multicellular organisms. In most well-known case studies, such as in the fruit fly *Drosophila melanogaster*, it has been shown that regulation among the genes that control early development and cell differentiation.

Genes are the blueprint of our bodies, a blueprint that creates the variety of proteins essential to any organism's survival. These proteins, which are used in countless ways by our bodies, are produced by genetic sequences, i. e. our genes, as described in the cell biology section, protein synthesis pages. Some other genes that will be functional during specialization determine the physical characteristics of the cell, i. e. long and smooth for a muscle cell or indented like a goblet cell.

(2) Three flows within cells Cells can be described by the processes that take place within them. The functions of the cell can be broken down into three basic flows: flow of information, flow of mass, flow of energy. None of these three flows operates in isolation from the other two. The flow of mass and flow of energy are extremely closely linked, as biological molecules are broken apart to make other biological molecules or to harvest energy.

1) **Flow of information**: The flow of information in the cell is described simplistically by the central dogma of molecular biology — "DNA makes DNA makes RNA makes protein" (Figure 1-8).

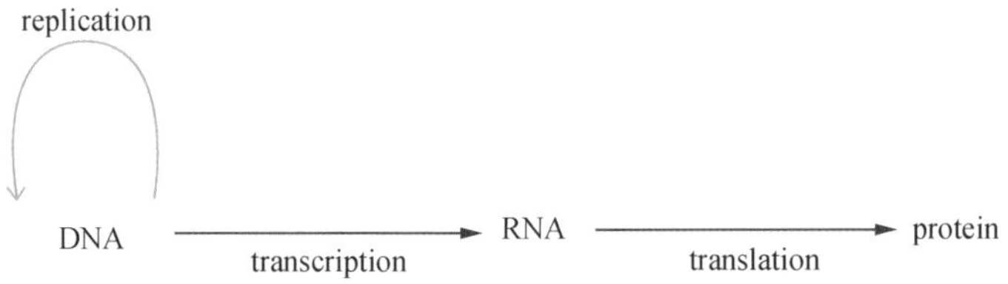

Figure 1-8 Central dogma of molecular biology

The central dogma of molecular biology contains: chromosomal DNA is used as a template to make more chromosomal DNA; chromosomal DNA is used as a template to make RNA; mRNA is used as a template from which proteins are made; other relationships will become apparent as we cover the course material.

2) **Flow of mass**: Simple molecules absorbed by cells are incorporated into larger molecules and (or) used to make other molecules. The characteristic particular to material transformation is the necessity, for the living cell to maintain a non-equilibrium state while flow of matter occurs in directions towards equilibrium driven by thermodynamic potentials. In contrast, in passive, irreversible flow of matter thermodynamic driving force results in flow until equilibrium is reached. The mammalian cell has solved this problem by controlled dissipation of thermodynamic driving force via enzyme networks and replenishment of driving force by oxidation of foodstuffs (Figure 1-9).

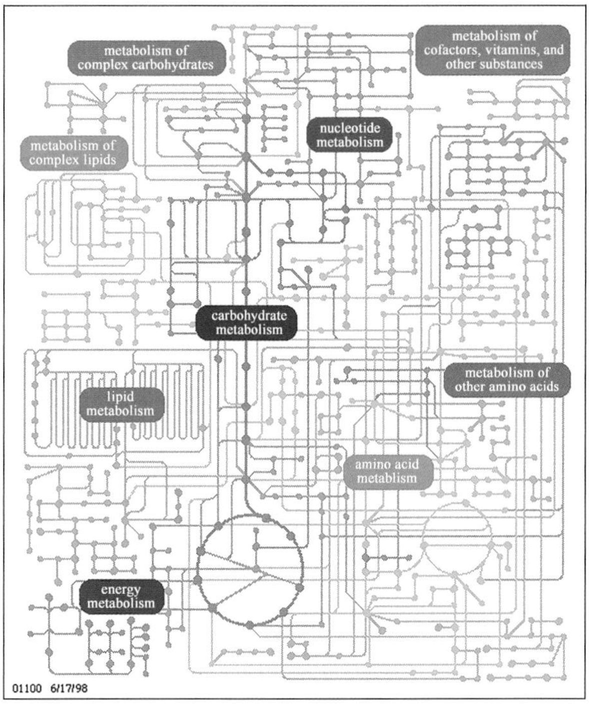

Figure 1-9 Metabolic pathway networks in a cell (From: Per Kraulis, 2003)

3) **Flow of energy**: Some processes (anabolic) are energy consuming. Other processes are energy producing (catabolic). Energy intermediates are the molecules, adenosine triphosphate (ATP) and adenosine diphosphate (ADP), which are used to allow endergonic reactions to occur in cells.

Biological reactions inside a cell are all catalyzed by enzymes. Enzymes are named on the basis of the reactions they catalyze. Within cells a number of phenomena regulate enzyme function. In some cases, irreversible inhibition may occur. Often, reversible inhibition occurs by either of two methods, competitive and noncompetitive inhibition. The latter occurs as a result of involvement of the allosteric site of the enzyme.

1.6.3 Systems biology of the cell

Currently, biologists tend to study the cell structure and behaviors as the ultimate objective to systemically understand the effect of the molecules on the whole cell, and the effect of the individual cell on the whole organism as well.

Molecular alterations in cells as elementary units of cell systems, tissues and organs may manifest as disease at the organism level. They are detectable as altered molecular cell phenotypes by single cell flow, image or chemical cytometry. Molecular cell phenotypes result from genotype and exposure influences within the heterogeneity of various cell types, but also within given cell types according to cell cycle or functional status.

(1) Molecular cytomics Molecular **cytomics** is an entry to cell systems biology, and a cell systems research, which aims at the understanding of the molecular architecture and functionality of cell systems (**cytomes**) by single-cell analysis in combination with exhaustive bioinformatic knowledge extraction.

The cytomics concept has been significantly advanced by a multitude of current developments. Amongst them are confocal and laser scanning microscopy, multiphoton fluorescence excitation, spectral imaging, fluorescence resonance energy transfer (FRET), fast imaging in flow, optical stretching in flow, and miniaturised flow and image cytometry within laboratories on a chip or laser microdissection, as well as the use of bead arrays.

The elucidation of the molecular pathways from the 20 ~ 40 thousands genes of the sequenced human genome via investigation of genetic networks and molecular pathways up to the cellular and organismal phenotypes is highly complex and time consuming.

The concept of a **human cytome project** systematizes this approach by analyzing in a first step the multitude of multiparametric cytometric data from large patient groups available at many clinical sites and establishing a structure for public databases for the standardized relational knowledge. In a second step molecular disease pathways are investigated and mathematically modeled by biomedical cell systems biology. Leukemia/lymphomas, cancers, rheumatism diseases, allergies and infections are of high medical interest while stem cell differentiation, cell cycle regulation, cell proteomics and cell organelle functionality are attractive for basic research.

Human cytome project will immediate medical use, facilitate access to the detection of new drug targets, increase research speed and the stimulation for advanced technological developments.

(2) Cellular genomics Cell **genomics** is the measurement of gene expression in single cells, including what is the transcriptional status of single cells and how their transcription patterns might change in response to external stimuli.

1) **Gene expression profile**: Gene expression is a complex process involving a cascade of events that starts with interactions between a multitude of transcription factors and gene promoter sequences, RNA polymerase binding and elongation, and RNA processing and maturation. The approach for gene-expression profiling in single cells is conceptually simple, but in practice it is technically challenging.

2) **The focuses of cell genomics**: The research of cellular genomics emphases following aspects: ①The spatial analysis of linkage between genetic loci as a means of detecting

translocations; ②The study of nuclear organization, namely whether active genes occupy different positions relative to inactive genes; ③The organization of active genes in relation to chromosome territories; ④The study of gene expression during different stages of differentiation and the cell cycle.

(3) Cellular proteomics The genome sequences of important model systems are available and the focus is now shifting to large-scale experiments enabled by this data. Following in the footsteps of genomics, we have functional genomics, proteomics, and even metabolomics, roughly paralleling the biological hierarchy of the transcription, translation, and production of small molecules.

1) **Cellular proteomics is a main part of proteomics**: Cellular proteomics is a main part of proteomics, and initially concerned with determining the structure, expression, localization, biochemical activity, interactions, and cellular roles of as many proteins as possible. There has been great progress owing to novel instrumentation, experimental strategies, and bioinformatics methods. The area of protein-protein interactions has been especially fruitful.

2) **Protein expression profiling in single cell**: An established and widely accessible strategy for protein profiling is two-dimensional gel electrophoresis(2DE), which displays changes in protein expression and post-translational modifications(PTMs) on the basis of protein staining intensities and electrophoretic mobility. An alternative strategy for protein profiling, variously referred to as multidimensional protein identification technology (MudPIT), multidimensional LC-MS/MS (multidimensional liquid chromatography coupled with tandem mass spectrometry), or "bottom up" shotgun proteomics. Several studies have successfully identified novel signal transduction targets by selectively activating or inhibiting pathways and screening molecular responses by the methods mentioned above, such as MAP kinase signaling in human erythroleukemia cells, and Rho GTPase signaling in a melanoma cell.

3) **Cellular proteomics assays**: Cellular proteomic systems are increasingly being exploited to assess the activity of proteins in a single cell, and this promise to provide unparalleled advances towards annotation of gene function and the identification of novel drug targets. Genes can be systematically evaluated for their participation in signaling pathways as well as cellular processes and physiological phenotypes. Then, integration of extracellular signals often involves the crosstalk between signal cascades that has been suggested to share some common traits with neural networks.

(4) Cellular epigenomics **Epigenetics** study the heritable changes in gene function that occur without a change in the DNA sequence. The **epigenome** is the sum of both the chromatin structure and the pattern of DNA methylation, which is the result of an interaction between the genome and the environment.

Epigenomics is a whole genome approach to epigenetics. Takes a whole-genome approach to study environmental or developmental epigenetic effects, primarily DNA methylation, on gene function. Thus, epigenomics focuses on those genes whose functions are determined by external factors.

An approach that views the imprinting, metabolic networks, genetic hierarchies in embryonic development, and epigenetic mechanisms of gene activation in a cell and other complex phenotypes from the genomic level down, rather than from the genetic level up, can provide powerful insights into the functional interrelationships of genes in health and disease.

(5) In silico cell "In silico cell" is also named as "e-cell", which is produced in or by means of a computer simulation. This must originally have been a computer scientist's joke,

but it now appears in print often enough that it has to be added to our list of modern latinisms. The e-cell system permits the construction of model structure, such as gene regulatory and (or) biochemical networks equivalent to a cell system or a part of the cell and perform both quantitative and qualitative analysis.

1) **"In silico cell" is the computational model of living cell:** Systems biology involves the use of computer simulations of cellular subsystems (such as the networks of metabolites and enzymes which comprise metabolism, signal transduction pathways and gene regulatory networks) to both analyze and visualize the complex connections of these cellular processes. Artificial life or virtual evolution attempts to understand evolutionary processes via the computer simulation of simple (artificial) life forms.

2) **"In silico cell" reflects the molecular networks within a cell:** Cellular components interact with each other to form networks that process information and evoke biological responses. A deep understanding of the behavior of these networks requires the development and analysis of mathematical models, which has become an attractive focal point for engineers, computer scientists, mathematicians, and systems biologists.

1.7 The technology of cell biology

Despite the considerable success in sequencing the human genome and other genomes, the function of many thousands of genes in humans remains unknown. Developments of techniques can screen thousands of genes simultaneously for their effect on the biological processes in living cells — cell division, signaling pathways and gene expression, for example and would result in major benefits to both the basic biomedical sciences and the pharmaceutical industry.

The details of how cells are able to grow, reproduce, differentiate and communicate with each other are highly important in modern biology. The investigation of these processes in living cells at the sub-cellular and molecular level requires a number of new technological developments.

1.7.1 Cell culture

Cell culture is one of the most widely used techniques in life sciences. This cell culture workshop will cover the basic techniques used for maintaining animal cells in culture: aseptic technique, counting cells, subculturing, cryopreservation (freezing) and thawing. Participants will obtain hands-on training in all techniques listed above. Lecture and discussion sessions will include the techniques mentioned above as well as the following topics: cell culture equipment, contamination and optimization of growth conditions.

1.7.2 Flow cytometry

Flow cytometry is a means of measuring certain physical and chemical characteristics of cells or particles as they travel in suspension one by one past a sensing point. In one way flow cytometers can be considered to be specialized fluorescence micro-scopes. The modern flow cytometer consists of a light source, collection optics, electronics and a computer to translate signals to data. In most modern cytometers the light source of choice is a laser which emits coherent light at a specified wavelength. We can measure physical characteristics such as cell size, shape and internal complexity and, of course, any cell component or function that can be detected by a fluorescent compound can be examined. So the applications of flow cytometry are numerous, and this has led to the widespread use of these instruments in the cellular biological and medical fields.

Flow cytometric applications include: phenotypic analysis, sterile sorting of transfectants, DNA analysis, assessment of apoptosis、functional studies.

1.7.3 Functional bio-imaging of cell

In the longer-term, additional multi-parameter imaging-measurement techniques could also be developed and applied to single living cells inside animals.

(1) Modern light microscopy Modern light microscopy has become a most powerful analytical tool for studying molecular processes in cells. Recent advances combining sample preparation, microscope design and image processing allow the generation of "multidimensional" image data, simultaneously reporting the three-dimensional distribution and concentrations of several different molecules within cells and tissues at multiple time points with sub-micron spatial resolution and sub-second temporal resolution. Molecular interactions and processes that were approached by biochemical analyses *in vitro* can now be monitored in live cells.

(2) Confocal microscopy Biological specimens are three-dimensional objects. In confocal microscopy, a pinhole aperture is inserted at the optical focal plane in the microscope. This passes only light going through the structure of interest and blocks all reflected light from the rest of the specimen, resulting in a clear image. Movable mirrors scan the area of good focus across the specimen in a raster pattern, creating a large, in-focus image of a thin slice of the specimen. The raster scan then moves vertically, and another slice is scanned. After multiple scans, computer processing of the images produces a three-dimensional image of the specimen(Figure 1-10).

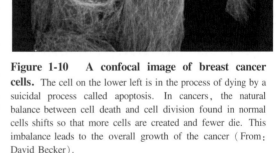

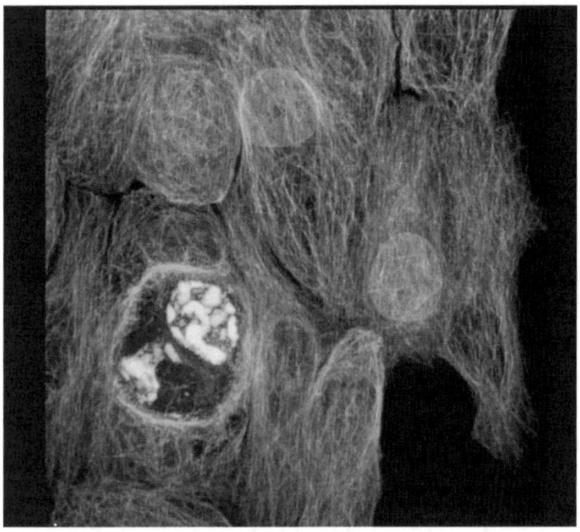

Figure 1-10 A confocal image of breast cancer cells. The cell on the lower left is in the process of dying by a suicidal process called apoptosis. In cancers, the natural balance between cell death and cell division found in normal cells shifts so that more cells are created and fewer die. This imbalance leads to the overall growth of the cancer (From: David Becker).

(3) Two-photon microscopy In two-photon microscopy fluorophores are excited simultaneously by two photons of twice the wavelength of the single, smaller-wavelength photons used in traditional confocal microscopy. Longer wavelength photons have lower energies and penetrate further into tissue than short wavelength photons, making the technique ideal for imaging deep into live tissue samples.

With two-photon imaging, lower energy photons illuminate the specimen. A dye molecule that absorbs two of the lower energy photons within a short period of time will receive energy above the threshold for fluorescence. To increase the chances that a dye molecule will absorb two photons within the short time period, the laser is pulsed causing all photons to arrive within a few seconds. The laser beam is focused on the area of interest, causing all of the photons to illuminate just the area of interest. These two techniques put all of the photons on target at the same time.

The technique could allow researchers in many fields of biology to track migrating cells in a form of real-time movie, which biologists have discovered are common in many types of tissue, ranging from nerves to lymph nodes. To date, such long-range migrations have been inferred from observations of chemically fixed tissue at different stages of development.

(4) Live cell image analysis Live-cell image analysis started with the earliest microscopists. Although most of these measurements were based on manual inspection and intervention, with the advent of fluorescence microscopy, many studies also involved quantitative imaging of living cells either using video or CCD cameras. CCD camera uses a small, rectangular piece of silicon rather than a piece of film to receive incoming light, this silicon piece is called a charge-coupled device (CCD). In the early years of live-cell microscopy, methods for segmentation and tracking of cells were rapidly developed and adapted from other areas. Nowadays, techniques for fully automated analysis and time - space visualization of time series from living cells involve either segmentation or tracking of individual structures, or continuous motion estimation.

A major requirement of an automated, real-time, computer vision-based cell tracking system is an efficient method for segmenting cell images. The usual segmentation algorithms proposed in the literature exhibit weak performance on live unstained cell images, which can be characterized as being of low contrast, intensity-variant, and unevenly illuminated.

By modern microscopy system, life scientists are able to conduct three-dimensional time-lapse studies with shorter exposure times and more frequent exposure opportunities, allowing them to capture a more complete picture of intracellular activity. This system now offers live-cell researchers the shortest exposure time of any microscopy system with the greatest molecular-level resolution available. Improved and more frequent image sampling over time dramatically increases the chance that a researcher might discover a new key structure or function occurring in the cell they are viewing.

1.7.4 Bioinformatics and cell biology

Bioinformatics is set to play a significant role in cell biology immediately and in the foreseeable future through the development of software to organize the multitude of data emerging from simultaneous time — course measurements and allow interpretation of that data in the context of cell function. An example is in the construction of 3D images. At present this software has been written with the medical imaging in mind and will need to be modified to suit the particular application.

1.7.5 RNA interference technology

RNA interference (RNAi) is a molecular biological technique that has been extensively used to deplete specific components in cells and has been invaluable in determining the requirements of a given protein in a particular cellular process. In the response, small segments of double stranded RNA direct specific message RNA cleavage resulting in little or no protein expression. The initial use of RNAi technology was published in 1998 where Mello and colleagues showed double-stranded RNA-mediated genetic interference in caenorhabditis elegans. Researchers are still exploring the pathway mechanisms and components, but it appears to be conserved in other organisms including plants and humans.

1.7.6 Antisense technique

The **antisense** technique is tools we are using in an attempt to selectively block the gene expression. The primary idea of antisense technique is to interfere with the information flow from gene to protein, and to do so in a very specific manner. Antisense oligodeoxynucleotides (ODN) have been used selectively to arrest translation of target mRNA into functional proteins in a wide variety of systems from cell-culture to animals. The action of antisense technique depends on the ability of short synthetic ODN, which has been taken up by the living cells, to hybridize selectively to the target mRNA, and thereby interfere with ribosomal activity, activate ribonuclease H-mediated degradation of mRNA, or both.

1.7.7 Coming technological developments

Understanding the subcellular localization of activity and the dynamics of the activity as well as being able to quantify these measurements represent significant technical challenges. In addition, it is often necessary to study a wide range of parameters simultaneously in living cells under normal physiological conditions. To assist the greater understanding of cell function, the following technological developments will need to occur over the next five years.

(1) New chemically based technologies ①The development of a range of new non-invasive biological assays for specific biochemical processes; ②The development of peptide tags for the localization of proteins in cells; ③The synthesis of new chemical compounds for monitoring specific cellular processes in living cells.

(2) New types of equipment ①More sensitive and versatile confocal microscopes; ②Cameras which are specific for particular assays combining for example, higher quantum efficiency, operation at faster speeds with broader wavelength sensitivity and low background noise; ③Cameras with greater versatility, which can be used for a variety of assays combining for example, higher quantum efficiency, operation at faster speeds with broader wavelength sensitivity and low background noise; ④Digital cameras able to record numerous simultaneous events; ⑤Detectors which are specific for the type of reporter or probe being used; ⑥Detectors need to work at the appropriate speed with the minimum of background noise; ⑦Specific new software to control the microscope and camera in order to collect and subsequently to analyze multi-parameter time course data.

1.8 Training the scientists of tomorrow

1.8.1 The principles in the course of cell biology

The intent of modern molecular cell biology is to interpret the properties of the whole organism through the structure of its own cell's component molecules. The cell biology also is the study of the life processes, which occur within the makeup of a cell and its molecules.

(1) Understanding the cell from molecular view The goals of the molecular cell biology lie in understanding the cell theory from molecular view, which by itself cannot explain the development and unity of the multi-cellular organism. A cell is not necessarily an independently functioning unit, and a plant or an animal is not merely an assembly of individual cells.

(2) Understand the cell from systemic view Molecular cell biology is not so self-inclusive as to eliminate the concept of the organism as more than the sum of its parts. But the study of a particular organism does require the investigation of cells, as both individuals and groups. A plant or an animal governs the division of its own cells; the correct cells must divide, become differentiated, and then be integrated into the appropriate organ system at the right time and place. Breakdown of this process results in a variety of abnormalities, one of which may be cancer. When a cell biologist studies the problem of the regulation of cell division, the ultimate objective is to understand the effect of the individual cell on the whole organism.

1.8.2 What dose cell biology show students

Cell biology will not only show students what we know about the cell, but also show that how we know and why we do about the cell.

(1) What we know about the cell The main content of the molecular cell biology course includes: ①The structure and function of the cell; ②The organelles of the cell; ③The mechanisms of important activities of the cell; ④The regulation of the activities; ⑤The advance in cellular research.

This course covers some modern topics of cell biology, such as the proliferation and differentiation of cells; single transduction in the cell; gene expression and regulation in the eukaryote cell; and the origin and evolution of cells. Thus cell biology is an important basic course for all students of life science.

(2) How we know what we know about the cell It is critical to present to students the experimental basis of our understanding — to show them how we know what we know. Hopefully this will demonstrate the dynamic nature of science and prepare them not only to engage actively in scientific research and teaching, but also to become educated members of a public that increasingly is asked to deal with complex issues such as environmental toxins, genetically modified foods, and human gene technology.

(3) Why we do what we do about the cell Of course, we want students to learn not only how we know what we know, but why we do what we do. As in other cell biology books, our coverage of basic cell biology, medical topics, biotechnology, human biology is integrated throughout. We know that these topics may be of particular interest to students.

The cell biologists focus recently on systems cell biology, epigenetics of the cell and functional cellular imaging. These researches lead to full understanding the cell activity and the cellular roles in organ as well as organism. With the development of research, more interests will produce further research projects to further our knowledge about the cell biology. The awareness of what we do and why we do is essential for a cell biologist of tomorrow.

1.8.3 Open heuristic teaching mode to cultivate innovative talents

To better inspire students' creative ability and increase learning enthusiasm and subjective initiative, some stories, such as the ones of Nobel Prize winners and famous Chinese scientists, are suggested to be introduced behind typical inspiring scientific discoveries while explaining the theories. The ultimate goal is to cultivate more innovative talents for the country.

Incorporating the heuristic teaching model and the cultivation of innovative talents into the teaching of cell biology has the following purposes:

(1) Let students understand the importance of learning knowledge and see further by standing on the shoulders of predecessors; stimulate students' interest and enthusiasm of science; cultivate students' perseverance, correct attitude, and good study habits.
(2) While imparting knowledge, teachers purposefully tell the process of representative scientific discoveries to inspire students to think critically and master subjective initiative and cultivate students' good thinking styles. The so-called original discoveries are all knowledge that was not discovered by the predecessors. Therefore, when learning knowledge, students should never regard the previous discoveries as absolute truth. Instead, they should always ask why, question existing discoveries or experimental phenomena, make bold hypotheses, and strive to find evidence to prove the hypothesis.
(3) Science and technology are double-edged swords and are beneficial to mankind, but improper use can also bring disasters. Therefore, scientific research must be ethical and conducted on the premise that it is beneficial to the world. Organize students to discuss through some examples, such as human cloning, gene-edited babies, stem cell transplantation, and other technologies that involve ethical issues.
(4) Through the scientific discovery stories of scientists, we can understand the importance of cooperation, especially now that disciplines intersect each other, and coordination is more needed. Therefore, it is important to establish teamwork. Only by overall planning, coordination, and division of labor can we achieve better results.
(5) The teacher must teach and educate people. Therefore, apart from teaching, young adults must be guided to establish a correct outlook on life and good morals, establish lofty ambitions, and understand that the purpose of learning is not only for professional development, but also to enhance the meaning of life. No matter what occupation they will pursue in the future, they must be enthusiastic, conscientious, and do the job wholeheartedly. While taking care of themselves and their family, they also contribute to the country and humanity.
(6) Through the study of cell biology, students understand that many diseases, such as cancer, cardiovascular and cerebrovascular diseases, and even immunity are closely related to personal psychology and living habits. If people want to be healthy, they must maintain an optimistic and open-minded attitude, and at the same time have a regular schedule and regulate their diet.

Summary

Cell biology is the modern biology of the cell. It concerns on the cell which is the units of life and a complex system. Current genomic and proteomic knowledge shapes the cell biology, leading to the expanse of the contents of cell biology from essential cell biology, to molecular cell biology and systemic cell biology. The development of modern cell biology depends on the progress of contemporary biotechnology. Cell biology is also an integrated view of cells at work, involving essential cell biology, molecular biology of cell and cell systems biology.

Development of techniques which can screen thousands of genes simultaneously for their effect on the biological processes in living cells — cell division, signaling pathways and gene expression, for example, would result in major benefits to both the basic biomedical sciences and the pharmaceutical industry.

Questions

1. What does cell biology concern?

2. Why we say the cell is a complex system?
3. What is cell theory?
4. What are the contemporary views on cell biology?
5. What cause cell diversity?
6. How many modern techniques are employed in cell biology research?

（吴超群，李　瑶修改）

Chapter 2

Cell membrane and cell surface

All cells are surrounded by a membrane layer called **cell membrane** or **plasma membrane**. Cell membranes consist of membrane lipid and proteins and are crucial for the life of the cell by defining the cell boundaries, therefore, maintaining the differences between the intra-and extracellular environment. On the surface of the cell plasma membrane, structures enriched with sugar groups form cell coats, also termed **glycocalyx** or **extracellular matrix** (**ECM**), which is functionally important for cell-cell communication and cell-environment communication.

Within a eukaryotic cell, membranes compartmentalize the cell into smaller subcompartments with specialized functions termed organelles, such as endoplasmic reticulum, vesicle, Golgi body (see chapter 4), mitochondria, and chloroplasts (see chapter 5). All these membranes form closed structures and a related net called cytoplasmic membrane system. The plasma membrane and cytoplasmic membrane system belong to biomembrane system. The most important characteristic of a biomembrane is selectively permeable (semi permeability). It forms an adjustable barrier between the cell and the extracellular environment or between inside and outside of organelles in eukaryotic cells. Prokaryotic cells have a very simple internal structure and contain no internal membrane-limited subcompartments. Most enzymes and metabolites are thought to diffuse freely within the single internal aqueous compartment. The replication of DNA and the production of ATP, take place at the plasma membrane.

In this chapter, the components, structures and functions of biomembrane will be described and the plasma membrane will be mainly focused on. Cell surface, extracellular matrix, cell wall, trans-membrane transport, and cell junction will be discussed in detail. Other structures with biomembrane, such as nuclear envelope, mitochondrion, chloroplast, endoplasmic reticulum, Golgi body and vacuole, will be discussed in pertinent chapters.

2.1 Components and structure of cell membranes

All cell membranes have a common structure: a thin film of lipids form a bilayer framework structure of biomembrane and protein molecules held together mainly by non-covalent interactions are embedded within this framework (Figure 2-1). Besides, a small quantity of saccharide chains covalently binds to the membrane proteins (glycoproteins) and lipids (glycolipids).

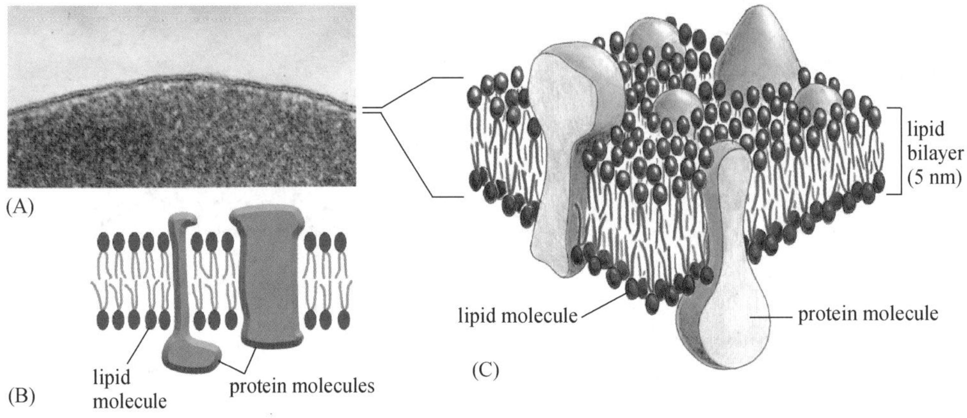

Figure 2-1 The structure of plasma membrane. (A) Structure of human red blood cell plasma membrane under electron microscope; (B) and (C) Schematic view of a plasma membrane (B: two dimension; C: three dimension) (From: Alberts B, et al. Molecular Biology of the Cell. 2008)

2.1.1 Lipids

Lipids contained in a typical biomembrane are phospholipids (include phosphoglycerides, sphingolipids) and steroids. All three classes of lipids are **amphipathic** molecules having a hydrophobic core called **polar head** and a hydrophobic core called **nonpolar tail**. This molecular character enables phospholipids to spontaneously aggregate with their hydrophobic tails buried in the interior and their hydrophilic heads exposed to water. Such bilayer structure is maintained by hydrophobic and van der Waals interactions between the lipid chains.

(1) Phosphoglycerides **Phosphoglycerides**, derivatives of glycerol 3-phosphate, are the most abundant class of lipids in most membranes. A typical phosphoglyceride molecule consists of a polar head attached to the phosphate group and a hydrophobic tail composed of two fatty acyl chains esterified to the two hydroxyl groups in glycerol phosphate (Figure 2-2). The two fatty acyl chains may differ in the number of carbons (commonly 16 or 18) and their degree of saturation (0, 1, or 2 double bonds). The length and the saturation degree of the fatty acid tails are important in regulating the fluidity of the mem-

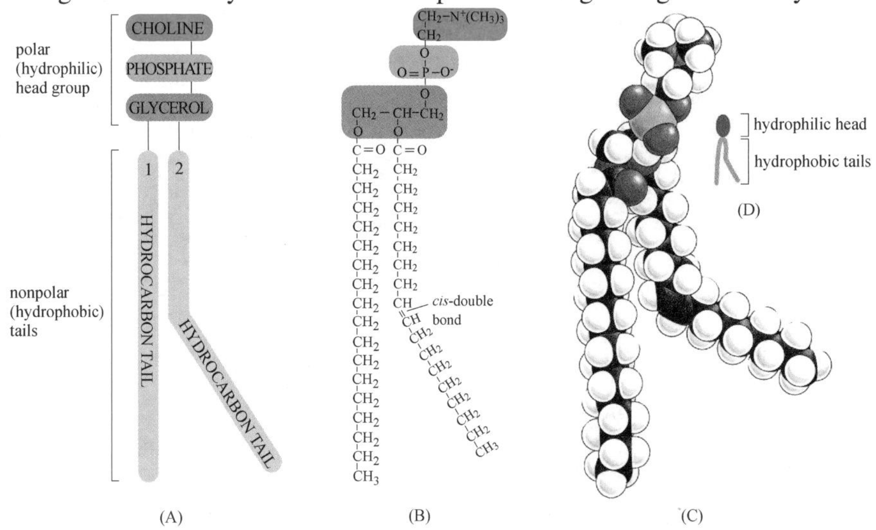

Figure 2-2 Structure of phosphatidylcholine, a typical phosphoglyceride molecule. (A) Scheme of molecular structure; (B) Formula; (C) Space-filling model; (D) Symbol (From: Alberts B, et al. Molecular Biology of the Cell. 2008)

brane. Phosphoglycerides are classified according to the nature of their head groups. In phosphatidylcholines, for example, the head group consists of choline, a positively charged alcohol, which is esterified to the negatively charged phosphate. In other phosphoglycerides, an OH-containing molecule such as ethanolamine, serine, or the sugar derivative inositol is linked to the phosphate group (Figure 2-3 A ~ C). The negatively charged phosphate group and the positively charged or the hydroxyl groups on the head group interact strongly with water.

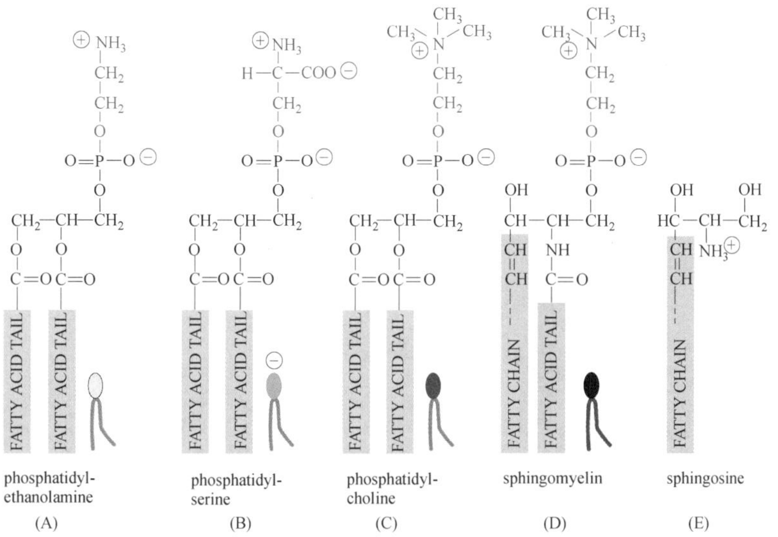

Figure 2-3 Major phospholipid molecules in the mammalian plasma membrane (From: Alberts B, et al. Molecular Biology of the Cell. 2008)

(2) Sphingolipids The second class of membrane lipid is **Sphingolipid**. Sphingolipid is derived from sphingosine and contains a long-chain fatty acid attached to the sphingosine amino group (Figure 2-3 D ~ E). In sphingomyelin, the most abundant sphingolipid, phosphocho-line is attached to the terminal hydroxyl group of sphingosine. Thus the overall structure of sphingom-yelin is quite similar to that of phosphatidylcholine and therefore sphingomyelin is a phospholipid.

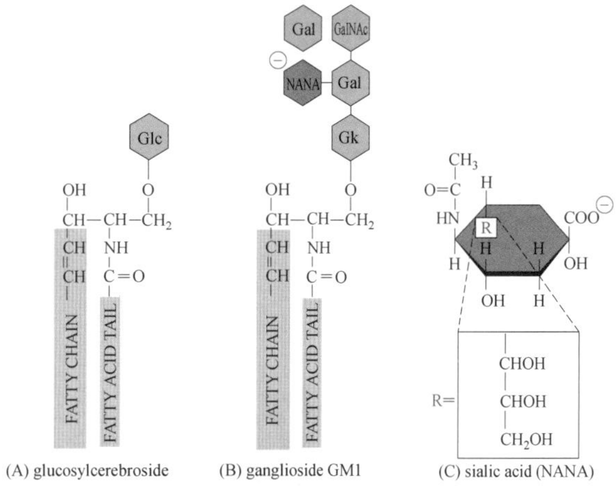

Figure 2-4 Structure of glycolipid molecules in plasma membrane

Other sphingolipids are amphipathic glycolipids whose polar head groups are sugars (Figure 2-4). Glycolipids, including glyceroglycolipid, constitute 2 ~ 10 percent of the total lipid in plasma membrane and are the most abundant lipids in nervous tissue.

(3) Cholesterol Eukaryotic plasma membranes contain especially large amounts of **cholesterol**, which is absent from most prokaryotic cells. Cholesterol is a sterol. Its basic structure is a rigid four-ring hydrocarbon with a hydroxyl substituent on one ring and a short

nonpolar hydrocarbon chain on the other side (Figure 2-5). Although the composition of cholesterol is almost entirely hydrocarbon, it is still amphipathic since its hydroxyl group can interact with water. The cholesterol molecules orient themselves in the bilayer with their hydroxyl group close to the polar head groups of adjacent phospholipid molecules. Their rigid, platelike steroid rings interacting with, and partly immobilizing the first few CH_2 groups of the hydrocarbon chains of the phospholipid molecules, and therefore make the lipid bilayer less deformable in this region and less permeable for small water-soluble molecules. At high concentrations, cholesterol prevents the hydrocarbon chains from coming together and crystallizing and therefore keeps the bilayer fluidity.

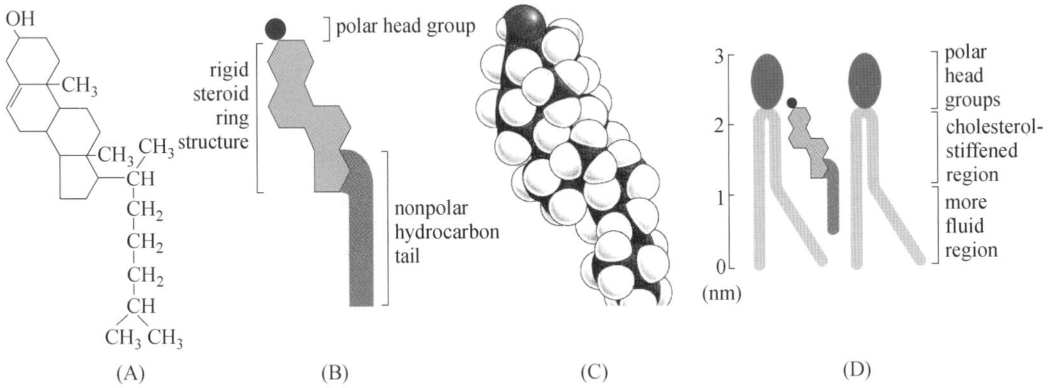

Figure 2-5 Cholesterol molecules in the mammalian plasma membrane. (A) Formula; (B) Symbol; (C) Space-filling model; (D) Interaction between phospholipids and cholesterols in plasma membrane (From: Alberts B, et al. Molecular Biology of the Cell. 2008)

2.1.2 Proteins

The lipid bilayer provides the basic structure of biological membranes, but the membrane proteins perform most of the membrane's functions. The lipid bilayer presents a unique two-dimensional hydrophobic environment for membrane proteins either inserted in or linked to the surface of the bilayer. Membrane proteins can be classified into three categories: integral proteins (buried within the bilayer, Figure 2-6①~④), lipid-anchored proteins (Figure 2-6⑤,⑥), and peripheral proteins (Figure 2-6⑦,⑧).

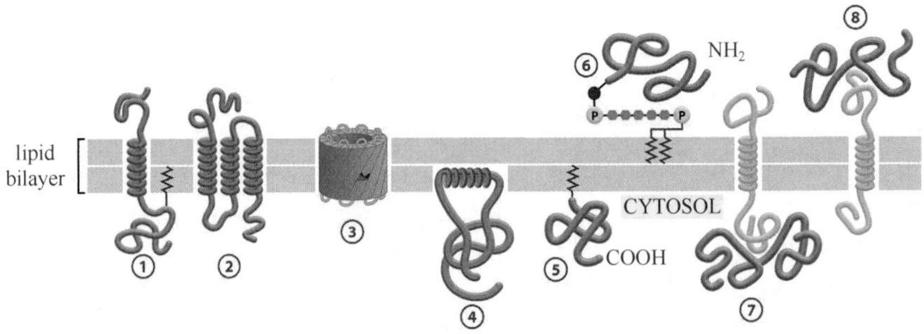

Figure 2-6 Diagram of proteins associated with the lipid bilayer (From: Albert B, et al. Molecular Biology of the cell. 2008)

(1) Integral proteins (transmembrane proteins) Integral proteins contain three different segments called domains: cytosolic domain, exoplasmic domain and transmembrane domain. The cytosolic and exoplasmic domains have hydrophilic exterior surfaces that interact with the aqueous solutions on the two faces of the membrane. Most transmembrane proteins are glycosylated with a complex branched sugar group attached to amino acid side chains which invariably are localized to the exoplasmic domains and generally bind to external signaling proteins, ions, and small metabolites (e. g., glucose, fatty acids), and to adhesion molecules on other cells or in the external environment. The cytosolic domain lying along the cytosolic face of the plasma membrane anchors cytoskeleton proteins and triggers intracellular signaling pathways. In most transmembrane proteins, their transmembrane regions contain many hydrophobic amino acids and form one or more α helices which are embedded in membranes. The structure of bacteriorhodopsin, a protein found in the membrane of certain photosynthetic bacteria, illustrates the general structure of proteins with α helix transmembrane region (Figure 2-7A). An alternative way for the peptide bonds in the lipid bilayer to satisfy their hydrogen-bonding requirements is to arrange the multiple transmembrane domains as β sheets that are rolled up into a closed barrel (a so-called β barrel). This form of multipass transmembrane structure is seen in OmpX, a porin protein (Figure 2-7B). Most multipass transmembrane proteins in eukaryotic cells and in the bacterial plasma membrane are constructed as transmembrane α helices whereas β barrel proteins are largely restricted to bacterial, mitochondrial, and chloroplast outer membranes. Compared to the β barrel, the helix transmembrane domains are rather flexible which can slide against each other, allowing conformational changes in the protein that can open and close ion channels, transport solutes, or transduce extracellular signals into the cell. In β barrel proteins, each β strand binds rigidly to its neighbors via hydrogen bonds, making conformational changes within the wall of the barrel unlikely. In addition, α helix can associate with the helix in another integral protein to form a coiled-coil dimmer, which is a common mechanism for creating dimeric membrane proteins.

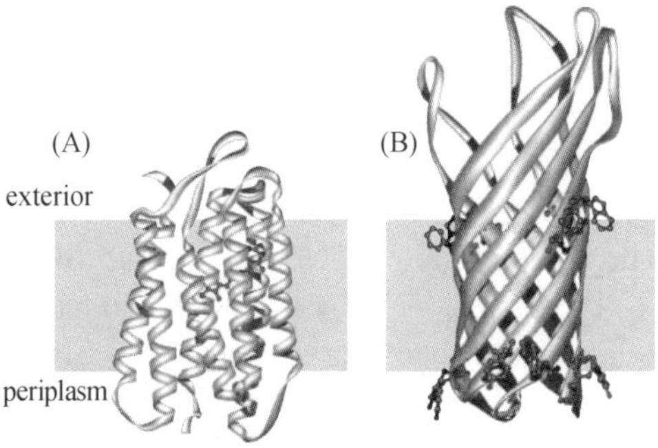

Figure 2-7 Structural basis of transmembrane proteins. (A) α helix model of bacteriorhodopsin (From: Luecke H, et al. J Mol Biol, 1999, 291: 899); (B) β barrel model of one subunit of OmpX (From: Schulz GE. Curr Opin Struc Biol, 2000, 10: 443)

(2) Lipid-anchored membrane proteins Lipid-anchored membrane proteins are bound covalently to one or more lipid molecules without their polypeptide chains entering the phospholipid bilayer: solely by a covalently bound lipid chain — either a fatty acid chain or a prenyl group — in the cytosolic monolayer (Figure 2-6 ⑤) or, via an oligosaccharide linker, to phosphatidylinositol in the noncytosolic monolayer — called a GPI anchor (Figure 2-6 ⑥).

(3) Peripheral membrane proteins Peripheral membrane proteins do not interact with the hydrophobic core of the phospholipid bilayer, but are usually bound indirectly to the membrane by interactions with integral proteins or directly by interactions with lipid head groups (Figure 2-6 ⑦,⑧).

2.1.3 Membrane carbohydrate

About 2 ~ 10 percent of membrane content is made up of carbohydrates. These carbohydrates occur as oligosaccharide chains covalently bound to membrane proteins (glycoproteins) and lipids (glycolipids). They also occur as the polysaccharide chains of integral membrane proteoglycan molecules found mainly outside the cell, as part of the extracellular matrix. The carbohydrate layer protects cells against mechanical and chemical damage and also helps to keep cells at a distance, preventing unwanted protein-protein interactions. The membrane carbohydrates help membrane proteins to form and maintain the correct packing and three-dimensional configures and to transfer to a correct position. The oligosaccharide side chains of glycoprotein and glycolipids usually contain fewer than 15 sugars, but are enormously diverse in their arrangement and are often branched and bonded together by various covalent linkages. Both the diversity and the exposed position of the oligosaccharides on the cell surface make them especially well suited to function in specific cell-recognition processes. For example, some glycolipids are important blood group determinants, and plasma membrane-bound lectins recognize specific oligosaccharides on cell-surface and mediate a variety of transient cell-cell adhesion processes in sperm-egg interaction, blood clotting, lymphocyte recirculation, and inflammatory response.

2.1.4 Structure characters of plasma membrane

The phospholipid molecules diffuse freely in biomembrane and form a bilayer framework structure with their hydrophobic nonpolar tails buried in the interior and their hydrophilic heads exposed to water. Depending on their shape, the cone-shaped lipid molecules can form spherical micelles with the tails inward whereas the cylinder-shaped lipids form double layer sheets with the hydrophobic tails sandwiched between the hydrophilic head groups. A mixture of two shapes can form **liposomes**, bilayer spherical vesicles with diameter spanning 25 nm to 1 μm (Figure 2-8).

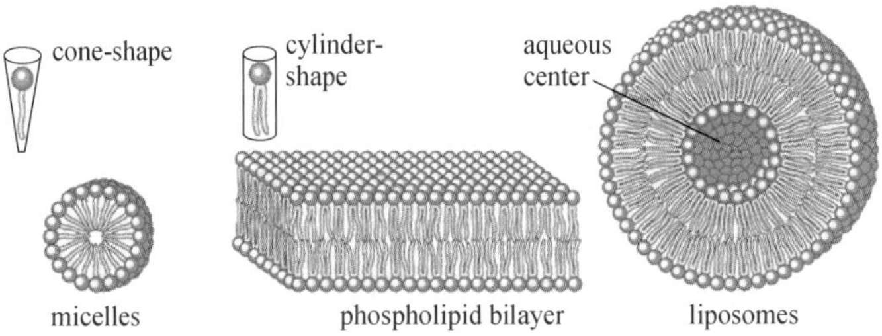

Figure 2-8 Arrangement of lipid molecules in an aqueous environment

(1) Fluid mosaic model of plasma membrane structure As to the structural organization of plasma membrane, several models have been proposed. Singer and Nicolson established the well-accepted **fluid mosaic model** in 1972. Fluid-mosaic model presents membranes as dynamic structures in which both lipids and associated proteins are mobile and capable of moving within the membrane to engage in interactions with other membrane molecules. The lipid bilayers form the basis of the membranes; a large number of proteins either span the bilayer or are attached to either side of the lipid membrane; the membranes are asymmetrical, in which lipid and protein molecules diffuse more or less easily.

(2) Lipid raft model **Lipid rafts model** has been recently established as a complementation for the former fluid mosaic model. In this model, some regions of the plasma membrane are especially rich in cholesterol, glycosphingolipid, and GPI-anchored membrane proteins, which are organized in small (10 ~ 200 nm) glycolipoprotein microdomains, termed lipid rafts (Figure 2-9). These specialized membrane microdomains compartmentalize cellular processes by serving as organizing centers for the assembly of signaling molecules, influencing membrane fluidity and membrane protein trafficking, and

regulating neurotransmission and receptor trafficking.

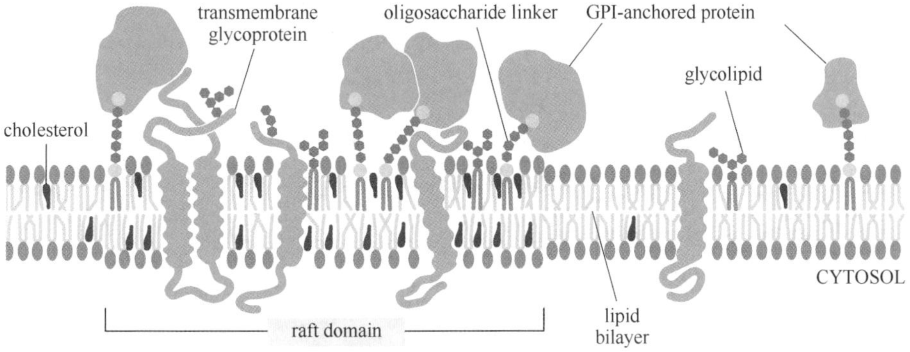

Figure 2-9 Structure of lipid raft in the plasma membrane(From: Alberts B, et al, Molecular biology of the cell. 2008)

(3) Membrane fluidity **Membrane fluidity**, referring to the viscosity of the lipid bilayer, is the physical state of the lipid molecules within the cell membrane and is essential for the membrane function. It allows molecules (e. g., integral proteins) to diffuse rapidly in the plane of the bilayer and interact with each other, thus many basic cellular processes such as cell movement, cell growth, cell division, secretion etc. can take place. Certain types of movement (e. g., diffusion and rotation) of molecules within the membrane are more frequent and rapid than the others (e. g., "flip-flop") (Figure 2-10). This is because, for "flip-flop" to occur, the hydrophilic head group of the lipid must overcome the internal hydrophobic sheet of the membrane, which is thermodynamically unfavorable.

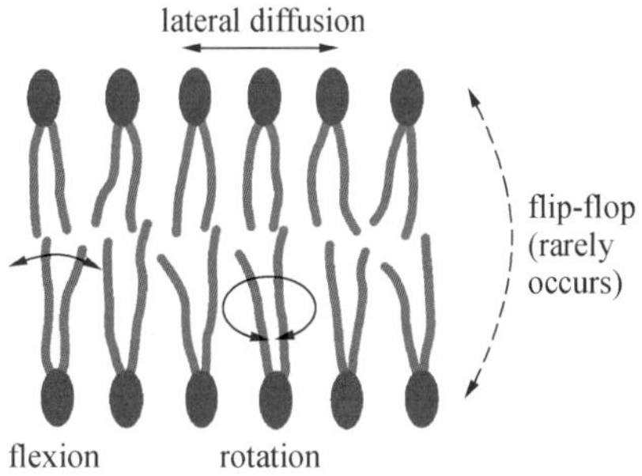

Figure 2-10 The possible movements of phospholipids in a membrane(From: Alberts B, et al. Molecular Biology of the Cell. 2008)

Temperature and membrane composition are the main determinants for membrane fluidity. At relatively warm temperature (e. g., 37℃), the lipid bilayer is in a relatively fluid state. When lowering the temperature slowly, there comes a point (transition temperature) when the lipid is converted to a frozen crystalline gel in which the movement of the phospholipid fatty acid chains is greatly limited. The length and the degree of unsaturation of hydrocarbon tails determine the viscosity and fluidity of the membrane. The longer the hydrocarbon chain is, the more likely the tails interact with one another and the less fluid the membrane is. As for unsaturation, the double bond in the chain creates a small kink in the hydrocarbon tail, which makes it more difficult for the tail to pack against one another. Therefore, the larger the proportion of unsaturated hydrocarbon tails is, the more fluid the membrane is. In most eukaryotic cells, the higher concentrations of cholesterol in the plasma membrane help to prevent the hydrocarbon chains from coming together and crystallizing and therefore keep the bilayer fluidity. Membrane fluidity needs to be maintained within certain limits for a cell to function normally. For example, when the temperature drops, the cells respond metabolically: enzymes remodel membranes to make the cell more cold-resistant by desaturating single bonds in fatty acyl chains to form double bonds. In addition, the cell changes the types of phospholipids being synthesized in favor of ones containing more unsaturated fatty acids.

Membrane protein mobility was first demonstrated using a cell fusion technique (Figure 2-11). In such studies, mouse and human cells were fused, and the locations of specific proteins of the plasma membrane were tracked using various fluorescent dye-labeled antibodies against certain mouse or human proteins. At the onset of fusion, the plasma membrane appeared half human and half mouse. As the time of fusion progresses, the membrane proteins from two species were seen to gradually move within the membrane into the opposite hemisphere and finally distribute uniformly in the entire hybrid cell membrane.

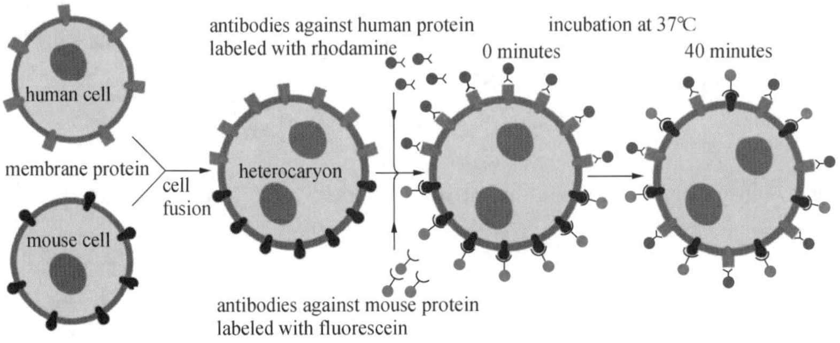

Figure 2-11 Membrane protein mobility revealed by cell fusion technique (From: Alberts B, et al. Molecular Biology of the Cell. 2008)

These early studies seem to show that integral membrane proteins were capable of moving freely within the membrane. However, other techniques such as fluorescence recovery after photobleaching (FRAP) show that there are restrictions on protein mobility (Figure 2-12). In FRAP assays, cells with a particular membrane protein labeled with a fluorescent probe are placed under the microscope, and irradiated by a laser beam that bleaches the fluorescent molecules in its path, leaving a circular spot on the cell surface lacking fluorescence. If the labeled proteins are mobile, the random movements of these molecules should produce a gradual reappearance of fluorescence in the irradiated blank circle. The rate of the fluorescence recovery indicates the diffusion rate of the mobile molecules. The extent of fluorescence recovery represents the percentage of the labeled molecules that are free to diffuse. The result of FRAP studies indicated that a significant portion of membrane proteins (30%~70%) was not free to diffuse back into the irradiated circle, indicating that the movement of certain proteins on cell membrane is restricted. In epithelial tissue, such as those that line the gut or the tubules of the kidney, certain plasma membrane enzymes and transport proteins are confined to the apical surface of the cells, whereas others are confined to the basal and lateral surfaces. Such restricted movement of proteins is controlled by a specific type of intercellular junction such as tight junction. The membrane proteins that form these intercellular junctions cannot be allowed to diffuse laterally in the interacting membranes.

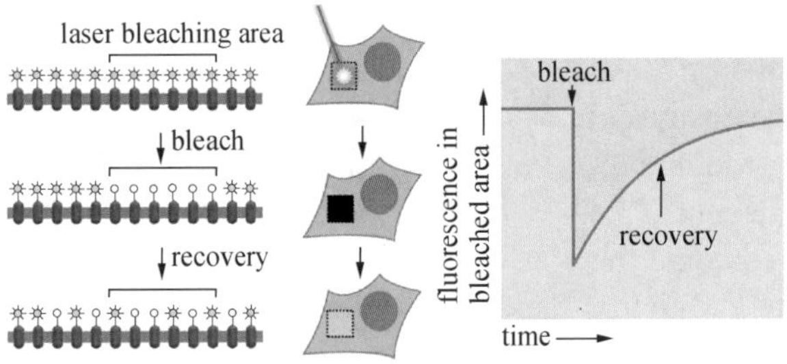

Figure 2-12 Membrane protein mobility revealed by FRAP technique

(4) Membrane asymmetry All cellular membranes have an internal face designated as the cytosolic face and an external face designated as the exoplasmic face. Therefore cell membranes are asymmetrical: the two halves of the bilayer often contain different types of phospholipids and glycolipids. In addition, the proteins embedded in the bilayer have a specific orientation.

The asymmetry of the lipid bilayer was found by taking advantage of lipid-digesting enzymes that cannot penetrate the plasma membrane, and are only able to digest lipids that reside in the external monolayer of the bilayer. When intact human red blood cells are treated with these enzymes, about 80% of the membrane's phosphatidylcholine (PC) is hydrolyzed, whereas only approximately 20% of the phosphatidylethanolamine (PE) and less than 10% of its phosphatidylserine (PS) are attacked. These data suggest that the external monolayer has a relatively higher concentration of PC and sphingomyelin (SM) and inner monolayer has higher concentration of PE and PS. Glycolipids are exclusively located in the outer surface monolayer. Lipid asymmetry is functionally important, especially in converting extracellular signals into intracellular ones. The phospholipid asymmetry of plasma membrane can be used to distinguish between live and dead cells. When animal cells undergo apoptosis, PS normally confined to the cytosolic monolayer rapidly translocates to the extracellular monolayer and signals neighboring cells such as macrophages to clean up the dead cells.

2.2 Transmembrane transport

The plasma membrane is a selectively permeable barrier between the cell and the extracellular environment to maintain a constant internal environment. The selective permeability ensures that essential molecules such as ions, glucoses, amino acids, and lipids readily enter the cell, metabolic intermediates remain in the cell, and waste components leave the cell. Few molecules can diffuse across a membrane, however, most molecules and all ions require a process, termed transmembrane transport, to across cellular membranes. This process is mediated by selective **membrane transport proteins** embedded in the phospholipid bilayer.

2.2.1 Overview of transmembrane transport

There are different mechanisms of transmembrane transport for various molecules and ions. To define a type of transmembrane transport, some characters below should be considered: ①Whether need the aid of specific transport proteins? ②If the direction of movement is against the concentration gradient of molecules or ions? ③Whether require the energy supplied by ATP hydrolysis or other types of energy, such as light? ④If it is driven by a movement of a cotransported ion down its gradient?

According to all of the above, the transmembrane transports are divided into two major types: **passive transport** and **active transport** (Figure 2-13). More detailed classifications and comparisons are listed in table 2-1. **Cotransport** is a special type of active transport, within which the movement of

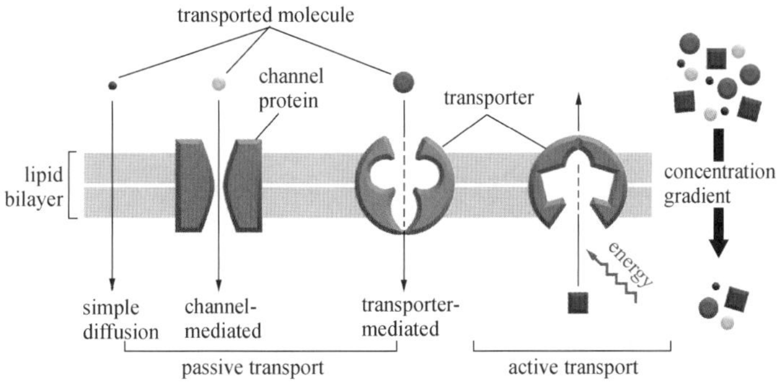

Figure 2-13 Comparison between passive and active transport
(From: Alberts B, et al. Molecular Biology of the Cell. 2008)

one type of ion or molecule against its concentration gradient is coupled with the movement of one or more different ions down its concentration gradient.

Table 2-1 Mechanisms for transporting ions and molecules across cell membranes

property	passive diffusion	facilitated diffusion	active transport	cotransport
requiring specific transport protein	no	yes	yes	yes
solute transported against its gradient	no	no	yes	yes
coupled to ATP hydrolysis	no	no	yes	no
driven by movement of an ion down its gradient	no	no	no	yes
examples	O_2, CO_2, steroid hormones, many drugs	glucose and amino acids (uniporters), ions and water (channels)	ions, small hydrophilic molecules, lipids (ATP-powered pumps)	glucose and amino acids (symporters), various ions and sucrose (antiporters)

There are two main classes of membrane transport proteins, transporters and channels (Figure 2-13). **Transporters** (also called carriers or permeases) bind the specific solute and undergo a series of conformational changes to transfer the bound solute across the membrane. Three types of transporters have been identified (Figure 2-14). **Uniporters** transport a single type of molecule (e. g. glucose and amino acids) down its concentration gradient via **facilitated diffusion**. In contrast, **antiporters** and **symporters** couple the movement of one type of ion or molecule against its concentration gradient with the movement of one or more different ions down its concentration gradient. These proteins are often called **cotransporters** and play roles in active transport.

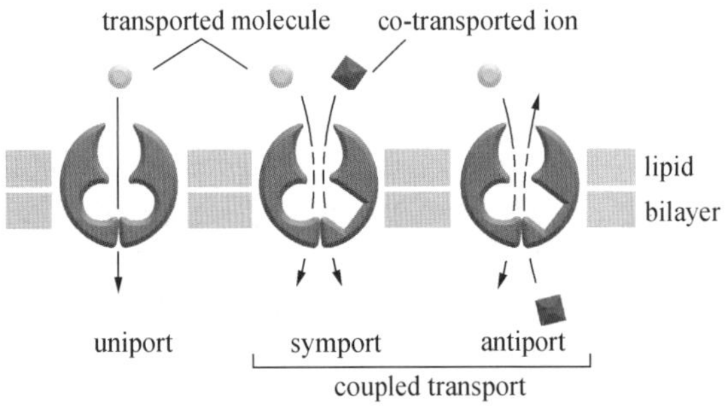

Figure 2-14 Three types of transporters (From: Alberts B, et al. Molecular Biology of the Cell. 2008)

Channels usually interact much more weakly with the solute and form aqueous pores that extend across the lipid bilayer; when pores open, these pores allow specific solutes (usually inorganic ions of appropriate size and charge) to pass through. Transport through channels occurs at a much faster rate than that mediated by transporters. Although water can diffuse across lipid bilayer, all cells contain specific channel proteins (called water channels or aquaporins) that greatly increase membrane permeability to water.

2.2.2 Passive transport

Passive transport includes both simple diffusion and facilitated diffusion (Figure 2-13). In this process, molecules move down their electrochemical concentration gradient and no metabolic energy is required. In the case of transport of a single uncharged molecule, the difference in the concentration on the two sides of the membrane drives passive transport and determines its direction. If the solute carries a net charge, both its concentration gradient and the electrical potential difference across the membrane influence its transport. The concentration gradient and the electrical gradient combine to form a net driving force, the **electrochemical gradient**. A few molecules, such as O_2, CO_2, and small, uncharged polar molecules, such as urea and ethanol, can diffuse across cellular membranes without the aid of transport proteins, a process called **simple diffusion**. In contrast, many polar molecules, such as, glucose, amino acids, ions and water, are transported across membrane by a protein-mediated movement, namely **facilitated diffusion**. The rate of facilitated diffusion within which the transported molecules never enter the hydrophobic core of the phospholipid bilayer is far higher than simple diffusion.

Transport action of facilitated diffusion can occur via a limited number of uniporter molecules. The best-understood uniporter, GLUT1 (glucose transporter) alternates between two conformational states: state A with the glucose-binding site facing the outside of the membrane and state B with the glucose-binding site facing inside (Figure 2-15). Through sequential conformational changes, the uniporter GLUT1 facilitates the unidirectional transport of glucose down its concentration gradient. Such transport is specific and there is a maximum transport rate that is achieved when the concentration gradient across the membrane is very large and each uniporter is working at its maximal rate.

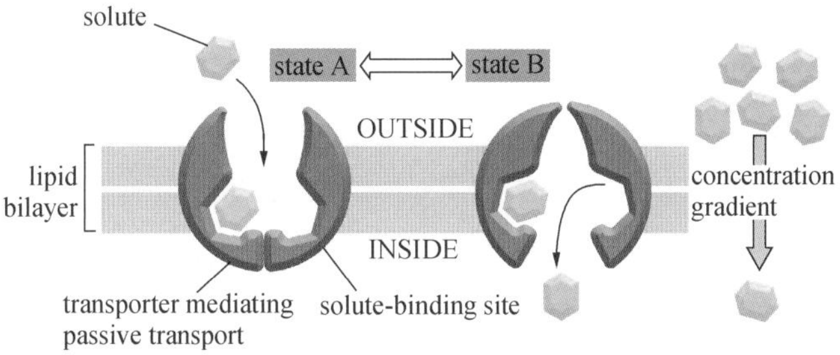

Figure 2-15 Uniporter mediated passive movement of a glucose solute (From: Alberts B, et al. Molecular Biology of the Cell. 2008)

Channel protein-assisted transport is another major type facilitated diffusion and transports water or ions and hydrophilic small molecules. Channel proteins form a hydrophilic passageway across the membrane through which multiple water molecules or inorganic ions, such as K^+, Na^+, Ca^{2+} and Cl^-, diffuse rapidly down their electrochemical gradients across the membrane at a very rapid rate (up to 10^6 ions/s). Above 100 kinds of ion channels have been discovered on various cell plasma membrane and are essential for many cell functions. In particular neurons and muscles have used many different ion channels to receive, conduct and transmit signals.

Ion channels have two distinct characters compared to simple aqueous pores. First they show specific ion selectivity. K^+ channel is probably the most common channel found in the plasma membrane of animal cells. They conduct K^+ 10,000-fold better than Na^+, yet the two ions are both featureless spheres and have similar diameters (0.133 nm and 0.095 nm, respectively). A single amino acid substitution in the pore of K^+ channel can result in a loss of ion selectivity and cell death.

The second distinct character of ion channels is that they are so gated that allows them to open briefly and then close again. Some ion channels are open much of the time and are referred to as **nongated** channels. Most ion channels open only in response to specific chemical or electrical signals which lead to the conformation changes of the channel protein and are referred to as **gated** channels, include **voltage-gated** channel (signal is a change in the voltage across the membrane), **mechanically-gated** channel (signal is a mechanical stress) and **ligand-gated** channel (Figure 2-16). The ligand can be an extracellular mediator-specifically a neurotransmitter (transmitter-gated channels) or an intracellular mediator such as an ion (ion-gated channels) or a nucleotide (nucleotide-gated channels). The ion channels also have an automatic inactivating mechanism. When cells receive continuous stimuli, the channel recloses rapidly and remains in this inactivated state until the membrane potential or chemical and electrical signals have returned to their initial value.

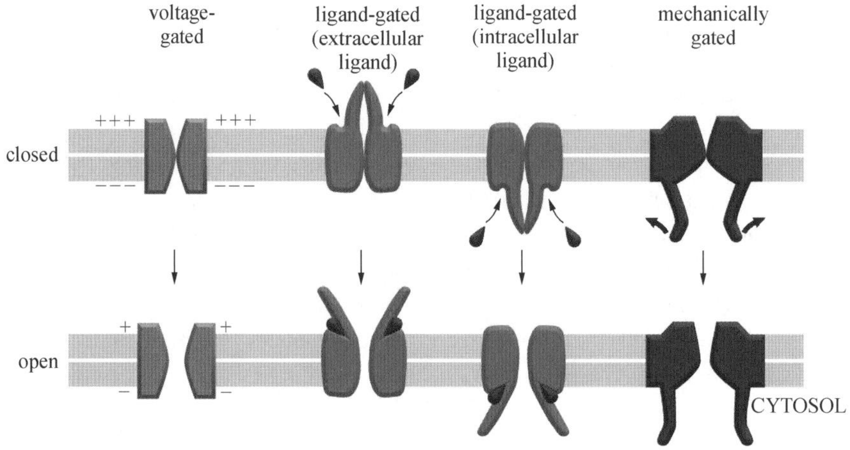

Figure 2-16 The gating of ion channels (From: Alberts B, et al. Molecular Biology of the Cell. 2008)

Prokaryotic and eukaryotic cells have water channels, or aquaporins. Aquaporins are especially abundant in cells that must transport water at particularly high rates, such as the epithelial cells of the kidney. To avoid disrupting ion gradients across membranes, they have to allow the rapid passage of water molecules while completely blocking the passage of ions. The crystal structure of an aquaporin reveals the water channels have a narrow pore that allows water molecules to traverse the membrane in a single file, following the path of carbonyl oxygens that line one side of the pore. Hydrophobic amino acids line the other side of the pore. The pore is too narrow for any hydrated ion to enter, and the energy cost of dehydrating an ion would be enormous because the hydrophobic wall of the pore cannot interact with a dehydrated ion to compensate for the loss of water.

2.2.3 Active transport

Like the membrane proteins mediatedly facilitated diffusion, active transports are mediated by specific membrane proteins. The difference is that active transports move ions or molecules against their concentration gradient and need the energy supplied by ATP hydrolysis, or by light energy, or by transporter coupled the uphill transport of one solute across the membrane to the downhill transport of another(Figure 2-17).

(1) ATP-driven pumps **ATP-driven pumps** are **ATPases** that use the energy of ATP hydrolysis to move ions or small molecules across a membrane against a chemical concentration gradient or electric potential gradient or both. All ATP-powered pumps are transmembrane proteins with one or more binding sites for ATP located on the cytosolic face of the membrane.

The general structures of the three classes of ATP-powered pumps are depicted in figure 2-18. Note that the types P-, F-, and V-pumps transport ions only, whereas the ABC transporters transport small molecules primarily.

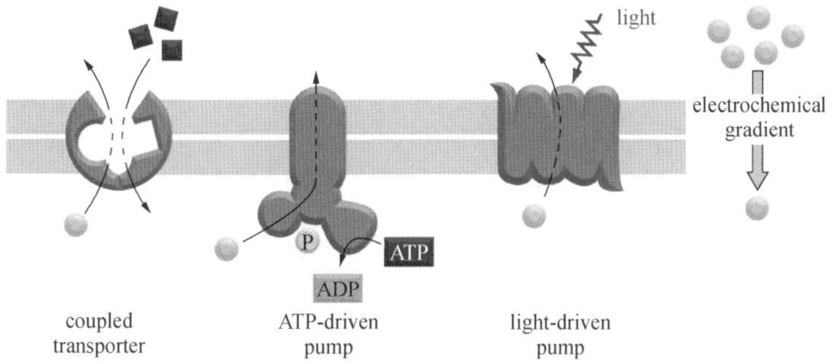

Figure 2-17 Three ways driving active transport (From: Alberts B, et al. Molecular Biology of the Cell. 2008)

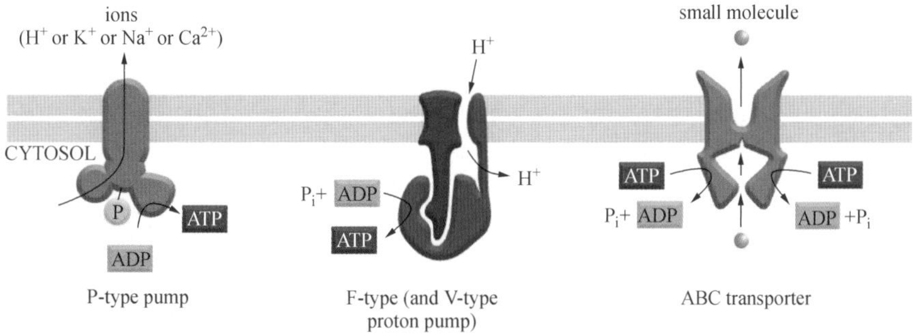

Figure 2-18 Three types of ATP-driven pumps (From: Alberts B, et al. Molecular Biology of the Cell. 2008)

1) **P-type ion pump** possesses two identical catalytic α subunits that contain an ATP-binding site, and most also contain two smaller β subunits that usually have regulatory functions. During the transport process, at least one of the α subunits is phosphorylated (hence the name P-type), and the transported ions are thought to move through the phosphorylated subunit.

This class includes the Na^+-K^+ ATPase in the plasma membrane. The pump operates as an ATP-driven antiporter, actively pumping each time 3 Na^+ out of the cell against its steep electrochemical gradient and pumping 2 K^+ in. Since three positively charged ions are pumped out of the cell and only two are pumped in, the Na^+-K^+ ATPase is electrogenic and maintains the low cytosolic Na^+ and high cytosolic K^+ concentrations in animal cells. It drives a net current across the membrane tending to create an electrical potential with the inside of the cell negative relative to the outside. The Na^+ gradient established across the membrane drives the transport of most nutrients into cells and is also crucial in regulating cytosolic pH and osmolarity.

Certain Ca^{2+} ATPases are P-type ion pumps and pump Ca^{2+} ions out of the cytosol into the external medium; others pump Ca^{2+} from the cytosol into the endoplasmic reticulum or into the sarcoplasmic reticulum, which is found in muscle cells. Another member of the P class, found in acid-secreting cells of the mammalian stomach, transports protons (H^+ ions) out of the cell and K^+ ions into the cell. The H^+ pump that maintains the membrane electric potential in plant, fungal, and bacterial cells also belongs to this class.

2) **V-type and F-type ion pumps** are similar in structure which is more complicated than that of P-type pumps. V-type and F-type pumps contain several different transmembrane and cytosolic subunits. All known V-type and F-type pumps transport only protons, in a process that does not involve a phosphoprotein intermediate. V-type pumps generally function to maintain the low pH of plant vacuoles and of lysosomes and other acidic vesicles in animal

cells by pumping protons from the cytosolic to the exoplasmic face of the membrane against a proton electrochemical gradient. F-type pumps are found in bacterial plasma membranes and in the inner membrane of mitochondria and chloroplasts. In contrast to V-type pumps, F-type pumps generally function to power the synthesis of ATP from ADP and Pi by movement of protons from the exoplasmic to the cytosolic face of the membrane down the proton electrochemical gradient.

3) **ABC (ATP-binding cassette) transporters** referred to as the **ABC superfamily**, include several hundred different transport proteins found in organisms ranging from bacteria to humans. Each ABC protein is specific for a single substrate or a group of related substrates, which may be ions, sugars, amino acids, phospholipids, peptides, polysaccharides, or even proteins. All ABC transport proteins share a structural organization consisting of four "core" domains: two transmembrane domains, forming the passageway through which transported molecules cross the membrane, and two cytosolic ATP-binding domains. ATP binding leads to dimerization of the two ATP-binding domains, and ATP hydrolysis leads to their dissociation. Such structural changes in the cytosolic domains are thought to be transmitted to the transmembrane segments, driving conformational changes that alternately expose substrate-binding sites on one side of the membrane and then on the other. As a consequence, small molecules are transported across the bilayer.

The bacterial ABC transporters are used for both import and export, whereas those identified in eukaryotes seem mostly specialized for export. ABC transporters are of great clinical importance. The first eukaryotic ABC transporters were discovered because of their ability to pump hydrophobic drugs out of the cytosol. One of these transporters is the multidrug resistance (MDR) protein, the overexpression of which in human cancer cells can make the cells simultaneously resistant to a variety of chemically unrelated cytotoxic drugs that are widely used in cancer chemotherapy. A similar phenomenon occurs with the antimalarial drug *chloroquine*. The protist *Plasmodium falciparum* develops a drug resistance to *chloroquine* by amplifying a gene encoding an ABC transporter that pumps out the *chloroquine*.

(2) Cotransporters Cotransporters mediate coupled reactions in which an energetically unfavorable reaction (i. e., uphill movement of molecules) is coupled to an energetically favorable reaction (see "symporter" and "antiporter" in figure 2-14). Cotransporters use the energy stored in an electrochemical gradient, which differs from ATP pumps that use energy directly from hydrolysis of ATP. The establishment of electrochemical gradients by Na^+-K^+ pumps or H^+ pumps consumes ATP, so, the energy supplied by ATP is indirectly utilized by cotransporter to operate the transport. This process sometimes is referred to as secondary active transport.

The cotransporters fall in two types, antiporters and symporters. **Symporters**, driven by ions, transfer molecules in or out of the cell or organelle across the membrane in the same direction. There are concurrent two binding sites on the surface of symporters, one for the ions and another for the molecules. For example, there are two types of glucose and amino acid carriers that enable gut epithelial cells to transfer glucose and amino acid across the gut lining: one is Na^+-driven symporters that mediate the glucose and amino acid to enter the cells against their concentration gradient; another is a passive transport carrier protein mediating the glucose and amino acid out of the cells down the concentration gradient (Figure 2-19). In epithelial cells which absorb nutrients from the gut, transporters are nonuniformly distributed in the plasma membrane and thereby facilitate the transcellular transport of the absorbed nutrients across the epithelial cell layer into blood. Na^+-linked symporters located in the absorptive domain of the plasma membrane actively transport nutrients into the cell, building up concentration gradients for these solutes across the cell membrane. Na^+ independent transport proteins in the basal and lateral domain allow the nutrients to leave the cell passively down their concentration gradients.

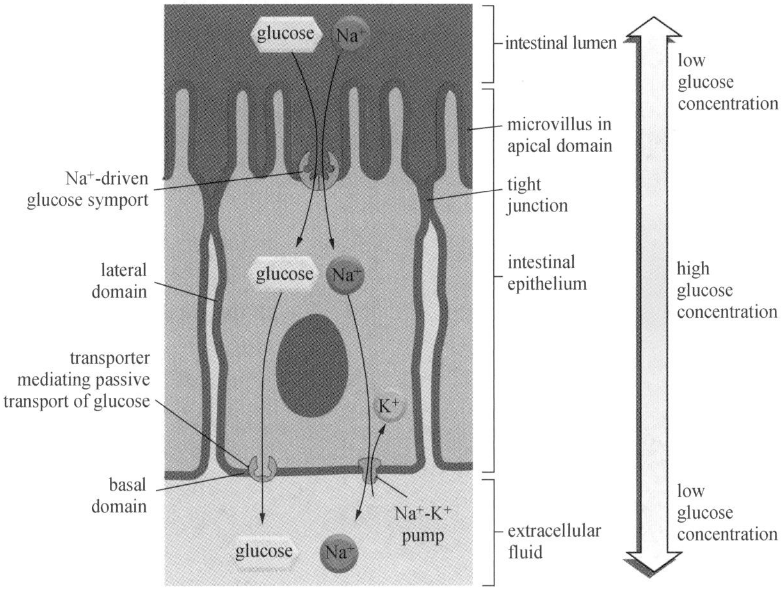

Figure 2-19 Two types of glucose and amino acid carriers enable gut epithelial cells to transfer glucose and amino acid across the gut lining (From: Alberts B, et al. Molecular Biology of the Cell. 2008)

Proteins within different intracellular compartments require an optimal pH to function correctly. Lysosomal enzymes in lysosomes function best at the low pH (5), whereas cytosolic enzymes in the cytosol function best at neutral pH (7.2). Most cells have one or more types of Na^+-driven antiporters in their plasma membrane that help to maintain the cytosolic pH at about 7.2. These transporters use the energy stored in the Na^+ gradient to pump out excess H^+. One of the antiporters that uses a Na^+-H^+ exchanger, couples an influx of Na^+ to an efflux of H^+. Another is a Na^+-driven Cl^--HCO_3^- exchanger that couples an influx of Na^+ and HCO_3^- to an efflux of Cl^- and H^+.

There are similarities and differences in carrier-mediated solute symport in animal and plant cells. An electrochemical gradient of Na^+, generated by Na^+-K^+ pump, is often used to drive the active transport of solutes across the animal cell plasma membrane, whereas in plant, bacteria and fungi cells, an electrochemical gradient of H^+, usually set up by H^+ ATPase, is often used for this purpose. The electrochemical H^+ gradient drives the active transport of many sugars and amino acids across the plasma membrane and into bacterial cells. One well-studied H^+-driven symporter is lactose permease, which transports lactose across the plasma membrane of *E. coli*.

2.3 Cell adhesion molecules and cell junction

In multicellular organisms, it is often necessary for adjacent cells within a tissue to be held together. Cells in tissues can adhere directly to one another (cell-cell adhesion) through specialized integral membrane proteins called **cell-adhesion molecules** (**CAM**) that often cluster into specialized cell junctions. Cells can also adhere indirectly (cell-matrix adhesion) through the binding of adhesion receptors in the plasma membrane to components of the surrounding **extracellular matrix** (ECM) that cells secreted into the spaces between them. Such connections create pathways for cell communication, allowing cells to exchange the signals that coordinate their functions. Attachments to other cells and to extracellular matrix control the orientation of the internal structure of each cell.

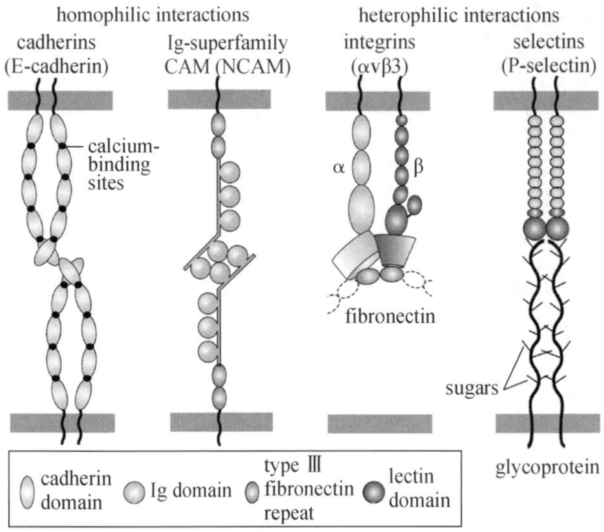

Figure 2-20 Major families of CAM (From: Lodish H, et al. Molecular Cell Biology. 2008)

2.3.1 Cell adhesion molecules

CAM can be classified into 4 main groups: **the cadherins**, **the integrins**, **the selectins** and **the immunoglobulin** (**Ig**) **superfamily** (Figure 2-20 and Table 2-2). A CAM on one cell can directly bind to the same kind of CAM on an adjacent cell (homophilic binding) or to a different class of CAM (heterophilic binding). CAM can be broadly distributed along the plasma membranes that contact other cells by forming Cell-cell adhesions. Cell-cell adhesions can be tight and long-lasting (such as cell junctions) or relatively weak and transient. The associations between nerve cells in the spinal cord or the metabolic cells in the liver exhibit tight adhesions. In contrast, immune cells in the blood show only weak, short-lasting interactions, allowing them to roll along and pass through a blood vessel wall to fight an infection within a tissue.

Table 2-2 CAM

family	ligands recognized	stable cell junction
cadherins	homophilic interactions	adherens junctions and desmosomes
integrins	extracellular matrix	focal adhesions and hemidesmosomes
	members of Ig superfamily	no
selectins	carbohydrates	no
Ig superfamily	integrins	no
	homophilic interactions	no

(1) Cadherins Cadherins are key molecules in cell-cell adhesion and cell signaling, and play a critical role during tissue differentiation. Cadherins take their name from their dependence on Ca^{2+} ions (calcium-dependent adhesion). In structure, they share cadherin repeats in N-terminus, which are the extracellular Ca^{2+}-binding domains. The C-terminal cytosolic domain of cadherins is linked to the actin cytoskeleton by many cytosolic adapter proteins. These linkages are essential for strong adhesion. For example, disruption of the interactions between cadherins and two common adapter proteins (α- or β-catenin) that link these cadherins to actin filaments dramatically reduces cadherin-mediated cell-cell

adhesion. Three classical cadherin family members are closely related in sequence and are named according to the main tissues in which they were found: E-cadherin is present on many types of epithelial cells; N-cadherin on nerve, muscle, and lens cells; P-cadherin on cells in the placenta and epidermis. There are also a large number of nonclassical cadherins with distantly related sequences.

Cadherins generally bind to one another homophilically: the head of one cadherin molecule binds to the head of a similar cadherin on an opposite cell. This selection enables mixed populations of different types of cells to sort out from one another according to the specific cadherins they express, which helps to control cell rearrangement during development. Epithelial-mesenchymal transition (EMT) and mesenchymal-epithelial transition (MET) are typical examples concerning the function of E-cadherin. The assembly of cells into an epithelium is a reversible process. By switching on the expression of cadherins, the dispersed and unattached mesenchymal cells can come together to form an epithelium. Conversely, epithelial cells can change their character, disassemble, and migrate away from their parent epithelium as separate individuals at least in part by inhibiting the expression of the cadherins. Such processes play important roles in normal embryonic development, as well as pathological events, such as carcinogenesis, during adult life.

(2) Integrins Integrins are membrane receptors that mediate mainly the attachment between cells and ECM. Via interactions with the ECM components, such as fibronectin, vitronectin, collagen, and laminin, integrins pass information about the chemical composition of ECM into the cell. However, integrins, particularly expressed by certain blood cells, work with other adhesion proteins such as cadherins, Ig superfamily, selectins and syndecans, to participate in heterophilic cell-cell interactions. Ligands binding to integrins also require the simultaneous binding of divalent cations (positively charged ions). Integrins typically exhibit low affinities for their ligands. However, the multiple weak interactions generated by the binding of hundreds or thousands of integrin molecules to their ligands on cells or in ECM, allow a cell to remain firmly anchored to its ligand-expressing target. Moreover, the weakness of individual integrin-mediated interactions facilitates cell migration.

Integrins are heterodimers containing two distinct chains, α and β subunits. The α and β subunits each penetrate the plasma membrane and possess small cytoplasmic domains. Like other cell-surface adhesive molecules, the cytosolic region of integrins interacts with adapter proteins that in turn bind to the cytoskeleton and intracellular signaling molecules. In addition to their adhesion function, integrins can mediate signaling involved in processes as diverse as cell survival, cell proliferation, and programmed cell death.

(3) Selectins Selectins are expressed on white blood cells, blood platelets, and endothelial cells. They bind heterophilically to carbohydrate groups on cell surfaces, mediate transient cell-cell adhesions in the bloodstream and control the binding of white blood cells to the endothelial cells lining blood vessels, thereby enabling the blood cells to migrate out of the bloodstream into sites of inflammation.

Selectins contain a conserved Ca^{2+}-dependent lectin domain, which is located at the distal end of the extracellular region of the molecule and recognizes oligosaccharides in glycoproteins or glycolipids. There are at least three types: L-selectin on leukocytes (white blood cells), P-selectin on platelets and endothelial cells that can be locally activated by inflammatory responses, and E-selectin on activated endothelial cells. Selectins do not act alone, but collaborate with integrins, which strengthen the binding of the blood cells to the endothelium.

(4) Ig superfamily (IgSF) Numerous transmembrane proteins characterized by the presence of multiple immunoglobulin domains in their extracellular regions constitute the IgSF of CAM. Such proteins mediate Ca^{2+}-independent cell-cell adhesion both homophilically and heterophilically. intercellular cell adhesion molecules (ICAM) or vascular cell adhesion molecules (VCAM) on endothelial cells are members of IgSF and both mediate heterophilic binding to integrins. Another IgSF member neural cell adhesion molecule (NCAM) expressed on neurons mediates homophilic binding. Most IgSF mediate the specific interactions of lymphocytes with cells required for an immune response (e.g., macrophages, other lymphocytes, and target cells). However, some IgSF members, such as VCAM and NCAM mediate adhesion between nonimmune cells.

2.3.2 Cell junctions

The physical attachment between cell and cell or between cell and matrix are diverse in structure. There are mainly three types of junctions: **tight junctions**, **anchoring junctions** (including **adherens junction**, **focal adhesion**, **desmosome**, **hemidesmosome**) and **gap junctions** (**plasmodesmata**, a specified gap junction in plant cells). These cell junctions not only transmit physical forces, but also define the diverse functions of different tissues (Figure 2-21).

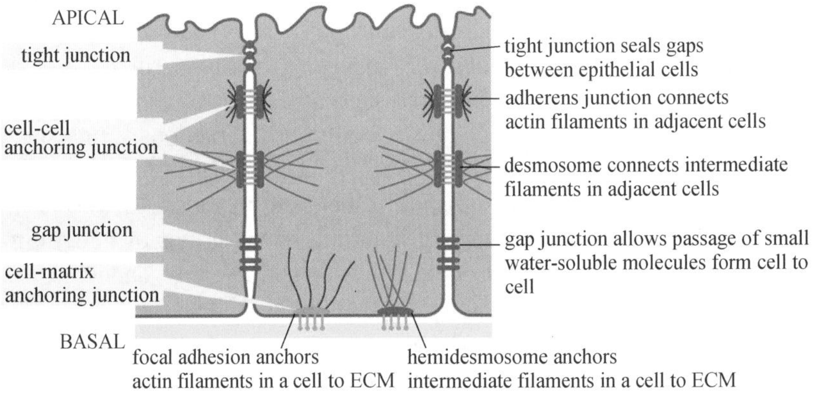

Figure 2-21 Typical cell junctions found in vertebrate epithelial tissues
(From: Alberts B, et al. Molecular Biology of the Cell. 2008)

(1) Tight junction Tight junctions are the closest known contacts between adjacent cells and are particularly common in epithelial cells such as those lining the small intestine. They are formed by interactions between strands of transmembrane proteins (occludin and claudins) on adjacent cells that continue around the entire circumference of the cell. The adjacent cells are pressed together so tight that no intercellular space left (Figure 2-22).

Tight junctions play at least three main roles: ①Form seals that prevent the free passage of molecules between the cells of epithelia; ②Prevent leakage of molecules across the epithelium through the gaps between cells; ③ Separate the apical and basolateral domains of plasma membrane by preventing the free diffusion of lipids and proteins between them, and help establishing and maintaining cell polarity.

(2) Anchoring junction Anchoring junction anchors cells together allowing the tissue to be stretched without tearing. There are four main types: adherens junction, desmosome junctions binding one epithelial cell to another, while focal adhesions and hemidesmosomes binding cells to the ECM.

1) **Cell to cell anchoring junction**: The adherens junctions and desmosome are cell-cell

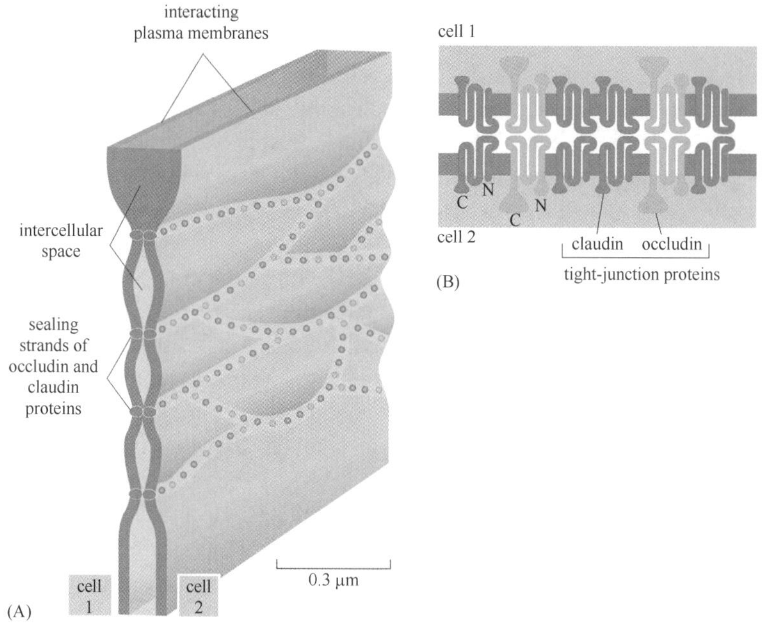

Figure 2-22 A structure model of tight junction (From: Alberts B, et al. Molecular Biology of the Cell. 2008)

junctions in which cadherins are linked to actin bundles or intermediate filaments, respectively (Figure 2-23). At an adherens junction, cadherin molecules bind to actin filaments via several linker proteins (catenin α, β) and form a continuous adhesion belt between each of the interacting cells. This belt is close beneath the apical face of the epithelium. The potentially contractile nature of actin network provides the motile force for folding epithelial cell sheets into tubes, vesicles and other related structures. Such epithelial movement is very important in embryonic development, such as the formation of neural tube structures which later develop into the central nervous system.

Desmosome junctions confer great tensile strength to the epithelia. Desmosomes contain two specialized cadherin proteins, desmoglein and desmocollin, whose cytosolic domains are distinct from those in the classical cadherins. The cytosolic domains of desmosomal cadherins interact with plakoglobin (similar in structure to β-catenin) and the plakophilins. These adapter proteins, which form the thick cytoplasmic plaques characteristic of desmosomes, in turn, interact with intermediate filaments.

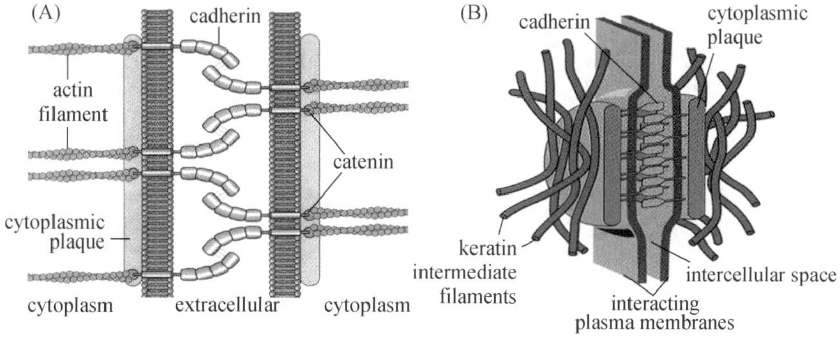

Figure 2-23 Structure components of cell to cell anchoring junctions. (A) Adherens junctions; (B) Desmosomes.

2) **Cell-matrix anchoring junction**: There are two types of cell-matrix anchoring junctions, focal adhesion and hemidesmosomes. Hemidesmosome junctions anchor the keratin intermediate filaments in an epithelial cell to the extracellular basal lamina through intergrin proteins in the basal plasma membrane of the epithelial cells. Such structures, just like half a desmosome, can attach and anchor the epithelial cells to the underlying tissue. In focal adhesions, integrins connect the extracellular matrix and cytoskeletal actin filaments mainly in nonepithelial cells and transduce adhesion-dependent signals for cell growth and cell motility.

(3) Gap junction Gap junctions bridge between adjacent cells to created a direct connection between the cytoplasm of one with the other.

1) **Gap junction in animal tissues**: In animal tissues, gap junctions are constructed by four-pass transmembrane proteins called connexins. Six connexins assemble to form a connexon with an open hydrophilic pore in its center. A connexon in one cell then aligns with the connexon in an adjacent cell forming a channel between the two cell interiors that connects the two cytoplasms (Figure 2-24A). Most cells in animal tissues-including epithelial cells, endothelial cells, smooth muscle and heart muscle cells communicate by gap junctions. Gap junctions allow ions and small molecules (molecular mass less than 1,000 daltons) including signal molecules cAMP and Ca^{2+} to pass freely between adjacent cells but prevent the passage of macromolecules, such as proteins, nucleic acids and polysaccharides. Gap junctions create metabolic and electrical coupling between the cells. For example, between heart muscle cells, gap junctions provide coupling that allows electrical signal of excitation to spread through the tissue, triggering coordinated contraction of the heart cells.

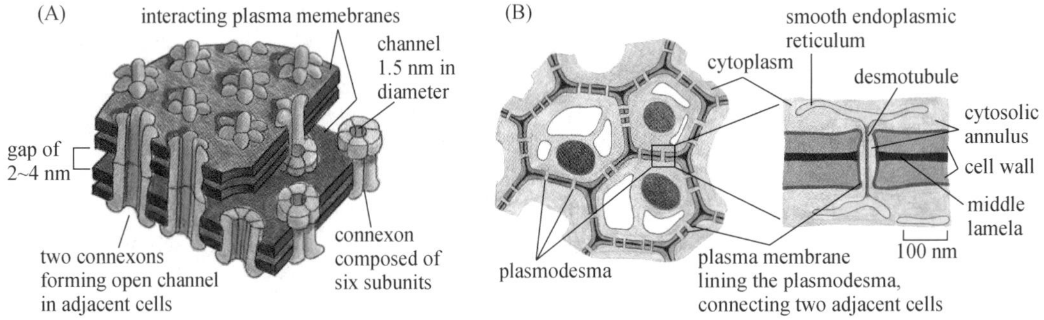

Figure 2-24 Structure models of gap junctions. (A) Gap junctions in animal cells; (B) Plasmodesma in plant cells.

2) **Plant cell adhesion and plasmodesmata**: Plant cells are imprisoned within rigid cell walls composed of an extracellular matrix rich in cellulose and other polysaccharides. Adhesion between plant cells is mediated by their cell walls. A specialized pectin-rich region of the cell wall called the **middle lamella** acts as mucilage to hold adjacent cells together. Plant cells have only one class of intercellular junctions, plasmodesmata (Figure 2-24B), which function analogously to animal cell gap junctions.

At each plasmodesma, the plasma membrane of one cell is continuous with that of the adjacent cell creating an open channel between the two cytoplasms. An extension of the smooth endoplasmic reticulum passed through the center of channel is a narrower cylindrical structure, the **desmotubule**, leaving a ring of surrounding cytoplasm through which ions and small molecules (molecular mass less than 800 daltons) can pass from cell to cell. In addition, plasmodesma can expand in response to appropriate signals, permitting the regulated passage of macromolecules, such as mRNA and virus RNA, between neighboring cells.

2.4 Extracellular matrix and cell wall

The ECM is a complex meshwork of proteins and polysaccharides secreted by cells into the spaces between them. The ECM plays important role in cell-cell signaling, wound repair, cell adhesion and tissue function.

2.4.1 Composition of the ECM

In plants, the ECM (cell wall) is primarily composed of cellulose surround each cell and is more important in cell organization which gives strength to the plants. In arthropods and fungi, the ECM is largely composed of chitin. In animal cells, the relative volumes of cells versus matrix vary greatly among different animal tissues and organs. In connective tissues, such as bone, cartilage or tendon, the extracellular matrix is plentiful and cells are bound to the matrix and sparsely distributed within it. However, many organs are composed of very densely packed cells with relatively little matrix. All epithelial cells as well as some other types (e.g., smooth muscle cells) are attached to a thin basal lamina (also known as the basement membrane, a specialized ECM structure).

In vertebrates, the ECM is made of three abundant components: **collagens**, structure proteins that often form fibers and provide mechanical strength and resilience; **proteoglycans**, a unique type of glycoproteins that cushion cells and bind a wide variety of extracellular molecules; and soluble **multiadhesive matrix proteins** (e.g., fibronectin and laminin), which bind to and cross-link cell-surface adhesion receptors and other ECM components.

(1) Structure proteins Collagens and elastins are the main structural proteins in ECM and serve as a structural scaffold. Collagens are the most abundant proteins in mammals and make up about 25% to 35% of the whole-body protein content. They have a wide variety of types. Bone utilizes type Ⅰ collagen (also the major collagen in skin), cartilage utilizes type Ⅱ, while basement membrane utilizes type Ⅳ. Collagens Ⅰ and Ⅱ form long fibrillar arrays, while type Ⅳ forms a mesh-type structure, reminiscent of a chain-link fence.

Collagens are trimeric proteins made from three polypeptides called collagen α chains, which are wrapped around one another to form a triple-stranded helical rod at approximately 300 nm long and 1.5 nm in diameter (Figure 2-25). The collagen is abundant in three amino acids: glycine, proline, and a modified form of proline called hydroxyproline, which make up the characteristic repeating motif Gly-X-Y. X and Y can be any amino acid but are often proline and hydroxyproline and less often lysine and hydroxylysine. Collagen molecules assemble into higher-order polymers called collagen fibrils, which are thin structures (10 ~ 300 nm in diameter) with many hundreds of micrometers long. Collagen fibrils have characteristic cross-striations every 67 nm, reflecting the regularly staggered packing of the individual collagen molecules in the fibril. After the fibrils have formed in the extracellular space, they are greatly strengthened by the formation of covalent cross-links between lysine residues of the constituent collagen molecules. Collagen fibrils often aggregated into collagen fibers several micrometers in diameter which are visible as cablelike bundles under the light microscope.

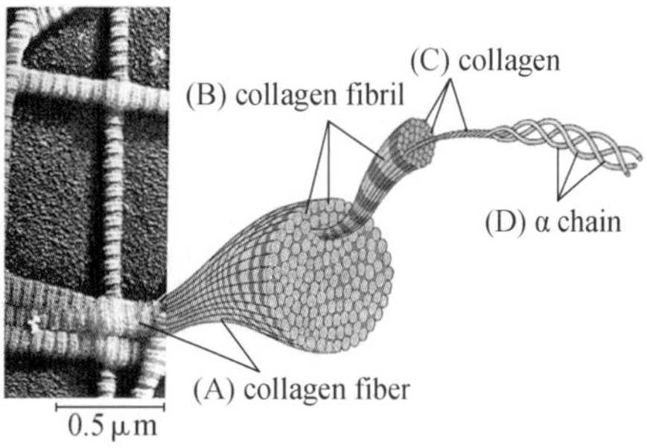

Figure 2-25 Structure of collagens

Elastins are ECM structure proteins and present in connective tissues. They are rich in hydrophobic amino acids such as glycine and proline, which form mobile hydrophobic regions bounded by crosslinks between lysine residues. These proteins are elastic and provide flexibility to many tissues such as skin, arteries, and lungs.

(2) **Proteoglycans** Proteoglycans are also glycoproteins but consist of much more carbohydrate than protein. They are huge clusters of carbohydrate chains, glycosaminoglycans (GAG), often attached covalently to a protein backbone. One GAG is composed of a repeating disaccharide (chondroitin sulfate, keratan sulfate, heparan sulfate, hyaluronic acid) with a -A-B-A-B-A-structure (A and B represent two different sugars) and is highly acidic due to the presence of both sulfate and carboxyl groups attached to the sugar rings. Proteoglycans can be assembled into huge complexes by linkage of the core proteins to a nonsulfated GAG, hyaluronic acid. The sulfated GAG are negatively charged and bind huge numbers of cations, which in turn bind large numbers of water molecules. Therefore, proteoglycans form a porous, hydrated gel-like packing material filling the extracellular space constructed by the adjacent collagen molecules and buffer compression forces. Together, collagens and proteoglycans give ECM strength and resistance to deformation.

(3) **Multiadhesive matrix proteins** Fibronectins are abundant multi adhesive matrix proteins that play a key role in migration and cellular differentiation. A human fibronectin molecule consists of two similar, but nonidentical, polypeptides joined by a pair of disulfide bonds located near the C-termini (Figure 2-26A). Each polypeptide is composed of a linear series of distinct modules, which are organized into several larger functional units. Each unit possesses binding sites for numerous components of the ECM such as collagens, proteoglycans, and other fibronectin. Through these binding sites fibronectins interact and link these diverse molecules into a stable, interconnected network. Fibronectins also possess binding sites, the tripeptide RGD sequence (Arg-Gly-Asp), for integrins which are ECM receptors located on the cell plasma membrane.

Laminins are a family of multi adhesive proteins in the basal lamina, a specialized ECM structure. Laminins consist of three different polypeptide chains linked by disulfide bonds and organized into a molecule resembling a cross with three short arms and one long arm (Figure 2-26B). Like fibronectin, extracellular laminins can bind to integrins and greatly influence a cell's potential for migration, growth, and differentiation. Laminins also bind to type Ⅳ collagen, proteoglycans, other laminin molecules, and other components of basement membranes to form a fibrous two-dimensional network. These networks give basement membranes both strength and flexibility.

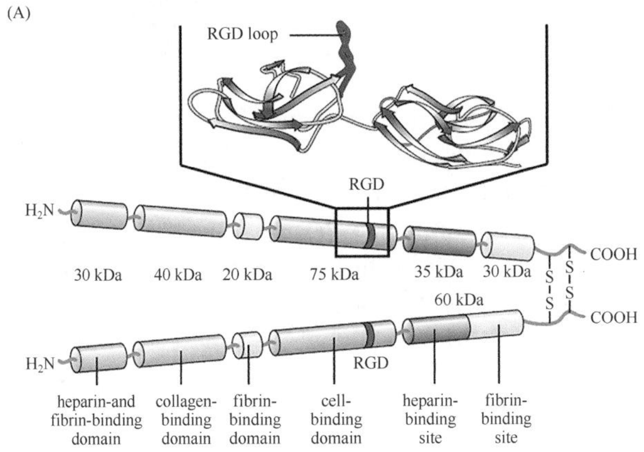

(B)

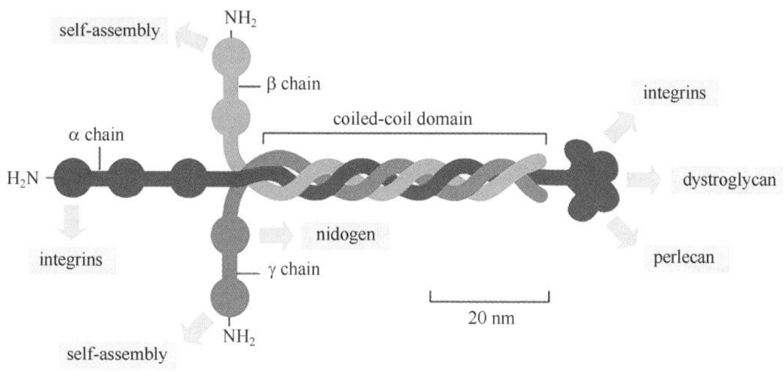

Figure 2-26 Structure of fibronectins (A) and laminins (B)

2.4.2 Connecting cells to the ECM

Most normal vertebrate cells cannot survive unless they are anchored to the extracellular matrix. This anchorage dependence is often lost when a cell turns malignant. The components of the ECM, such as fibronectin, laminin, proteoglycans, and collagen are capable of binding to receptors situated on the cell surface (Figure 2-27). The integrin is the most important family of receptors that attach cells to their extracellular microenvironment.

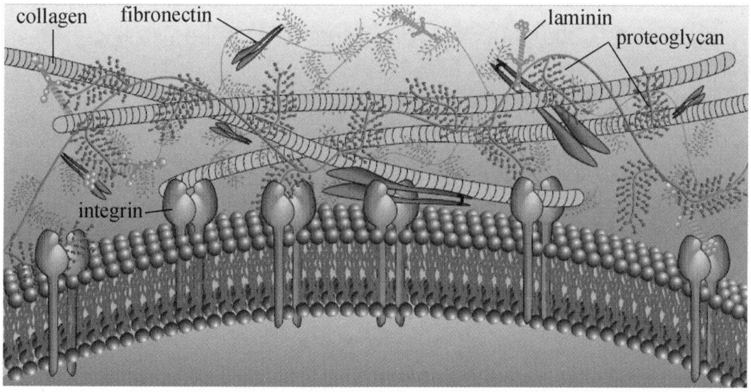

Figure 2-27 Connection between cells and the ECM

Integrins are part of a large family of cell adhesion receptors that are involved in cell-ECM and cell-cell interactions. Integrins play a key structural role in cells, binding to various types of ECM proteins via their extracellular domain and to the actin filaments of the cytoskeleton via their intracellular portion. Therefore integrins are important initiators to transduce ECM signal into the cells.

The basement membrane is a specialized ECM structure as a thin sheet of fibers that underlies the epithelium, which lines the cavities and surfaces of organs including skin, or the endothelium, which lines the interior surface of blood vessels. Most of the ECM components in the basal lamina are synthesized by the cells that rest on it. Four ubiquitous protein components are found in basal lamina: laminins, type IV collagen, entactin (also called nidogen) and perlecan. Type IV collagens are trimeric molecules with both rod-like and globular domains that form a two-dimensional network. Entactin is a rod-like molecule that cross-links type IV collagen and laminin and helps incorporate other components into the ECM. Perlecan is a large multidomain proteoglycan that binds to and cross-links many ECM components and cell-surface molecules. The primary function of the basement membrane is to anchor the epithelium to its loose connective tissue underneath. As a mechanical barrier, the basement membrane prevents malignant cells from invading the deeper tissues. Early stages of malignancy that are thus limited to the epithelial layer by the basement membrane are called carcinoma in situ.

2.4.3 Cell wall

Prokaryotic cells and plant cells both have a rigid cell wall made up of polysaccharides. The cell wall provides and maintains the shape of these cells, and serves as a protective barrier.

(1) The peptidoglycan cell wall All bacteria except mycoplasmas have a cell wall containing peptidoglycan, also called murein. Peptidoglycan is a vast polymer consisting of sugars and amino acids that forms a mesh-like layer outside the plasma membrane of bacteria (but not archaea). A peptidoglycan monomer consists of two joined amino sugars, N-acetylglucosamine (NAG) and N-acetylmuramic acid (NAM). Peptidoglycan serves a structural role in the bacterial cell wall, giving structural strength, as well as counteracting the osmotic pressure of the cytoplasm. Peptidoglycan is also involved in binary fission during bacterial cell reproduction. The peptidoglycan layer is thicker in Gram-positive bacteria (20 to 80 nm) than in Gram-negative bacteria (7 to 8 nm). The cell wall protects against lysis by osmotic swelling, and it is a target of host antibacterial proteins such as lysozyme and antibiotics such as penicillin. Penicillin interferes with the production of peptidoglycan, subsequently the cell wall formation, by binding to bacterial enzymes known as transpeptidases.

(2) The plant cell wall The plant cell walls were the first cellular structures to be observed with a light microscope that performing numerous vital functions. The cell wall is located outside the plasma membrane and gives the enclosed cell its characteristic polyhedral shape. Cell walls protects the cell against damage from mechanical abrasion and pathogens, and serve as a type of "skeleton" for the entire plant. Like the ECM at the surface of an animal cell, a plant cell wall mediates cell-cell interactions and is a source of signals that alter the activities of the cells that it contacts.

Cell wall is a secretion of the cell with sugar as 90% of its composition. Cellulose is the most abundant organic molecule providing the fibrous component of the cell wall, and proteins and pectin provide the matrix. Like starch, cellulose is a polysaccharide with glucose as its monomer. The matrix of the cell wall is composed of three types of macromolecules: ①Hemicelluloses are branched polysaccharides whose backbone consists of one sugar, such as glucose, and side chains of other sugars, such as xylose. Hemicellulose bind to the surfaces of cellulose microfibrils, cross-linking them into a resilient structural network. ② Pectins are a heterogeneous class of negatively charged polysaccharides containing galacturonic acid. Like the GAG in ECM, pectins hold water and thus form an extensive hydrated gel that fills in the spaces between the fibrous elements. Proteins mediate dynamic activities. For example, the expansins, facilitate cell growth.

Cell walls arise as a thin cell plate that forms between the plasma membranes of newly formed daughter cells following cell division. In addition to providing mechanical support and protection from foreign agents, the cell wall of a young, undifferentiated plant cell must be able to grow in conjunction with the enormous growth of the cell it surrounds. The walls of growing cells are called primary walls, and they possess extensibility that is lacking in the thicker secondary walls present around many mature plant cells. The transformation from primary to secondary cell wall occurs as the wall increases in cellulose content and, in most cases, incorporates a phenol-containing polymer called lignin. Lignin provides structural support and is the major component of wood.

Summary

All cells are surrounded by cell membranes. Prokaryotes cells contain no internal membrane limited subcompartments, while in the eukaryotes a cell was partitioned into smaller subcompartments termed organelles, such as endoplasmic reticulum, vesicle, Golgi body, mitochondria, and chloroplasts. Each organelle is surrounded by one or more membranes. All these membranes are formed by biomembrane. The basic compositions of all biomembranes are lipids, proteins and a small quantity of saccharide. The most important character of biomembrane is selectively permeable (semipermeability), which makes the biomembrane to form a barrier between the cell and the extracellular environment, or between inside and outside of organelles. This barrier controls the movement of molecules or ions between the two sides of the membrane.

The cell-cell interactions are mediated by four types of cell adhesion proteins, cadherins, integrins, selectins and Ig-like CAM. The cadherins link the cytoskeletons of adjacent cells at stable cell-cell junctions, and integrins link cell-matrix junctions. There are three classes of junctions in animal cells: tight junction, gap junction and anchoring junction. Plant cells have only one class of intercellular junctions — plasmodesmata. The ECM is the material found around cells. Biochemically the ECM of vertebrates is composed of complex mixtures of proteins, proteoglycans and in the case of bone, mineral deposits. The interaction between a cell and ECM maintains the normal growth and response of the cell.

Questions

1. When viewed by electron microscopy, the lipid bilayer is often described as looking like a railroad track. Explain how the structure of the bilayer creates this image.
2. Biomembranes contain many different types of lipid molecules. What are the three main types of lipid molecules found in biomembranes? How are the three types similar, and how are they different?
3. Explain the following statement: The structure of all biomembranes depends on the chemical properties of phospholipids, whereas the function of each specific biomembrane depends on the specific proteins associated with that membrane.
4. Name the three groups into which membrane-associated proteins may be classified. Explain the mechanism by which each group associates with a biomembrane.
5. What types of integral proteins would you expect to reside in the plasma membrane of an epithelial cell that might be absent from that of an erythrocyte? How do such differences relate to the activities of these cells?
6. Compare the features distinguishing active transport from passive transport, uniport transport from passive diffusion.
7. Compare the arrangement of integral membrane proteins in a tight junction versus a gap junction.
8. Which types of cell junctions contain actin filaments? Which contain intermediate filaments? Which contain integrins? Which contain cadherins?

9. How are integrins able to link the cell surface with materials that make up the ECM? What is the significance of the presence of an RGD motif in an integrin ligand?
10. What are the roles of each component that makes up a plant cell wall?

(张　儒)

Chapter 3

Cytoplasm, ribosomes and RNAs

All cells have four things in common. They all have a cell membrane, DNA and RNA, cytoplasm and ribosomes. They are all very important in the function of the cell. Ribosomes are scattered in the cytoplasm or attached to the endoplasmic reticulum in cytoplasm.

3.1 Structure and functions of cytoplasm

Cytoplasm is the thick, gel-like semitransparent fluid that is found in both plant and animal cell. It surrounds the nucleus of a cell and is bounded by the cell membrane. Cytoplasm was discovered in 1835. Discovery of different organelles found in cytoplasm can be attributed to different scientists, but no single scientist can be credited for discovering cytoplasm.

3.1.1 Structure of cytoplasm

Cytoplasm contains about 80% water and is usually clear in color. It is more like a viscous gel than a watery substance, but it liquefies when shaken or stirred. Cytoplasm is the substance of life that serves as a molecular soup in which all of the cell's organelles are suspended and held together by a fatty membrane.

Cytoplasm is formerly referred to as protoplasm. It contains three parts: the cytosol, the organelles and the cytoplasmic inclusions. The most basic part of cytoplasm is the cytosol, and it takes up most of the space of the cell.

(1) Cytosol The cytosol is the portion not within membrane-bound organelles, and is a gel, with a network of fibers dispersed through water. It is a complex mixture of cytoskeleton filaments, ribosome, dissolved molecules, and water that fills much of the volume of a cell. The inner, granular mass is referred to as the endoplasm and the outer, clear and glassy layer is called the cell cortex or the ectoplasm. Cytosol accounts for almost 70% of the total cell volume.

The dissolved molecules in cytosol include dissolved nutrients and waste products, including small organic molecules, such as sugars, amino acids, fatty acid. It also contains dissolved macromolecules, such as RNAs and proteins that the cell uses for growth and reproduction. The cytosol also contains many salts and is an excellent conductor of electricity, which therefore creates a medium for the vesicles, or mechanics of the cell.

The cytosol functions to suspend and to hold into place the organelles within the cell. The cytosol helps materials move around the cell by moving and churning through a process called cytoplasmic streaming.

(2) Organelles The organelles are microscopic semi-organs that facilitate a number of im-

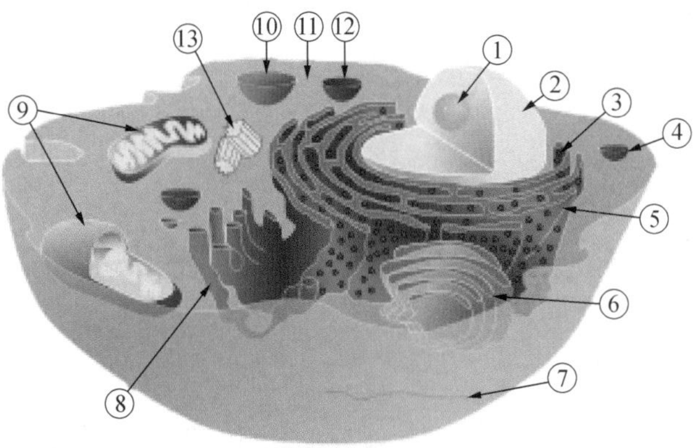

Figure 3-1 Schematic showing the cytoplasm, with major organelles of a typical animal cell. ①nucleolus; ②nucleus; ③ribosomes; ④ vesicle; ⑤ rough endoplasmic reticulum (ER); ⑥Golgi apparatus; ⑦cytoskeleton; ⑧smooth ER; ⑨mitochondria; ⑩vacuole; ⑪cytosol; ⑫lysosome; ⑬centrioles within centrosome.

portant metabolic reactions, such as synthesizing or breaking down macromolecules, producing energy and cell mitosis or meiosis. The major membrane-bounded organelles that are suspended in the cytosol are Golgi bodies, mitochondria, endoplasmic reticulum, vacuoles, lysosomes, and chloroplasts in plant cells (Figure 3-1), each organelle is bounded by a fatty membrane, and has some specific functions. There are also non-membrane structures including cytoskeleton and ribosome suspended in the cytosol. Without the function of these organelles, cells would wither and die, and life would not be possible.

Cytoplasm is the home of the cytoskeleton, a network of cytoplasmic filaments that are responsible for the movement of the cell and give the cell its shape. The cytoplasm, as seen through an electron microscope, appears as a three-dimensional lattice of thin protein-rich strands. These lattices serve to interconnect and support the other organelles or macromoleculars in the cytoplasm. All the contents of the cells of prokaryotes are contained within the cytoplasm. Within the cells of eukaryotes the contents of the cell nucleus are separated from the cytoplasm, which is called the nucleoplasm.

(3) Cytoplasmic inclusions The inclusions are small particles of insoluble substances suspended in the cytosol. For example, proteasomes exists inside all eukaryotes and archaea, and in some bacteria.

A huge range of inclusions exist in different cell types, and range from crystals of calcium oxalate or silicon dioxide in plants, to granules of starch, glycogen, or polyhydroxybutyrate, which can store energy. Lipid droplets are spherical droplets composed of lipids and proteins that are used in both prokaryotes and eukaryotes as a way of storing fatty acids and sterols. Lipid droplets make up much of the volume of adipocytes, which are specialized lipid-storage cells, but they are also found in a range of other cell types.

3.1.2 Functions of cytoplasm

In general, the function of cytoplasm is mostly mechanical in nature. It provides support to the internal structures by being a medium for their suspension. It maintains the shape and consistency of the cell. It stores many chemicals that are inevitable for life. It is in this cytoplasm, that vital metabolic reactions, like, anaerobic glycolysis and protein synthesis, take place.

(1) Metabolic pathways and biochemical reactions Cytoplasm is the site of many vital biochemical reactions crucial for maintaining life. It is within the cytoplasm that many of the most basic, and most important, facets of biology take place, such as many metabolic pathways including glycolysis, and processes such as cell division, expansion and growth.

Cytoplasm provides a medium in which the organelles can remain suspended. Besides, cytoskeleton found in cytoplasm gives the shape to the cell, and facilitates its movement. It also assists the movement of different elements found within the cell. Movement of calcium ions in and out of the cytoplasm is thought to be a signaling activity for metabolic processes.

The enzymes found in the cytoplasm breaks down the macromolecules into small parts so that it can be easily used by the other organelles. For example, mitochondria cannot use glucose present in the cell, unless it is broken down by the enzymes into pyruvate. They act as catalysts in glycolysis, as well as in the synthesis of fatty acid, sugar and amino acid. Cell reproduction, protein synthesis, anaerobic glycolysis, cytokinesis are some other vital functions that are carried out in cytoplasm.

(2) Ubiquitin-proteasome system The proteasomes included in cytoplasm, is a cylindrical complex containing a "core" of four stacked rings around a central pore. Each ring is composed of seven individual proteins. The inner two rings are made of seven β subunits that contain three to seven protease active sites. The outer two rings each contain seven α subunits whose function is to maintain a "gate" through which proteins enter the barrel. Proteins are tagged for degradation with a small protein called ubiquitin. The tagging reaction is catalyzed by enzymes called ubiquitin ligases. Once a protein is tagged with a single ubiquitin molecule, this is a signal to other ligases to attach additional ubiquitin molecules. The α subunits of proteasome are controlled by binding to "cap" structures or regulatory particles that recognize polyubiquitin tags attached to protein substrates. The overall system of ubiquitination and proteasomal degradation is known as the ubiquitin-proteasome system.

The main function of the proteasome is to degrade unneeded or damaged proteins by proteolysis, a chemical reaction that breaks peptide bonds. Enzymes that carry out such reactions are called proteases. Proteasomes are part of a major mechanism by which cells regulate the concentration of particular proteins and degrade misfolded proteins. The degradation process yields peptides of about seven to eight amino acids long, which can then be further degraded into amino acids and used in synthesizing new proteins.

The proteasomal degradation pathway is essential for many cellular processes, including the cell cycle, the regulation of gene expression, and responses to oxidative stress. The importance of proteolytic degradation inside cells and the role of ubiquitin in proteolytic pathways was acknowledged in the award of the 2004 Nobel Prize in Chemistry to Aaron Ciechanover, Avram Hershko and Irwin Rose.

(3) The functions of organelles Organelles present in cytoplasm have their specific vital functions, the smooth operation of all functions depend on the existence of cytoplasm, as it provides the medium for carrying out these vital processes.

(4) Maintenance and movement One of the most important functions of cell cytoplasm is to maintain the shape of the cell and suspension of organelles. Another function that cytoplasm sometimes involves is allowing cell movement. By squeezing organelles to a particular part of the cell, cytoplasm can cause the cell to move within the blood flow. In human, this allows white blood cells to get the parts of the body where they need to be in order to operate. In basic organisms, such as the amoeba, this provides their only means of transportation. In plants, this process, called cytoplasmic streaming, allows them to optimize cell organelles for the collection of sunlight necessary for photosynthesis.

(5) Storage Cytoplasm acts as a storage space for the chemical building blocks of the body, storing protein, lipid, oxygen and other substances until they can be used by the organelles, and storing the waste byproducts of metabolic reactions, such as carbon, until they can be disposed of. These stores are the cytoplasmic inclusions.

3.2 Ribosome

The ribosome is a ribonucleoprotein particle composed of RNA and protein, and serves as the site of protein synthesis. Specifically, the ribosome carries out the process of translation, decoding the genetic information encoded in messenger RNA, one amino acid at a time, into newly synthesized polypeptide chains.

Ribosomes were found in the cytoplasm of prokaryotic and eukaryotic cells. Some ribosomes occur freely in the cytosol whereas others are attached to the nuclear membrane or to the endoplasmic reticulum (ER) giving the latter a rough appearance, hence, the name rough ER (rER).

Ribosomes are sometimes referred to as organelles, but the use of the term organelle is often restricted to describing sub-cellular components that include a phospholipid membrane, which ribosomes, being entirely particulate. For this reason, ribosomes may sometimes be described as "non-membranous organelles".

The structure and function of the ribosomes and associated molecules, known as the translational apparatus, has been of research interest since the mid-twentieth century and is still a very active research field today. In mid-1950s, ribosomes were first observed as dense particles or granules by George Palade using electron microscope. The term "ribosome" was proposed by scientist Richard B. Roberts in 1958 to designate ribonucleoprotein particles in sizes ranging from 35 to 100 S. The word "ribosome" comes from ribo nucleic acid and the Greek: soma (meaning body). For his discovery, George won a Nobel Prize with Albert Claude and Christian de Duve in 1974. Venkatraman Ramakrishnan, Thomas A. Steitz and Ada E. Yonath have been awarded the the 2009 Nobel Prize in Chemistry for their landmark work revealing the detailed structure and mechanism of the ribosome.

3.2.1 Ribosome components

Ribosomes are roughly spherical. With a diameter of ~20 nm, they can be seen only with the electron microscope. They can make up 25% of the dry weight of cells (e. g., pancreas cells) that specialize in protein synthesis (A single pancreas cell can synthesize 5 million molecules of protein per minute).

Each ribosome consists of two subunits of many proteins and RNAs of substantial length. The smaller subunit binds to the mRNA pattern, while the formation of peptide bonds occurs in the large subunit where the acceptor-stems of the tRNAs are docked. When a ribosome finishes reading an mRNA molecule, these two subunits split apart.

Ribosomes from bacteria, archaea and eukaryotes differ in their size, composition and the ratio of protein to RNA. Prokaryotic ribosomes are around 20 nm in diameter and are composed of 65% ribosomal RNA and 35% ribosomal proteins. Eukaryotic ribosomes are between 25 nm and 30 nm in diameter, the ratio of rRNA to protein is about 1. The unit of measurement is the Svedberg unit, a measure of the rate at which the particles are spun down in the ultracentrifuge, rather than size and accounts for why fragment names do not add up (70 S is made of 50 S and 30 S).

Prokaryotes have 70 S ribosomes, each consisting of a small and a large subunit. The small subunit of the prokaryotic ribosome sediments at 30 S. It is composed of 16 S rRNA subunit (consisting of ~1,540 nucleotides, ~500 kDa) bound to 21 proteins (S1 ~ S21). The large subunit of the prokaryotic ribosome sediments at 50 S. It is composed of

5 S rRNA subunit (~120 nucleotides, 39 kDa), 23 S rRNA subunit (~2,900 nucleotides, 946 kDa) and 34 proteins(L1 ~ L34) (Figure 3-2).

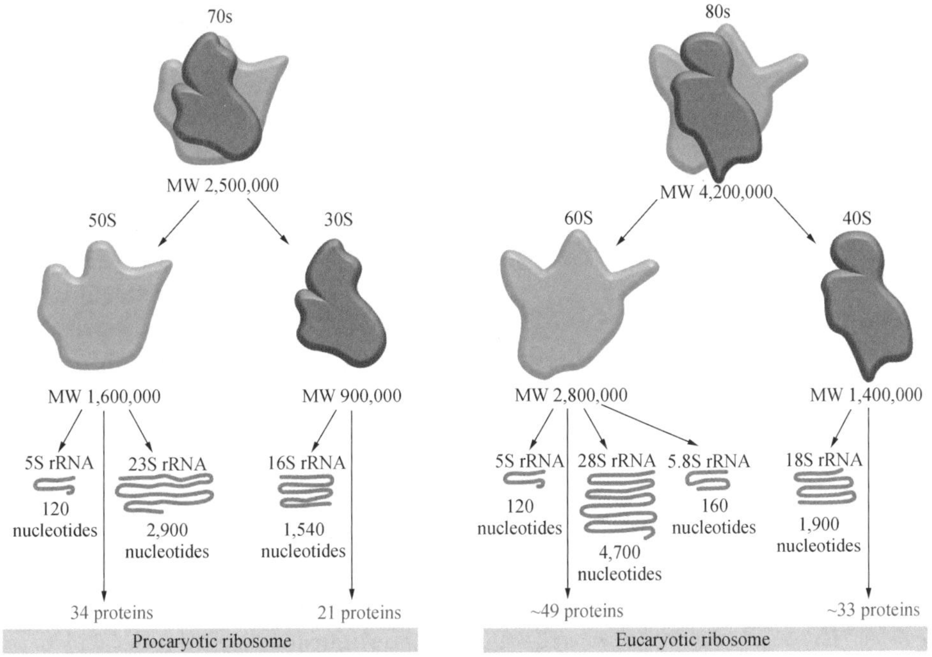

Figure 3-2 Composition of ribosomes of prokaryote and eukaryote (From: Alberts B, et al. 2008)

Other macromolecules in a functioning ribosome include three transfer RNA molecules, messenger RNA, and the nascent protein chain. The total molecular mass is several million Daltons.

The ribosome of eukaryote sediments at 80 S with two subunits sediment at 40 S and 60 S, it is generally about another million Daltons larger than the prokaryotic one(Figure 3-2). The 40 S subunit has an 18 S rRNA (1,900 nucleotides) and 33 proteins. The large subunit is composed of 5 S rRNA (120 nucleotides), 28 S rRNA (4,700 nucleotides), 5.8 S rRNA (160 nucleotides) subunits and about 49 proteins.

The ribosomes found in chloroplasts and mitochondria of eukaryotes also consist of large and small subunits bound together with proteins into 70 S particle, which are similar to those of bacteria(Table 3-1).

Table 3-1 Comparison of ribosome structure in bacteria, eukaryotes, and mitochondria

	bacterial (70 S)	eukaryotic (80 S)	mitochondrial (55 S)
Large subunit	50 S	60 S	39 S
rRNAs(1 of each)	23 S (2,900 nts)	25 S (4,700 nts)	16 S (1,560 nts)
	5 S (120 nts)	5 S (120 nts)	
		5.8 S (160 nts)	
Proteins	34	49	48
Small subunit	30 S	40 S	28 S
rRNA	16 S (1,542 nts)	18 S (1,900 nts)	12 S (950 nts)
Proteins	21	33	29

3.2.2 Ribosome structure

Despite differences in components, the basic operations of bacterial, eukaryotic, and mitochondrial ribosomes are very similar. Much of the RNA is highly organized into various tertiary structural motifs, for example pseudoknots that exhibit coaxial stacking (Figure 3-3). The extra RNA in the larger ribosomes is in several long continuous insertions, such that they form loops out of the core structure without disrupting or changing it.

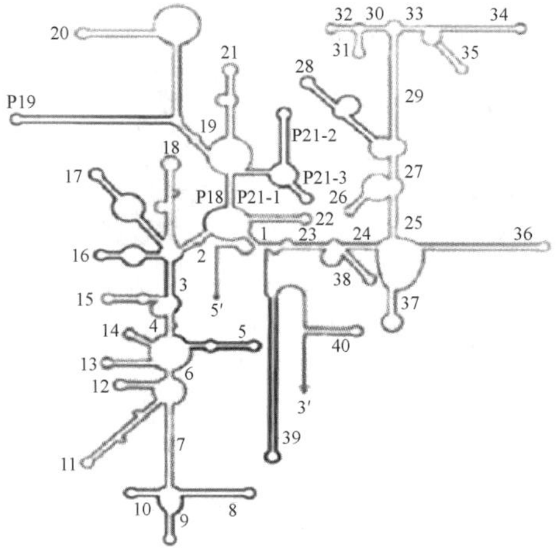

Figure 3-3 The second structure of *E. coli* 16 S rRNA (From: Cited Kleinsmith, et al. 1995)

The small subunit of the ribosome is the main site of decoding, directing the interaction of the mRNA codon with the anticodon stem-loops of the proper tRNA. The formation of peptide bonds occurs in the large subunit where the acceptor-stems of the tRNAs are docked. However, in the active ribosome the two subunits are in contact via bridges, and the actions in one subunit affect the other as the process of translation advances through the stages of initiation, elongation, and termination.

Affinity label for the tRNA binding sites on the *E. coli* ribosome allowed the identification of A and P site proteins most likely associated with the peptidyl transferase activity; labeled proteins are L27, L14, L15, L16, L2; at least L27 is located at the donor site. Additional research has demonstrated that the S1 and S21 proteins, in association with the 3′ end of 16 S ribosomal RNA, are involved in the initiation of translation.

Ribosomes have been classified as ribozymes, because the rRNA contributes directly to forming the site of both decoding and catalysis of peptide bond synthesis. Crystallographic work has shown that no ribosomal protein was observed close enough to the site of peptide bond synthesis. This suggests that the proteins in the subunit probably provide the scaffolding needed to maintain the three-dimensional structure of the RNA.

Though similarity, the ribosomes of prokaryotes, eukaryotes, and mitochondria differ in many details of their structure. The differences in structure between bacterial and eukaryotic allow some antibiotics to kill the infected bacteria by inhibiting their ribosomes without any detriment to a eukaryotic host's. Due to the differences in their structures, the bacterial 70S ribosomes are vulnerable to these antibiotics while the eukaryotic 80S ribosomes are not. Even though mitochondria possess ribosomes similar to the bacterial ones, mitochondria are not affected by these antibiotics because they are surrounded by a double membrane that does not easily admit these antibiotics into the organelle.

A number of antibiotics act by inhibiting translation, these include anisomycin, cycloheximide, chloramphenicol, tetracycline, streptomycin, erythromycin, and puromycin, among others.

3.2.3 Function of ribosome

Ribosome is the translation factory. Translation is the process in which mRNA, produced by transcription, is decoded by the ribosome to produce a specific polypeptide in the cytoplasm, that will later fold into an active protein. The basic process of protein production is addition of one amino acid at a time to the end of a protein.

The choice of amino acid type to add is determined by an mRNA molecule. The mRNA comprises a series of codons (three nucleotide subsequence) that dictate to the ribosome the sequence of the amino acids needed to make the protein. Each of those triplets codes for a specific amino acid. The successive amino acids added to the chain are matched to successive nucleotide triplets in the mRNA. In this way the sequence of nucleotides in the template mRNA chain determines the sequence of amino acids in the generated amino acid chain. Addition of an amino acid occurs at the C-terminus of the peptide and thus translation is said to be amino-to-carboxyl directed.

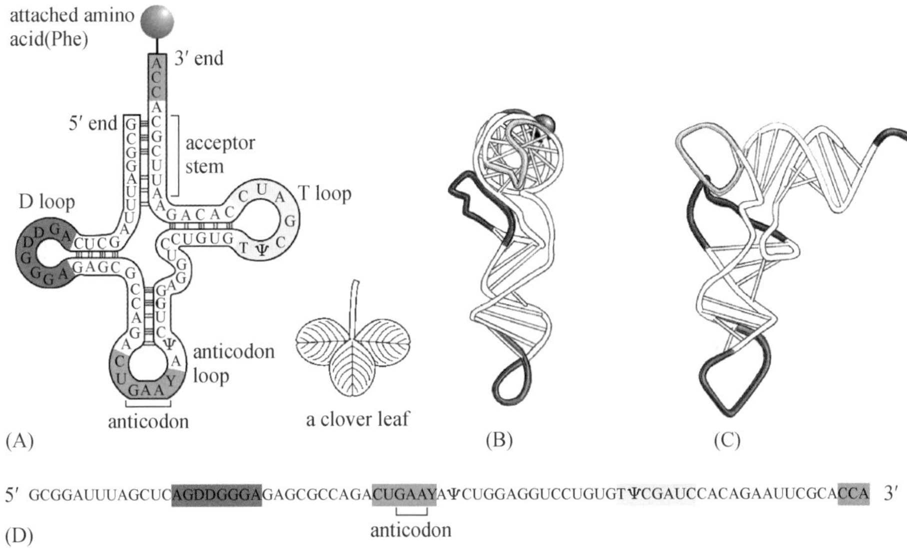

Figure 3-4 Tertiary structure of tRNA (From: Alberts B, et al. 2008)

Transfer RNA(tRNAs) are small noncoding RNA chains (74~93 nucleotides) that transport amino acids to the ribosome. tRNA is shaped like a clover leaf with three loops. It contains an amino acid attachment site on one end and a special section in the middle loop called the anticodon site. The anticodon is an RNA triplet complementary to the mRNA codon that codes for their cargo amino acid (Figure 3-4). Aminoacyl tRNA synthetase catalyzes the bonding between specific tRNAs and the amino acids that their anticodon sequences call for. The amino acid is joined by its carboxyl group to the 3′ OH of the tRNA with an ester bond. The product of this reaction is an aminoacyl-tRNA molecule, which is termed "charged".

This aminoacyl-tRNA travels inside the ribosome, where mRNA codons are matched through complementary base pairing to specific tRNA anticodons. The ribosome facilitates decoding by inducing the binding of tRNAs with complementary anticodon sequences to that of the mRNA. Two subunits of ribosome can bind together to an mRNA chain and use it as a template for determining the correct sequence of amino acids in a particular protein. Amino acids are selected, collected and carried to the ribosome by tRNA containing a complementary anticodon on one end and the appropriate amino acid on the other end, which enter one part of the ribosome and bind to the mRNA chain. The attached amino acids are then linked together by another part of the ribosome as the mRNA passes through.

More than one ribosome may move along a single mRNA chain simultaneously, forming what is called a **polyribosome** or **polysome**, each "reading" its sequence and producing a corresponding protein molecule. Because of the relatively large size of ribosomes, they can only attach to sites on mRNA 35 nucleotides apart.

In Bacteria, translation occurs in the cell's cytoplasm, where the large and small subunits of the ribosome are located, and bind to the mRNA. In Eukaryotes, Ribosomes are classified as being either "free" or "membrane-bound". Free and membrane-bound ribosomes differ only in their spatial distribution, they are identical in structure. Free ribosomes can move about anywhere in the cytosol, but are excluded from the cell nucleus and other organelles. Proteins that are formed from free ribosomes are released into the cytosol and used within the cell. Since the cytosol contains high concentrations of glutathione and is, therefore, a reducing environment, proteins containing disulfide bonds, which are formed from oxidized cysteine residues, cannot be produced in this compartment.

In many cases, the entire ribosome/mRNA complex causing it to bind to the outer membrane of the rough endoplasmic reticulum (ER) and release the nascent protein polypeptide inside for later vesicle transport and secretion outside of the cell. The ribosome making this protein can become "membrane-bound".

Whether the ribosome exists in a free or membrane-bound state depends on the presence of an ER-targeting signal sequence on the protein being synthesized, so an individual ribosome might be membrane-bound when it is making one protein, but free in the cytosol when it makes another protein.

The ribosome contains three RNA binding sites, they are the aminoacyl site (abbreviated A), the peptidyl site (abbreviated P) and the exit site (abbreviated E). With respect to the mRNA, the three sites are oriented 5′ to 3′ E-P-A, because ribosomes moves toward the 3′ end of mRNA. The A site binds the incoming tRNA with the complementary codon on the mRNA(except for the first aminoacyl tRNA, fMet-$tRNA^{fMet}$, which enters at the P site). The P site is where the peptidyl tRNA is formed in the ribosome. The E site binds the uncharged tRNA before it exits the ribosome.

3.2.4 Translation on ribosomes

Protein synthesis is accomplished through a process called translation, the process of converting the mRNA codon sequences into an amino acid polypeptide by ribosomes. Translation proceeds in four phases: **initiation**, **elongation**, **termination** and **recycling.**

(1) Initiation In prokaryote The initiation of translation involves the assembly of the initiation complex of the translation system which are: the small ribosomal subunits, the mRNA, the first (formyl) aminoacyl tRNA, three initiation factors(IF1, IF2, and IF3) and GTP, which help the assembly of the initiation complex.

During translation, a small ribosomal subunit attaches to a mRNA molecule. At the same time an initiator tRNA (aminoacyl-tRNA containing the amino acid N-formylmethionine, fMet-$tRNA^{fMet}$) molecule recognizes and binds to an AUG codon on the same mRNA molecule. A large ribosomal subunit then joins the newly formed initiation complex. The initiator tRNA resides in one binding P site of the ribosome, leaving the A site open. Protein synthesis begins at a start codon AUG near the 5′ end of the mRNA.

The ribosome is able to identify the start codon by use of the Shine-Dalgarno sequence of the mRNA(Kozak box in eukaryotes). The 30 S subunit binds to the mRNA template at a purine-rich region (the Shine Dalgarno sequence) upstream of the AUG initiation codon. The Shine Dalgarno sequence is complementary to a pyrimidine rich region on the 16 S rRNA component on thc 30 S subunit. During the formation of the initiation complex, these complementary nucleotide sequences pair to form a double stranded RNA structure that binds the mRNA to the ribosome in such a way that the initiation codon is placed at the P site.

(2) Initiation of eukaryotic translation

1) Cap-dependent initiation: Initiation of eukaryotic translation usually involves the interaction of certain key proteins with a special tag bound to the 5′ end of an mRNA molecule, the 5′ cap, named cap-dependent initiation. The protein factors bind the 40 S subunit of ribosome, and these initiation factors hold the mRNA in place. The eukaryotic initiation factor 3 (eIF3) is associated with the small ribosomal subunit, keeping the large ribosomal subunit from prematurely binding. eIF3 also interacts with the eIF4F complex which consists of three other initiation factors: eIF4A, eIF4E and eIF4G. eIF4G is a scaffolding protein which directly associates with both eIF3 and the other two components. eIF4E is the cap-binding protein. It is the rate-limiting step of cap-dependent initiation, and is often cleaved from the complex by some viral proteases to limit the cell's ability to translate its own transcripts. This is a method of hijacking the host machinery in favor of the viral (cap-independent) messages. eIF4A is an ATP-dependent RNA helicase, which aids the ribosome in resolving certain secondary structures formed by the mRNA. There is another protein associated with the eIF4F complex called the Poly(A)-binding protein (PABP), which binds the poly-A tail of most eukaryotic mRNA molecules. This protein plays a role in circularization of the mRNA during translation. This pre-initiation complex (the 40 S subunit and tRNA) accompanied by the protein factors move along the mRNA chain towards its 3′ end, scanning for the "start" codon (typically AUG) on the mRNA. The initiator tRNA charged with Met(Met-$tRNA^{Met}$) forms part of the ribosomal complex and thus all proteins start with this amino acid (unless it is cleaved away by a protease in subsequent modifications). The Met-$tRNA^{Met}$ is brought to the P site of the small ribosomal subunit by eukaryotic initiation factor 2 (eIF2). It hydrolyzes GTP, and signals for the dissociation of several factors from the small ribosomal subunit which results in the association of the 60 S subunit. Then the complete ribosome (80 S) starts translation elongation.

Regulation of protein synthesis is dependent on phosphorylation of initiation factor eIF2. When large numbers of eIF2 are phosphorylated, protein synthesis is inhibited. This would occur if there is amino acid starvation or there has been a virus infection. However, naturally a small percentage of this initiation factor is phosphorylated. Another regulator is 4EBP which binds to the initiation factor eIF4E on the 5′ cap of mRNA stopping protein synthesis. To oppose the effects of the 4EBP, growth factors phosphorylate 4EBP reducing its affinity for eIF4E and permitting protein synthesis.

2) cap-independent initiation: There exits cap-independent initiation in eukaryotic translation. The best studied example of the cap-independent mode of translation initiation in eukaryotes is the internal ribosome entry site (IRES) approach. IRES are a RNA structure that allow for translation initiation in the middle of a mRNA sequence as part of the process of protein synthesis.

What differentiates cap-independent translation from cap-dependent translation is that cap-independent translation does not require the ribosome to start scanning from the 5′ end of the mRNA cap until the start codon. The ribosome can be trafficked to the start site by ITAFs (IRES trans-acting factors) bypassing the need to scan from the 5′ end of the untranslated region of the mRNA. This method of translation has been recently discovered, and has found to be important in conditions that require the translation of specific mRNAs, despite cellular stress or the inability to translate most mRNAs. Examples include factors responding to apoptosis, stress-induced responses.

(3) **Elongation** Elongation of the polypeptide involves addition of amino acids to the carboxyl end of the growing chain, which is catalyzed by peptidyl transferase (ribozyme). The growing protein exits the ribosome through the polypeptide exit tunnel in the large subunit. Elongation is dependent on elongation factors.

During elongation, each additional amino acid is added to the nascent polypeptide chain in a four-step microcycle. In step1, a correct aminoacyl-tRNA is positioned in the A site of the ribosome, and a spent tRNA dissociates from the E site. In step 2, a new peptide bond is formed. In step 3, the large subunit translocates relative to the mRNA held by the small subunit, thereby shifting the two tRNAs to the E and P sites of large subunit. In step 4, the small subunit translocates carrying its mRNA a distance of three nuceotides through the ribosome. (Figure 3-5). This four-step cycle is repeated and each time an amino acid is added the peptide.

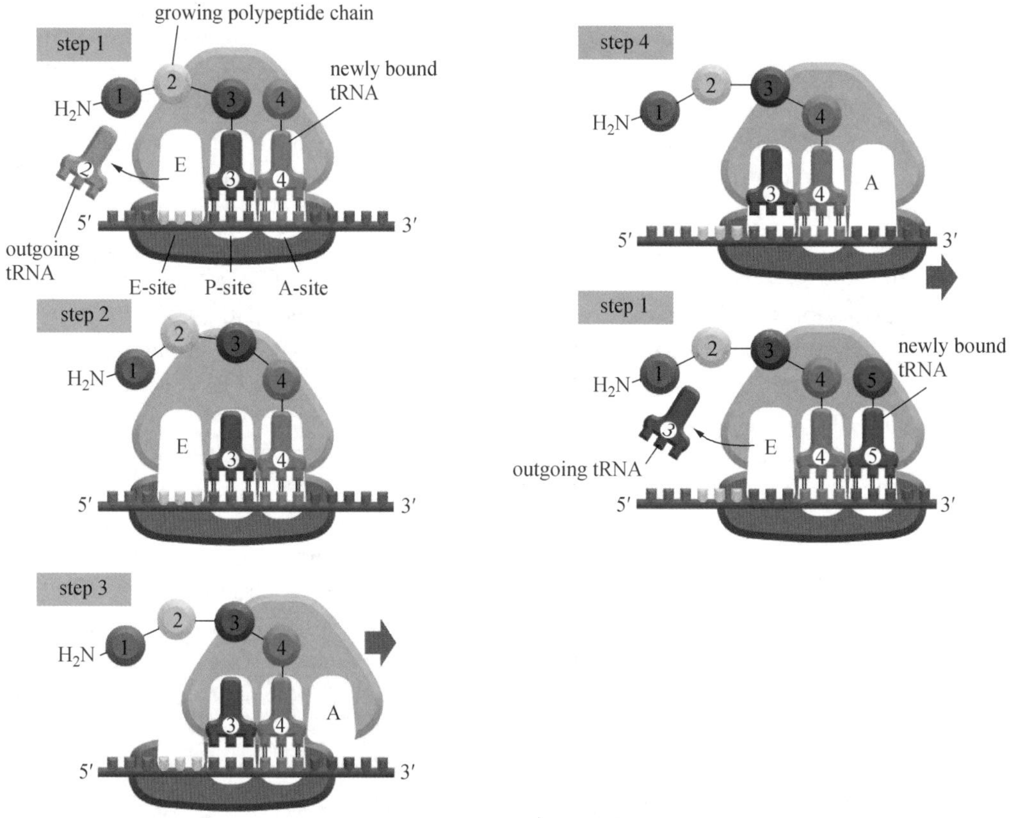

Figure 3-5 Four-step cycle of translating procedures (From: Alberts B, et al. 2008)

The initiator fMet-tRNA occupies the P site in the ribosome, causing a conformational change which opens the A site for the new aminoacyl-tRNA to bind. This binding is facilitated by elongation factor-Tu (EF-Tu), a small GTPase.

When an aminoacyl-tRNA recognizes the next codon sequence on the mRNA, it attaches to the open A site. A peptide bond forms connecting the amino acid of the tRNA in the A site and the amino acid of the charged tRNA in the P site. This process, known as peptide bond formation, is catalyzed by a ribozyme (the 23 S ribosomal RNA in the 50 S ribosomal subunit). Now, the A site has the newly formed peptide, while the P site has an uncharged tRNA.

In translocation stage, the uncharged tRNA in the E site leaves the ribosome via E site, the peptidyl-tRNA in the A site along with its corresponding codon moves to the P site and the codon after the A site moves into the A site. The A binding site becomes vacant again until another aminoacyl-tRNA that recognizes the new mRNA codon takes the open position.

After the new amino acid is added to the chain, the energy provided by the hydrolysis of a GTP bound to the translocase EF-G (in prokaryotes) and eEF-2 (in eukaryotes), moves the ribosome down one codon towards the 3' end. The energy required for translation of proteins is important. For a protein containing n amino acids, the number of high-energy phosphate bonds required to translate it is $4n$-1. The rate of translation varies; it is significantly higher in prokaryotic cells (up to 17~21 amino acid residues per second) than in eukaryotic cells (up to 6~9 amino acid residues per second).

The translation machinery works relatively slowly compared to the enzyme systems that catalyze DNA replication. The ribosome will translate the mRNA molecule as more aminoacyl-tRNA bind to the A site, until it reaches a termination codon on the mRNA (UAA, UGA, or UAG).

(4) Termination Termination of the polypeptide occurs when one of the three termination codons moves into the A site. No tRNA can recognize or bind to this codon. Instead, they are recognized by proteins called release factors, namely RF1 (recognizing the UAA and UAG stop codons) or RF2 (recognizing the UAA and UGA stop codons). These factors trigger the hydrolysis of the ester bond in peptidyl-tRNA and the release of the newly synthesized protein from the ribosome. A third release factor RF3 catalyzes the release of RF1 and RF2 at the end of the termination process (Figure 3-6).

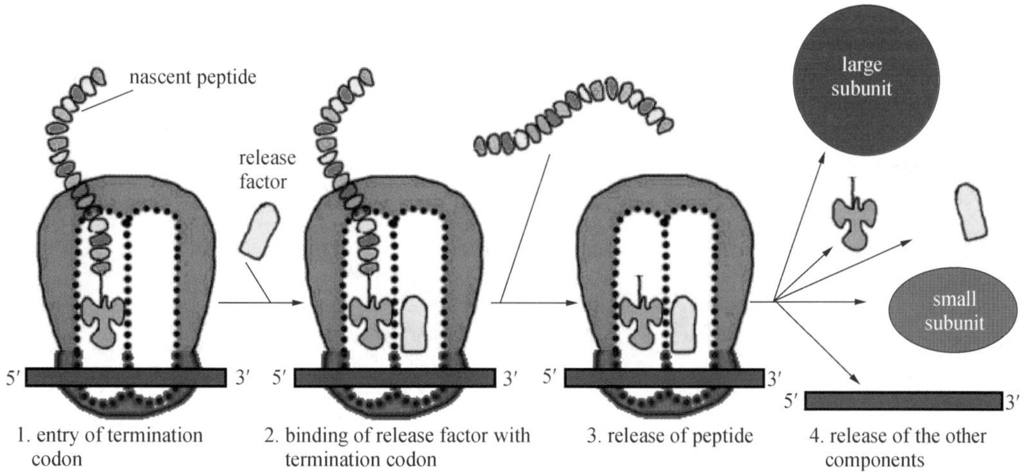

Figure 3-6 Termination of peptide synthesis

(5) Recycling After translational termination, mRNA and P site deacylated tRNA remain associated with ribosomes in post-termination complexes (post-TCs), which must therefore be recycled by releasing mRNA and deacylated tRNA and by dissociating ribosomes into subunits. Hence, ribosome recycling step is responsible for the disassembly of the post-termination ribosomal complex. Once the nascent protein is released in termination, ribosome recycling factor and elongation factor G (EF-G) function to release mRNA and tRNAs from ribosomes and dissociate the ribosome into two subunits. IF3 then replaces the deacylated tRNA releasing the mRNA. All translational components are free for additional rounds of translation.

During prokaryotic termination, RF1 and RF2 promote peptide release, whereas RF3 mediates release of RF1/RF2 from post-termination ribosomes and dissociates after hydrolyzing GTP, yielding post-termination complexes (post-TCs) that comprise 70 S ribosomes, mRNA, and P site deacylated tRNA. Recycling of post-TCs requires EF-G, RRF (ribosome recycling factor), and initiation factor IF3. EF-G and RRF dissociate post-TCs into free 50 S subunits and 30 S subunits bound to mRNA and P site deacylated tRNA, and IF3 induces release of tRNA from 30 S subunits, after which mRNA dissociates spontaneously.

In eukaryotic recycling, initiation factors eIF3, eIF1, eIF1A, and eIF3j, a loosely associated subunit of eIF3, can promote recycling of eukaryotic post-TCs. eIF3 is the principal factor that promotes splitting of post-termination ribosomes into 60S subunits and tRNA-and mRNA bound 40S subunits. Its activity is enhanced by eIF 3j, 1, and 1A. eIF1 also mediates release of P site tRNA, whereas eIF3j ensures subsequent dissociation of mRNA.

Proteins have a variety of functions. Some will be used in the membrane of the cell, while others will remain in the cytoplasm or be transported out of the cell. Many copies of a protein can be made from one mRNA molecule. This is because several ribosomes can translate the same mRNA molecule at the same time, forming polysomes.

3.2.5 Post-translational modification

Protein undergoes several modifications before becoming a fully functioning protein. These processes are called post-translational modification (PTM). PTM increases the functional diversity of the proteome by the covalent addition of functional groups or proteins, proteolytic cleavage of regulatory subunits or degradation of entire proteins. These modifications include phosphorylation, glycosylation, ubiquitination, nitrosylation, methylation, acetylation, lipidation and proteolysis.

3.3 Ribozyme

A ribozyme is defined as an RNA molecule capable of catalyzing a chemical reaction in the absence of proteins.

3.3.1 Discovery and activity of ribozyme

Before 1980, all known enzymes were proteins. But then it was discovered that some RNA molecules can act as enzymes; that is, catalyze covalent changes in the structure of substrates (most of which are also RNA molecules). Catalytic RNA molecules are called **ribozyme(RZ).** Ribozyme means ribonucleic acid enzyme. It may also be called an RNA enzyme or catalytic RNA.

The first ribozyme was discovered in the 1980s by Thomas R. Cech, who was studying RNA splicing in the ciliated protozoan. The term ribozyme was first introduced by Kelly Kruger et al. in 1982. In 1989, Thomas R. Cech and Sidney Altman won the Nobel Prize in chemistry for their "discovery of catalytic properties of RNA".

Ribozyme catalyzes a chemical reaction with its tertiary structure. It contains an active site that consists entirely of RNA. Many natural ribozymes cleave one of their own phosphodiester bonds (self-cleaving ribozymes), or bonds in other RNAs. Some have been found to catalyze the aminotransferase activity of the ribosome. Although most ribozymes are quite rare in the cell, their roles are sometimes essential to life. For example, the functional part of the ribosome, the molecular machine that translates RNA into proteins, is fundamentally a ribozyme. A recent test-tube study of prion folding suggests that an RNA may catalyze the pathological protein conformation in the manner of a chaperone enzyme. Ribozymes have been shown to be involved in the viral concatemer cleavage that precedes the packing of viral genetic material into virions.

3.3.2 Examples of naturally occurring ribozymes

(1) Peptidyl transferase 23S rRNA The three-dimensional structure of the large (50 S) subunit of a bacterial ribosome clearly shows that formation of the peptide bond that links each amino acid to the growing polypeptide chain is catalyzed by the 23S RNA molecule in the large subunit.

(2) RNase P Almost all living things synthesize an enzyme — called ribonuclease P (RNase P) — that cleaves the head (5′) end of the precursors of tRNA. In bacteria, ribonuclease P is a heterodimmer containing a molecule of RNA and one of protein. Separated from each other, the RNA retains its catalysis ability with less efficiency than the intact dimmer. Recent findings reveal that the human nucleus RNase P is also required for the normal and efficient transcription of various ncRNAs transcribed by RNA polymerase III, such as tRNA, 5S rRNA, SRP RNA and U6 snRNA genes.

(3) Spliceosomes Spliceosomes is a complex of snRNA and protein subunits participating in converting pre-mRNA into mRNA by excising the introns and splicing the exons of most nuclear genes. They are composed of 5 kinds of small nuclear RNA (snRNA) molecules and a large number of protein molecules. It is the snRNA — not the protein — that catalyzes the splicing reactions. There are two main groups of self-splicing RNAs: the group Ⅰ catalytic intron and group Ⅱ catalytic intron.

(4) Leadzyme Leadzyme is a small, artificially-made RZ which can cleave RNA in the presence of lead. Leadzyme is able to cleave RNA in the presence of lead. The structure of leadzyme has been determined by X-ray crystallography. Although initially created in vitro, it has been proposed that a naturally occurring leadzyme occurs in the 5 S rRNA and further that this may be an important mechanism in lead toxicity.

(5) Hammerhead ribozyme Hammerhead ribozyme is an RNA molecule that self-cleaves via a small conserved secondary structural motif termed a hammerhead because of its shape. Most hammerhead RNAs are subsets of two classes of plant pathogenic RNAs: the satellite RNAs of RNA viruses and the viroids. Viroids are RNA molecules that infect plant cells as conventional viruses do, but are far smaller (246 nucleotides), and are naked; that is, they are not encased in a capsid. The self-cleavage reactions are part of a rolling circle replication mechanism. The hammerhead sequence is sufficient for self-cleavage and acts by forming a conserved three-dimensional tertiary structure.

There are more ribozymes, including HDV ribozyme, Hairpin ribozyme, Mammalian CPEB3 ribozyme, VS ribozyme and glmS ribozyme.

3.3.3 Artificial ribozymes and deoxyribozymes

Since the discovery of ribozymes that exist in living organisms, there has been interest in the study of new synthetic ribozymes made in the laboratory. Scientists have produced ribozymes in the laboratory that are capable of catalyzing their own synthesis under very specific conditions, such as an RNA polymerase ribozyme and self-cleaving RNAs. Some of the artificial ribozymes had novel structures, while some were similar to the naturally occurring hammerhead ribozyme.

Deoxyribozymes or DNA enzymes or catalytic DNA, rDNAzymes are artificially DNA molecules that have the ability to perform a chemical reaction, such as catalytic action. In contrast to the RNA ribozymes, which have many catalytic capabilities, in nature DNA is only associated with gene replication and nothing else. The reasons are that DNA lacks the 2'-hydroxyl group of RNA, which diminishes its chemical reactivity and its ability to form complex tertiary structures, and that nearly all biological DNA exists in the double helix conformation in which potential catalytic sites are shielded. In comparison to proteins built up from 20 different monomers, both RNA and DNA have a much more restricted set of monomers to choose from which limits the construction of interesting catalytic sites. For these reasons DNAzymes exist only in the laboratory. The ability of ribozymes to recognize and to cut specific RNA molecules makes them exciting candidates for human therapy.

3.4 Non-coding RNA

A non-coding RNA (ncRNA) is a functional RNA molecule that is not translated into a protein. The DNA sequence from which a ncRNA is transcribed is often called a RNA gene.

3.4.1 Classification

Non-coding RNAs belong to several groups and include highly abundant and functionally important RNAs such as tRNA and rRNA, as well as RNAs such as small RNA, long ncRNAs and circRNA(Table 3-2). Recent transcriptomic and bioinformatic studies suggest the existence of thousands of ncRNAs. Although many of the newly identified ncRNAs have not been validated for their function, it is possible that many are functional. In this chapter, we introduce several important non-coding RNAs.

Table 3-2 The main RNA species in the cell

name	definition	function	size(nt)
mRNAs	messenger RNA	messenger RNA for encoding protein	large difference in length, mainly: 2~5 kb
rRNAs	ribosomal RNA	catalyze protein synthesis	in human: 120, 160, 1868, 5024
tRNAs	transfer RNA	as a bridge between mRNA and amino acids in protein synthesis	70~90
snRNAs	small nuclear RNA	play a role in nuclear processing, such as pre-mRNA processing	100~300
snoRNAs	small nucleolar RNA	process and modify rRNA	60~300
scaRNAs	small Cajal body-associated RNA	found in cajal bodies, used to modify snRNA	200~300
miRNAs	microRNAs	used to degrade mRNA and block translation	21~22
siRNAs	small interfering RNA	directly degrade mRNA and build compressed chromatin structure	20~25
piRNAs	piwi-interacting RNA	small RNAs in germ cells maintain the silence of transposons, resist transposons, and play a role in gametogenesis	23~30
lncRNAs	long noncoding RNA	multiple functions, such as regulating gene transcription, etc.	large difference in length, usually over 200
circRNAs	circular RNA	same as lncRNA, has multiple functions	large difference in length

3.4.2 Small RNA

Among non-coding RNAs, small RNAs have burst on the scene as ubiquitous, versatile repressors of gene expression in plants, animals and many fungi. Small RNAs are defined by their length (20~30 nucleotides) and their association with Argonaute family proteins (AGO family proteins), and they are typically classified into three classes: siRNA, microRNA (miRNA) and PIWI-interacting RNA (piRNA). They can control mRNA stability or translation, or target epigenetic modifications to specific regions of the genome. Small RNAs and enzymes that provide their biogenesis and functioning are involved in the organism development and coordination of biological processes, including metabolism, maintaining genome integrity, immune and stress responses. Small RNAs and RNA-mediated silencing pathways have established a new paradigm for understanding eukaryotic gene regulation and revealed novel host defenses to viruses and transposons (Figure 3-7).

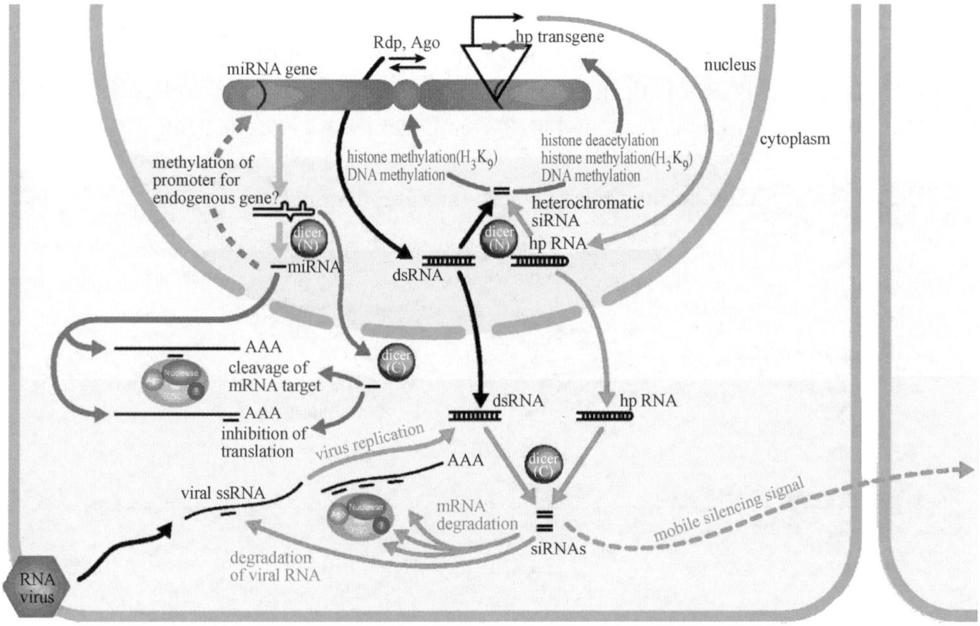

Figure 3-7 The small RNA world (From: Cell Science, 2003)

siRNA are usually generated via processing of longer double-stranded RNA (dsRNA) precursors by an RNase Ⅲ-like enzyme termed Dicer. Dicer acts in complexes with other proteins, including members of the Argonaute family and possibly HEN1, to produce small RNA. Both nuclear (N) and cytoplasmic (C) Dicer activities are found, but both forms may not be present in all organisms.

In general, siRNA can be derived from all regions of perfect duplex RNAs. Perfect duplex RNAs can be produced by transcription of a "hairpin" (hp) transgene, which produces a corresponding hpRNA. Plants make two functionally distinct size classes of small RNA. A shorter class, 21~22 nt, has been implicated in mRNA degradation, and a longer size class, 24~26 nt, in directing DNA methylation and in systemic silencing.

miRNA are small RNAs containing about 23 noncoding nucleotides that downregulate most protein-coding transcripts, miRNAs are involved in nearly all developmental and pathological processes in animals and plants. The roles of miRNAs were found in many physiological processes, such as development, cell death process, and cell signaling. Besides, emerging studies have shown that miRNAs contribute to the pathogenesis of different kinds of diseases, such as neurological disease, cardiovascular disease, metabolic syndrome, and cancer.

pri-miRNA are usually transcribed either from their own genes or from introns by RNA polymerase Ⅱ. A single pri-miRNA may contain from one to six miRNA precursors. These hairpin loop structures are composed of about 70 nucleotides each. The double-stranded RNA structure of the hairpins in a pri-miRNA is recognized cleaved by the enzyme Drosha to form pre-miRNA (precursor-miRNA) with imperfect duplex RNA, 70 ~ 200 nt in length. pre-miRNA hairpins are exported from the nucleus and is cleaved by the RNase Ⅲ enzyme Dicer in cytoplasm. This endoribonuclease interacts with the 3′ end of the hairpin and cuts away the loop joining the 3′ and 5′ arms, yielding an imperfect miRNA/miRNA duplex about 22 nucleotides in length. Although either strand of the duplex may potentially act as a functional miRNA, only one strand is usually incorporated into the RNA-induced silencing complex (RISC) where the miRNA and its mRNA target interact (Figure 3-8). Its modification by RNA editing, RNA methylation, uridylation and adenylation; Argonaute loading; and RNA decay. Non-canonical pathways for miRNA biogenesis, including those that are independent of Drosha or Dicer, are also emerging.

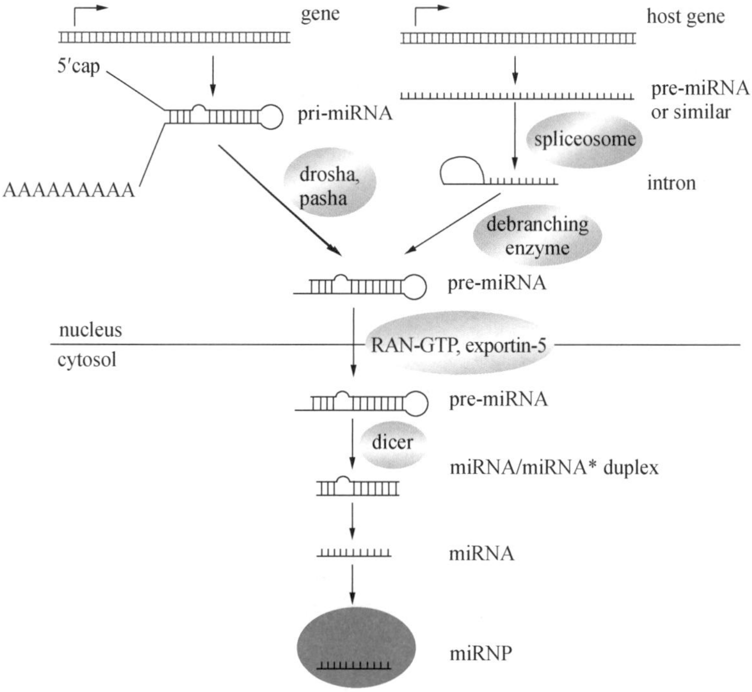

Figure 3-8 The procedure of miRNA synthesis and maturation

In RNA silencing, miRNA functions as a guide by base pairing with its target mRNAs, whereas AGO proteins function as effectors by recruiting factors that induce translational repression, mRNA deadenylation and mRNA decay. miRNA-binding sites are usually located in the 3′ untranslated region (UTR) of mRNAs. The domain at the 5′ end of miRNAs that spans from nucleotide position 2 to 7 is crucial for target recognition and has been termed the "miRNA seed". The downstream nucleotides of miRNA (particularly nucleotide 8 and less importantly nucleotides 13~16) also contribute to base pairing with the targets. More than 60% of human protein-coding genes contain at least one conserved miRNA-binding site, and, considering that numerous non-conserved sites also exist, most protein-coding genes may be under the control of miRNAs.

3.4.3 Biological roles of lncRNA

One of the most profound discoveries from the sequencing of the human genome is that over 85% of the genome is transcribed, yet less than 2% encodes protein-coding genes. Long non-coding RNAs (lncRNAs) represent the largest group of non-coding RNAs produced from the genome. lncRNAs are defined as transcripts greater than 200 nucleotides in length, lacking protein-coding potential. Most of their functions are unclear. From those that have been characterized, it is clear that lncRNAs can function through a variety of mechanisms to regulate gene expression both at the transcriptional and post-transcriptional levels.

lncRNAs can mediate their functions through interactions with proteins, RNA, DNA, or a combination of these. Furthermore, the function of lncRNAs can often be dictated by their localization, sequence and/or secondary structure. There are many categories and sub-categories of lncRNAs, but some of the major classifications include: antisense, bi-directional, enhancer-associated, intergenic lncRNAs (lincRNAs), pseudogenes. lncRNA function cannot be determined simply based on the lncRNA classification. Recently, some lncRNAs have been demonstrated to actually encode small peptides indicating that these genes are misclassified as noncoding, although it is possible that they could also have functions as a noncoding RNA in addition to their peptide coding capacity. It is therefore, important to have a logical methodology to study the biological importance of these genes.

lncRNAs play regulatory roles in many biological processes and diseases.

(1) Transcriptional regulation A lncRNA can either be a scaffold for transcription factors to enhance activation, or a scaffold for chromatin remodeler proteins to open or close chromatin (*trans* regulation). lncRNAs can regulate the expression of their neighboring coding genes (*cis* regulation) as an enhancer RNA forming DNA-RNA triplex.

(2) Post transcriptional regulation There are different ways that lncRNAs regulate mRNA expressin, by affecting either the stability, change the splicing activity, editing of modifications, or even the capping of the mature mRNA. A lncRNA can affect the stability of a mRNA transcript. A lncRNA can function as a miRNA sponge which indirectly de-represses the expression of a mRNA that would be targeted by the miRNAs. Lastly, a lncRNA can modulate the translation of a mRNA by binding to ribosomes or mRNA transcripts during translation.

In short, lncRNA can play its role by interacting with DNA, RNA and protein. lncRNA can regulate gene expression at the transcriptional level and post-transcriptional level, and can participate in the formation and activity regulation of protein complexes to perform its functions.

3.4.4 Biological roles of circRNA

Circular RNAs(circRNA) is a large class of non-coding RNA newly identified in recent years. Unlike linear RNAs terminated with 5′ caps and 3′ tails, circRNA are characterized by covalently closed loop structures with neither 5′ to 3′ polarity nor polyadenylated tail. CircRNAs are largely generated from exonic or intronic sequences, and reverse complementary sequences or RNA-binding proteins (RBPs) are necessary for circRNA biogenesis. The majority of circRNAs are conserved across species, are stable and resistant to RNase R, and often exhibit tissue/developmental-stage-specific expression. Mature circRNAs mainly are localized in the cytoplasm, but exon-intron circRNAs mainly appear to have nuclear origin due to their role in transcriptional regulation.

circRNAs are pre-mRNA derived and their biogenesis during back splicing is spliceosomal or catalyzed by group I and II ribozymes. circRNAs originate mainly from protein-coding exons but some from 3'-UTR, 5'-UTR, intronic and intergenic regions as well. circRNA can be derived from both canonical and non-canonical cleavage processes. So far, three types of circRNAs have been identified: exonic circRNAs (ecircRNAs), intronic RNAs (ciRNAs) and exon-intron circRNAs (ElciRNAs). Exonic circRNAs account for the largest class.

circRNAs play important roles as microRNA (miRNA) sponges, regulators of gene transcription, mRNA trap, and alternative splicing, RBP sponges and protein/peptide translators, regulators of parental gene expression. In recent years, a variety of circRNAs have been found to have translational functions. Internal ribosome entry site (IRES)- and N6-methyladenosines (m6A)-mediated cap-independent translation initiation have been suggested to be potential mechanism for circRNA translation.

Emerging evidence indicates that circRNAs might play important roles in diseases. Similar to miRNAs and lncRNAs, circRNAs are becoming a new research hotspot in the field of RNA and could be widely involved in the processes of life.

Summary

Cytoplasm is the thick, gel-like semitransparent fluid that is found in both plant and animal cell. It contains three parts, the cytosol, the organelles and cytoplasmic inclusions. Cytoplasm is the site of many vital biochemical reactions crucial for maintaining life. Cytoplasm can maintain the shape of the cell and suspension of organelles. Cytoplasm also acts as a storage space for the chemical building blocks of the body.

The ribosome is a ribonucleoprotein particle composed of RNA and protein, and serves as the site of protein synthesis. Each ribosome consists of two subunits. The ribosome contains three RNA binding sites, the site A, P and E. Translation is the process in which mRNA is decoded by the ribosome to produce a specific polypeptide in the cytoplasm, it proceeds in four phases: initiation, elongation, translocation and termination.

Ribozyme means ribonucleic acid enzyme. Ribozyme catalyzes a chemical reaction with its tertiary structure. It contains an active site that consists entirely of RNA. Many natural ribozymes cleave one of their own phosphodiester bonds, or bonds in other RNA. Some have been found to catalyze the aminotransferase activity of the ribosome. There are synthetic ribozymes made in the laboratory.

A non-coding RNA is a functional RNA molecule that is not translated into a protein. The term small RNA is often used for short ncRNA. ncRNA belong to several groups and include highly abundant and functionally important RNA such as tRNA and rRNA, as well as RNA such as small RNA and the long ncRNA. ncRNA are involved in many cellular processes, such as translation, RNA splicing, gene regulation, genome defense and chromosome structure.

Questions

1. Give short definitions of the following terms: Cytoplasm, Proteasome, Ribosome, Ribozyme, Deoxyribozyme, small RNA, miRNA.

2. Please compare the components and the structure of ribosome between prokaryotes and eukaryotes.
3. Please expatiate the translation procedure.
4. Give 5 examples of naturally occurring ribozymes.
5. What's the function of non-coding RNA?
6. Explain the structures and functions of cytoplasm.

(李 瑶)

Chapter 4

Endomembrane system, protein sorting and vesicle transport

A typical eukaryotic cell carries out thousands of different chemical reactions, many of which are mutually incompatible. As result cells have developed strategies for isolating and organizing their chemical reactions. In eukaryotic cells, different metabolic processes, and the proteins required to perform them, are confined within different membrane-enclosed compartments. The major membrane-enclosed compartments of a typical eukaryotic cell include nuclei, mitochondria, chloroplasts, endoplasmic reticulum (ER), Golgi apparatus, lysosomes, endosomes and peroxisomes. These organelles are surrounded by the cytosol, which is enclosed by the plasma membrane (Figure 4-1). The main functions of the membrane enclosed compartments of a eukaryotic cell are listed in table 4-1.

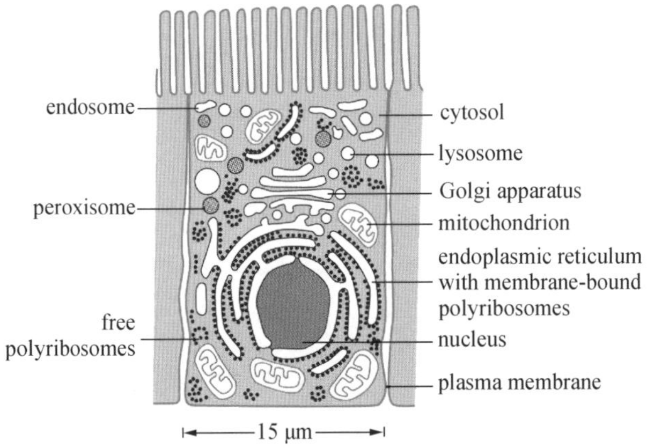

Figure 4-1 A cell from the lining of the intestine contains the basic set of organelles found in most animal cells (From: Alberts B, et al. Molecular Biology of the Cell. 2008)

Table 4-1 The main functions of the membrane-enclosed compartments of a eukaryotic cell

compartment	main function
cytosol	containing many metabolic pathways, protein synthesis
nuclei	contain main genome, DNA and RNA synthesis

Continued

compartment	main function
endoplasmic reticulum	synthesis of most lipids, synthesis of proteins for distribution to many organelles and to the plasma membrane
Golgi apparatus	modification, sorting, and packaging of proteins and lipids for either secretion or delivery to another organelle
lysosomes	intracellular degradation
endosomes	sorting of endocytosed material
mitochondria	ATP synthesis by oxidative phosphorylation
chloroplasts(in plant cells)	ATP synthesis and carbon fixation by photosynthesis
peroxisomes	oxidation of toxic molecules

4.1 Overview of endomembrane system and protein sorting

Membrane-enclosed organelles are thought to have arisen in evolution at least in two ways. Mitochondria and chloroplasts are thought to have originated in a way of endosymbiosis. The similarity of their genomes and some of their proteins to those of bacterial strongly suggests that mitochondria and chloroplasts evolved from bacteria that were engulfed by primitive eukaryotic cells with which they initially lived in symbiosis. The nuclear membranes and the membrane of the ER, Golgi apparatus, endosomes, and lysosomes are believed to have originated by invagination of the plasma membrane. These membranes, and the organelles they enclose, are all part of the **endomembrane system** which forms a dynamic, integrated network in which materials are shuttled back and forth from one part of the cell to another. In the present chapter, we will introduce the structure and functions of the endoplasmic reticulum, Golgi complex, endosomes, lysosomes, peroxisomes and vacuoles. Nucleus is the most prominent organelle in eukaryotic cells which stores genetic information for a cell and will be introduced in detail in a separated chapter.

A typical mammalian cell contains up to 10,000 different kinds of proteins, a yeast cell, about 5,000. For a cell to function properly, each of its numerous proteins must be localized to the correct cellular membrane or aqueous compartment. Directing newly made proteins to their correct organelle is therefore necessary for a cell to be able to grow, divide, and function properly. The process of directing each newly made polypeptide to a particular destination is referred to as protein targeting, or **protein sorting**.

Except for a few mitochondrial and chloroplast proteins that are synthesized on ribosomes inside these organelles, virtually all proteins in the cell are encoded by nuclear DNA and synthesized on ribosomes in the cytosol. After these proteins are made, they are imported into different organelles by three mechanisms (Figure 4-2).

(1) Gated transport through nuclear pores Proteins that function the inside nucleus are transported through the nuclear pores from the cytosol. The pores function as selective gates that actively transport specific macromolecules, but also allow free diffusion of smaller molecules. Such a process will be discussed in the chapter of nucleus.

(2) Transport across membranes Most proteins of mitochondria and chloroplasts, and all proteins of the ER and peroxisomes are transported across the organelle membrane by protein translocators located in the membrane. In this case, the transported protein molecular must usually unfold in order to snake through the membrane. This process will be discussed in detail within the current chapter in ER section.

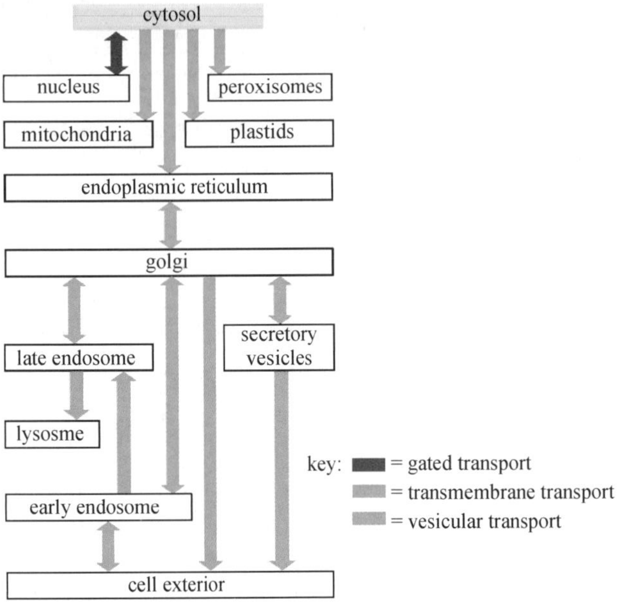

Figure 4-2 Three mechanisms that proteins are imported into organelles (From: Alberts B, et al. Molecular Biology of the Cell. 2008)

(3) Transport by vesicles

Proteins moving from the ER onward and from one compartment of the endomembrane system to another are transported by transport vesicles, which become loaded with a cargo of proteins from the interior space, or lumen, of one compartment, as they pinch off from its membrane. The vesicles subsequently discharge their cargo into a second compartment by fusing with its membrane. This kind of vesicular transport will be introduced in the last part of the current chapter.

The protein **localization** is determined by its **sorting signal**. The typical sorting signal is a continuous stretch of amino acid sequence of about 15 ~ 60 amino acids long. The properties of signal sequences that direct proteins from the cytosol to organelles are shown in table 4-2. The signal sequences of mitochondria, chloroplast and endoplasmic reticulum proteins are usually located in the N-terminus and are often removed from the find protein once the sorting decision has been executed. The signal sequence that directs a protein from the cytosol into the nucleus is called **nuclear localization signal** (**NLS**), which is located on the internal of a protein and will not be removed after transportation.

Table 4-2 Some typical signal sequences

target organelle	signal location	signal removal	nature of signal
ER	N-terminal	(+)	"core" of 6 ~ 12 mostly hydrophobic amino acids, often preceded by one or more basic amino acids
ER retention	C-terminal	(-)	KDEL
mitochondrion	N-terminal	(+)	3~5 nonconsecutive Arg or Lys residues, often with Ser and Thr, no Glu or Asp residues
chloroplast	N-terminal	(+)	no common sequence motifs; generally rich in Ser, Thr, and small hydrophobic amino acid residues and poor in Glu and Asp residues
peroxisome	C-terminal	(-)	usually Ser-Lys-Leu at extreme C-terminus
nucleus	internal	(-)	one cluster of 5 basic amino acids, or two smaller clusters of basic residues separated by 10 amino acids

In order to gain insight into the functions of different cell components, techniques to break up cells and isolate particular types of organelles were pioneered in the 1950s and 1960s by Albert Claude and Christian de Duve. In these techniques, cells are ruptured by homogenization and the cytoplasmic membranes become fragmented and the fractured edges of the membrane fragments fuse to form spherical vesicles less than 100 nm in diameter. Vesicles derived from different organelles have different properties and can be separated from one another by an approach called subcellular fractionation. Membranous vesicles derived from the endomembrane system (primarily the endoplasmic reticulum and Golgi complex) form a heterogeneous collection of similar sized vesicles referred to as **microsomes**. A rapid (and crude) preparation of the microsomal fraction of a cell is depicted in figure 4-3. Once isolated, the biochemical composition and possible functions of various fractions can be determined. For example, vesicles derived from different parts of the Golgi complex contained enzymes that added different sugars to the end of a growing carbohydrate chain of a glycoprotein or glycolipid. In the past few years, sophisticated proteomic technology has been used to identify the proteins present in cell fractions. Once a particular organelle has been isolated, the proteins can be extracted, separated, and identified by mass spectrometry. Hundreds of proteins can be identified simultaneously, providing a comprehensive molecular portrait of any organelle that can be prepared in a relatively pure state.

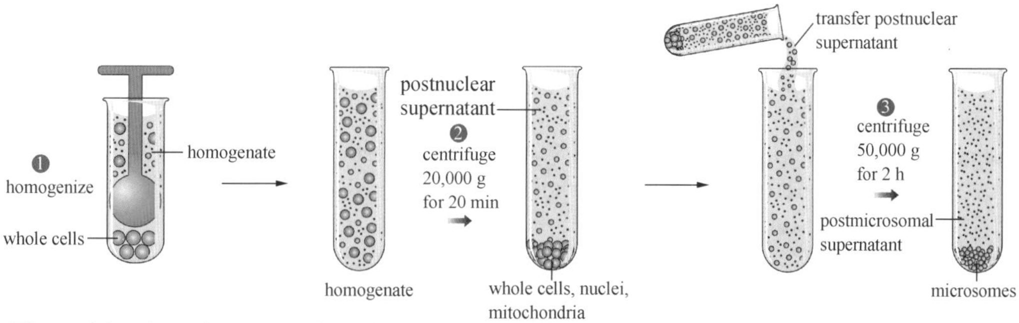

Figure 4-3 A rapid preparation of the microsomal fraction of a cell (From: Karp G, et al. Cell and Molecular Biology. 2009)

4.2 Endoplasmic reticulum

Endoplasmic reticulum (ER) comprises a system of membranes that enclose a space, or lumen, which is separated from the surrounding cytosol and forms a network of tubules, vesicles and sacs (**cisternae**) that are interconnected and often extends throughout most of the cell (Figure 4-4). The main functions of ER are the synthesis of many membrane lipids and proteins. Large areas of ER have ribosomes attached to the cytosolic surface and are designated **rough ER (RER)**. The RER is continuous with the outer membrane of the nuclear envelope and is typically composed of a network of flattened sacs. The attached ribosomes are actively synthesizing proteins that are delivered into the ER lumen or inserted into ER membrane. The **smooth ER (SER)** is scanty in most cells but is highly developed for performing particular functions in others. The membranous elements of SER are highly curved and tubular, forming an interconnecting system of pipelines curving through the cytoplasm lacking ribosome association. SER is the site of steroid hormone synthesis in cells of the adrenal gland. A variety of organic molecules, including alcohol, are detoxified by SER in liver cells. In many eukaryotic cells

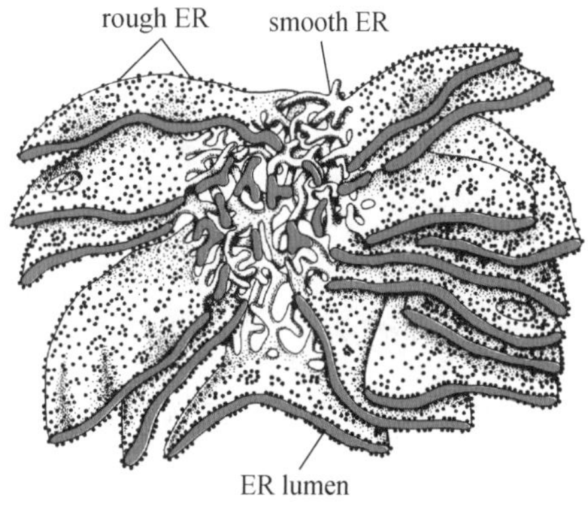

Figure 4-4 Rough ER and smooth ER (From: Alberts B, et al. Molecular Biology of the Cell. 2008)

the SER also serves as sequesters of Ca^{2+} from the cytosol which is involved in the rapid response to many extracellular signals.

4.2.1 RER and protein synthesis

Certain polypeptides are synthesized on ribosomes attached to the cytosolic surface of the RER membranes. These include: ①Soluble proteins that reside within compartments of the endomembrane system; ②Secreted proteins; ③Integral membrane proteins. The ER serves as the starting point of such biosynthetic pathways: ①Water-soluble proteins are completely translocated across the ER membrane and released into the ER lumen. Once inside the ER or in the ER membrane, they will be ferried by transport vesicles from organelle to organelle and, in some cases, secreted from organelle to the cell exterior. ②Prospective transmembrane proteins are only partly translocated across the ER membrane and destined to ER itself, Golgi apparatus, endosomes, lysosomes, as well as cell plasma membrane.

(1) Synthesis of water-soluble proteins on RER Synthesis of all proteins begins on "free" ribosomes in the cytosol. What drives the translocation of ER-targeted proteins in the ER lumen or on ER membrane during protein synthesis? In the early 1970s, Günter Blobel, David Sabatini and Bernhard Dobberstein of Rockefeller University demonstrated that the sequence of amino acids in the N-terminal portion of secreted proteins (**signal sequence**), which is the first part to emerge from the ribosome during protein synthesis, directs the polypeptide and ribosome to the ER membrane (known as the **signal hypothesis**) (Figure 4-5). The polypeptide moves into the cisternal space of the ER through a protein-lined, aqueous channel (so called **translocator**) in the ER membrane. It was proposed that the polypeptide moves through the membrane as it is being synthesized. A signal peptidase closely associated with the translocator clips off the signal sequence during translation, and the mature protein is released into the lumen of the ER immediately after synthesis. This proposal has been proved by a large body of experimental evidence and Blobel's original concept that proteins contain built-in "address codes" has been shown to apply in principle to virtually all types of protein trafficking pathways throughout the cell.

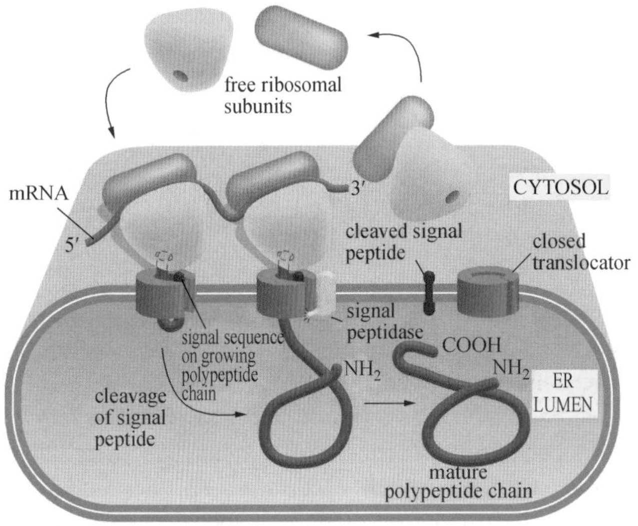

Figure 4-5 The signal hypothesis (From: Alberts B, et al. Molecular Biology of the Cell. 2008)

The ER signal sequence located at the N-terminus of the polypeptide varies greatly in amino acid sequence. However, each ER signal sequence contains eight or more nonpolar hydrophobic amino acids at its center. After the ER signal sequence is translated on "free" ribosomes in the cytosol, it is recognized by a **signal-recognition particle (SRP)**, a complex consisting of six different polypeptide chains bound to a single small 7S RNA molecule. The crystal structure revealed that the signal-sequence-binding site of the SRP protein forms a large hydrophobic pocket lined by methionines. Since methionines have unbranched, flexible side chains, the pocket is sufficiently plastic to accommodate hydrophobic signal sequences of different sequences, sizes, and shapes. Binding of an SRP to the signal sequence of the polypeptide chain slows down protein synthesis until the ribosome and its bound SRP bind to an SRP receptor, an integral membrane protein complex embedded in the rough ER membrane. After binding to its receptor, the SRP is released and protein synthesis recommences, with the polypeptide now being threaded into the lumen of the ER through a translocation channel in the ER membrane (Figure 4-6).

The transient pause of protein translation caused by SRP binding presumably gives the ribosome enough time to bind to the ER membrane before completion of the polypeptide chain, thereby ensuring that proteins, especially those secreted and lysosomal hydrolases, are not released into the cytosol before reaching their destination. In contrast to the post-translational import of proteins into mitochondria and chloroplasts, SRP-mediated translation does not require chaperone proteins to keep the newly synthesized protein unfolded during translocation to the ER membrane.

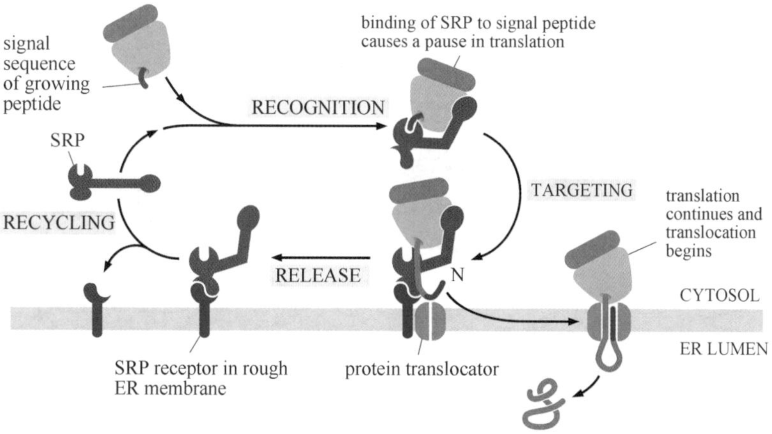

Figure 4-6 ER signal sequences and SRP direct ribosomes to the ER membrane (From: Alberts B, et al. Molecular Biology of the Cell. 2008)

In addition to directing proteins to the ER via SRP recognition, the signal sequence can be recognized by a specific site inside the translocation channel and trigger the opening of the pore in the protein translocator, where it serves as a start-transfer signal (or **start-transfer peptide**). Such dual recognition may help to ensure that only appropriate proteins enter the lumen of the ER. The signal peptide remains bound to the channel while the rest of the protein chain slides through the membrane as a large loop. At some stage during translocation, the signal sequence is cleaved off by a signal peptidase located on the luminal side of the ER membrane; the signal peptide is then released from the translocation channel and rapidly degraded into amino acids. Once the C-terminus of the protein has passed through the membrane, the protein is released into the ER lumen.

The core of the translocator is Sec61 complex containing three subunits that are highly conserved from bacteria to eukaryotic cells. The X-ray crystal structure of Sec61 complex showed the α helices contributed by the largest subunit form a central pore through which the polypeptide chain traverses the membrane. The pore is gated by a short helix to keep the translocator closed when it is idle. The short helix moves aside to open the pore when it is engaged in passing a polypeptide chain. In eukaryotic cells, Sec61 complexes may provide binding sites for the ribosome and other accessory proteins that help polypeptide chains fold as they enter the ER. The structure of the Sec61 complex suggests that the pore can also open along a seam on its side to allow lateral access of a translocating peptide chain into the hydrophobic core of the membrane. Such process is important both for the release of a cleaved signal peptide into the membrane and for the integration of membrane proteins into the bilayer membrane.

As we have seen, translocation of proteins into mitochondria, chloroplasts, and peroxisomes occurs post-translationally, after the protein has been made and released into the cytosol, whereas translocation across the ER membrane usually occurs during translation (**co-translation**). Such co-translational transfer process creates two spatially separate populations of ribosomes, membrane-bound and free ribosomes (Figure 4-7). They are structurally and functionally identical but differ only in the proteins they are making at any given time. Many ribosomes can bind to a single mRNA molecule and form a **polyribosome** directed to the ER membrane by the signal sequences on multiple growing polypeptide chains. When the translation is finished, the individual ribosomes can return to the cytosol whereas the mRNA remains attached to the ER membrane by a changing population of ribosomes that are transiently held at the membrane by the translocator.

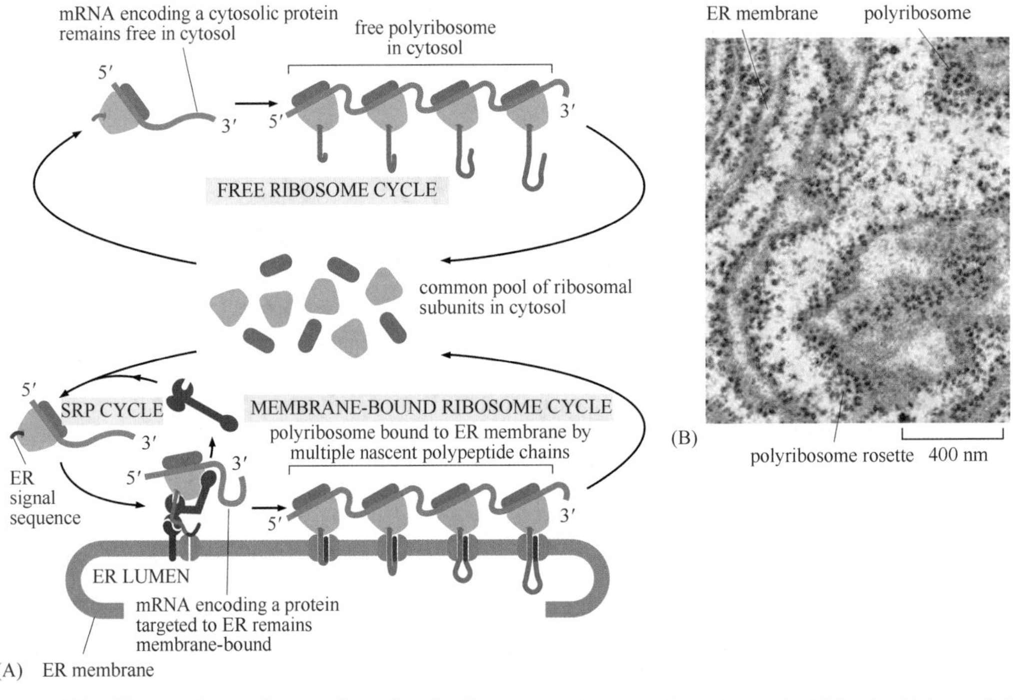

Figure 4-7 Free and membrane-bound polyribosomes. (From: Alberts B, et al, Molecular biology of the cell. 2008)

It was found that some completely synthesized proteins are first released into the cytosol and then imported into the ER as well. Such post-translational translocation is especially common across the yeast ER membrane. To function in such a process, completely synthesized proteins need to bind to chaperone proteins in cytosol to prevent folding, and accessory proteins associated with the sec 61 complex are needed to feed the polypeptide chain into the pore and drive translocation. These accessory proteins use a small domain on the lumenal side of the ER membrane to deposit a Hsp70-like chaperone protein (called BiP) onto the polypeptide chain as it emerges from the pore into the ER lumen.

(2) Synthesis of transmembrane proteins on RER Transmembrane proteins translocate to the ER and remain embedded in the ER membrane. Each transmembrane protein has a unique orientation with respect to the membrane's phospholipid bilayer. Either the N-terminal segment is located on the exoplasmic face and the C-terminal segment on the cytosolic face, or in the reverse orientation. During the transport to their final destinations, the topology of a membrane protein is preserved. Thus, the orientation of these membrane proteins is established during biosynthesis on the ER membrane. Transmembrane proteins contain one to several membrane-spanning hydrophobic segments, which form **transmembrane α helices** anchoring the protein in the phospholipid bilayer.

1) **Translocation of single-pass transmembrane proteins**: There are three ways in which single-pass transmembrane proteins become inserted into the ER. In the simplest case as for the insulin receptors, the N-terminal signal sequence initiates translocation of single-pass transmembrane proteins to ER similarly as described in the water-soluble protein translocation. However, the transfer process is paused when it meets an additional hydrophobic segment in the polypeptide chain (**stop-transfer signal**). After the ER signal sequence has been released from the translocator and has been cleaved off, this stop-transfer signal is released laterally from the translocation channel and drifts into the plane of the lipid bilayer, where it forms an α-helical membrane-spanning segment that anchors the protein in the membrane. As a result, the translocated protein ends up as a transmembrane protein with the N-terminus on the luminal side of the lipid bilayer and the C-terminus on the cytosolic side (Figure 4-8). In the other two cases, the start-transfer signal sequence is an internal hydrophobic sequence and functions as both an ER signal sequence and membrane-anchor sequence. This **internal start-transfer sequence** directs the insertion of the nascent polypeptide chain into the ER membrane. After release from the translocator, the internal start-transfer sequence remains in the lipid bilayer as a single membrane-spanning α helix (Figure 4-9). Internal start-transfer sequences can bind to the translocation apparatus in either of two orientations which in turn determines the orientation of transmembrane proteins. The orientation of the start-transfer sequence depends on the distribution of nearby charged amino acids. The negatively charged end of the signal sequence inserted adjacent to the ER lumen whereas the positively charged end pointed to the cytosol.

2) **Translocation of multipass transmembrane proteins**: Many important proteins, such as ion pumps, ion channels, and transporters, span the membrane multiple times. Each membrane-spanning α helix in these multipass transmembrane proteins is used as signal sequences to start and stop the protein transfer. Combinations of start-transfer and stop-transfer signals deter-

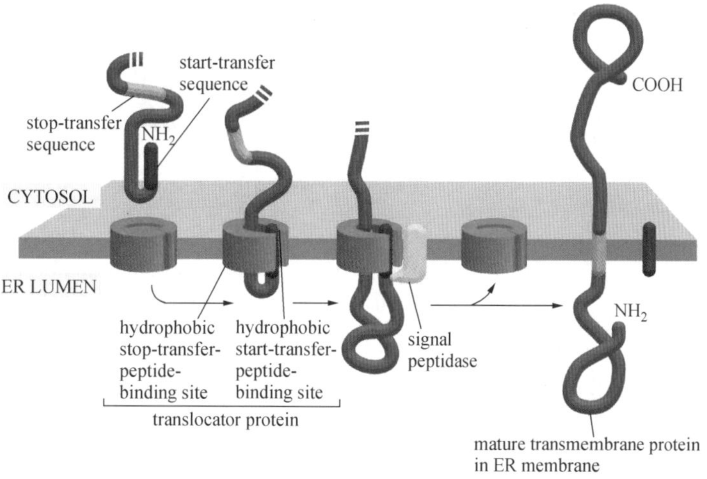

Figure 4-8 Integration of a single-pass transmembrane protein into the ER membrane (From: Alberts B, et al. Molecular Biology of the Cell. 2008)

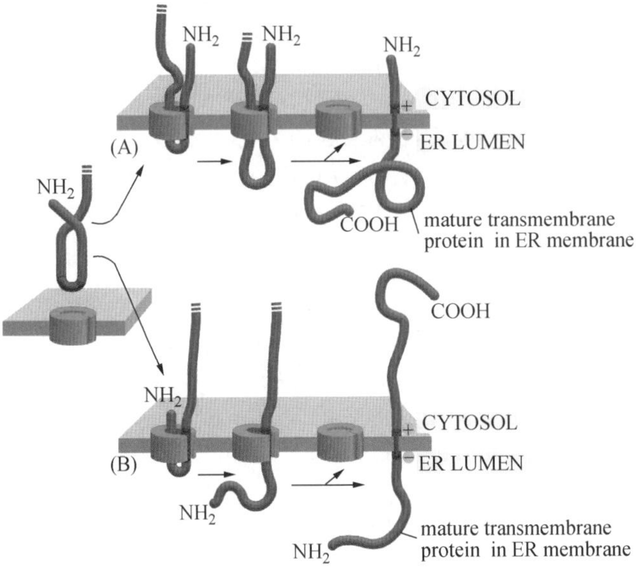

Figure 4-9 Integration of a single-pass transmembrane protein with an internal signal sequence into the ER membrane (From: Alberts B, et al. Molecular Biology of the Cell. 2008)

mine the topology of multipass transmembrane proteins. In this case, hydrophobic signal sequences are thought to work in pairs: an internal start-transfer sequence serves to initiate translocation, which continues until a stop-transfer sequence is reached; the two hydrophobic sequences are then released into the bilayer, where they remain as membrane-spanning α helices (Figure 4-10).

In complex multipass proteins, in which many hydrophobic α helices span the bilayer, additional pairs of stop and start sequences come into play: one sequence reinitiates translocation further down the polypeptide chain, and the other stops translocation

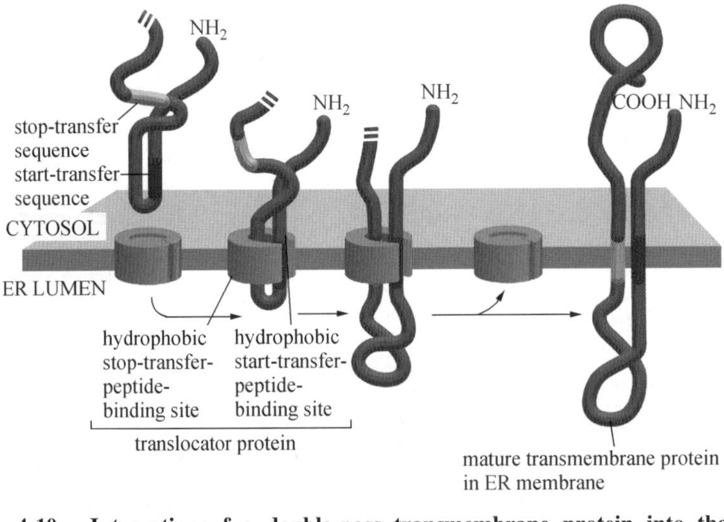

Figure 4-10 Integration of a double-pass transmembrane protein into the ER membrane (From: Alberts B, et al. Molecular Biology of the Cell. 2008)

and causes polypeptide release, and so on for subsequent starts and stops. Thus, multipass membrane proteins are stitched into the lipid bilayer as they are being synthesized (Figure 4-11).

Whether a given hydrophobic signal sequence functions as a start-transfer or stop-transfer sequence must depend on its location in a polypeptide chain. The differences between start-transfer and stop-transfer sequences come from their relative order in the growing polypeptide chain. It seems that the SRP begins scanning an unfolded polypeptide

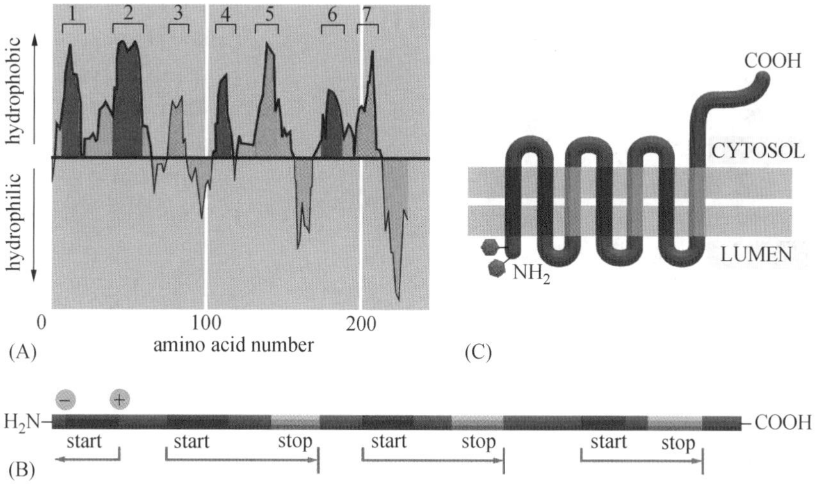

Figure 4-11 Integration of a multi-pass transmembrane protein rhodopsin into the ER membrane (From: Alberts B, et al. Molecular Biology of the Cell. 2008)

chain for hydrophobic segments from its N-terminus toward the C-terminus. By recognizing the first appropriate hydrophobic segment to emerge from the ribosome, the SRP sets the "reading frame" for membrane integration: after the SRP initiates translocation, the translocator recognizes the next appropriate hydrophobic segment in the direction of transfer as a stop-transfer sequence. A similar scanning process continues until all of the hydrophobic regions in the protein have been inserted into the membrane.

(3) Newly synthesized proteins are covalently modified in the ER The RER is a major protein processing place. To meet its obligations, the RER lumen is packed with molecular chaperones that recognize and bind to unfolded or misfolded proteins and give them the opportunity to attain their correct three-dimensional structures.

1) **Formation of disulfide bonds**: Disulfide bonds are formed by the oxidation of pairs of cysteine side chains on the newly synthesized proteins, a reaction catalyzed by protein disulfide isomerase (PDI), that resides in the ER lumen. The disulfide bonds help to stabilize the tertiary and quaternary structures of those proteins. Disulfide bonds do not form in the cytosol, because of the reducing environment there. Soluble cytosolic proteins, synthesized on free ribosomes, lack disulfide bonds and utilize other interactions to stabilize their structures.

2) **Glycosylation**: Another important modification of proteins in the ER is protein **glycosylation**. Oligosaccharide side chains linked to amino (NH_2) group of an asparagine in a protein (***N*-linked**) are the most common type of linkage found on glycoproteins. It is the major form of protein glycosylation in the ER. Oligosaccharides linked to the hydroxyl group on the side chain of a serine, threonine, or hydroxylysine amino acid, called *O*-linked glycosylation, are carried out mainly in the Golgi apparatus. In *N*-linked glycosylation, a preformed, branched oligosaccharide containing a total of 14 sugars is transferred to the NH_2 group of an asparagine side chain on the protein immediately after the target asparagine emerges in the ER lumen during protein translocation (Figure 4-12). This ubiquitous 14-residue high-mannose precursor of *N*-linked oligosaccharides is originally formed on dolichol, a lipid that is firmly embedded in the ER membrane and acts as a carrier for the oligosaccharide. Only asparagine residues in the tripeptide sequences Asn-X-Ser and Asn-X-Thr (where X is any amino acid except proline)

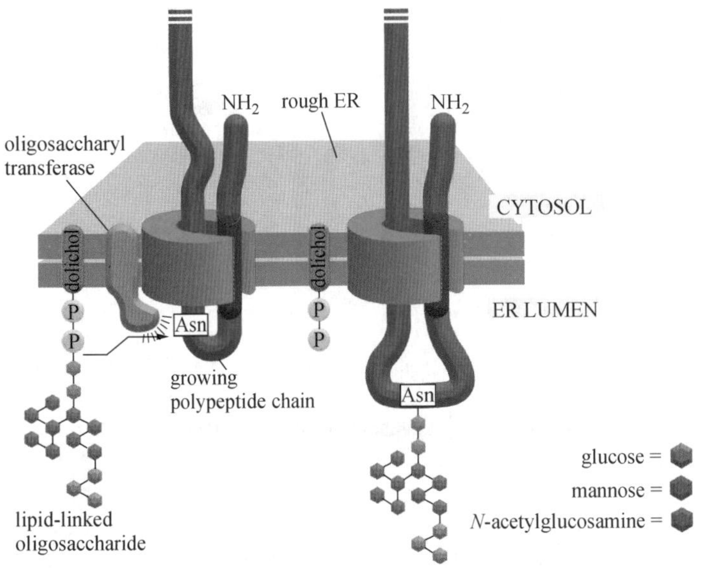

Figure 4-12 *N*-linked protein glycosylation in the ER (From: Alberts B, et al. Molecular Biology of the Cell. 2008)

are the glycosylation sites.The addition of the 14-sugar oligosaccharide to proteins in the ER is only the first step in a series of further modifications. Despite their initial similarity, the *N*-linked oligosaccharides on mature glycoproteins are remarkably diverse. Modifications to *N*-linked oligosaccharides are completed in the Golgi complex.

The oligosaccharides on proteins serve various functions, depending on the protein. They can protect the protein from degradation, hold the protein in the ER until it is properly folded, or help guide it to the appropriate organelle by serving as a transport signal for packaging the protein into appropriate transport vesicles. When displayed on the cell surface, oligosaccharides form part of the carbohydrate layer and can function in the recognition of one cell by another.

Oligosaccharides are used as tags to mark the state of protein folding. Two ER chaperone proteins, calnexin and calreticulin which require Ca^{2+} for their activities, bind to oligosaccharides on incompletely folded proteins and retain them in the ER. Like other chaperones, they prevent incompletely folded proteins from becoming irreversibly aggregated. Calnexin and calreticulin recognize *N*-linked oligosaccharides that contain a single terminal glucose, and therefore bind proteins only after two of the three glucoses on the precursor oligosaccharide have been removed by ER glucosidases. When the third glucose has been removed, the protein dissociates from its chaperone and can leave the ER. Another ER enzyme, a glucosyl transferase that keeps adding a glucose to those oligosaccharides that are attached to unfolded proteins. Thus, an unfolded protein undergoes continuous cycles of glucose removal (by glucosidase) and glucose addition (by glycosyl transferase), maintaining an affinity for calnexin and calreticulin until it has achieved its fully folded state (Figure 4-13).

Despite all the help from chaperones, many protein molecules translocated into the ER fail to achieve their properly folded or oligomeric state. Such proteins are exported from the ER back into the cytosol and are degraded. Chaperone proteins are probably necessary to keep the polypeptide chain in an unfolded state before and during transport. A source of energy and a translocator might be required. The *N*-linked oligosaccharides, which serve as timers that measure how long a protein has spent in the ER. The slow trimming of a particular mannose on the core-oligosaccharide tree by an enzyme (a mannosidase) in the ER is thought to create a new oligosaccharide structure that the retrotranslocation apparatus recognizes. Proteins that fold and exit from the ER faster than the action of the mannosidase would therefore escape degradation. Once a misfolded protein has been retrotranslocated

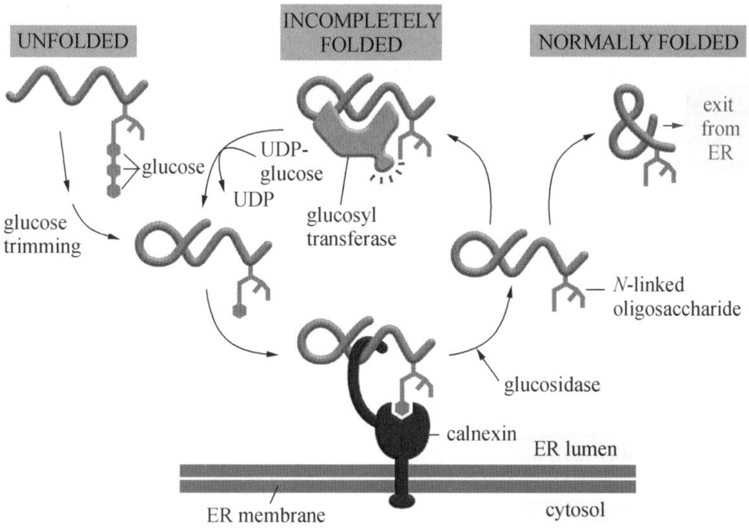

Figure 4-13 The role of *N*-linked glycosylation in ER protein folding
(From: Alberts B, et al. Molecular Biology of the Cell. 2008)

into the cytosol, an *N*-glycanase removes all its oligosaccharide chains. The deglycosylated polypeptide is rapidly ubiquitylated by ER-bound ubiquitin-conjugating enzymes and is then fed into proteasomes and degraded.

3) **Glycosylphosphatidyl-inositol anchor of membrane proteins**: Several cytosolic enzymes catalyze the covalent addition of a single fatty acid chain or prenyl group to selected proteins. The attached lipids help to direct these proteins to cell membranes. A related process is catalyzed by ER enzymes, which covalently attach a glycosylphosphatidyl-inositol (GPI) anchor to the C-terminus of some membrane proteins destined for the plasma membrane. This linkage forms in the lumen of the ER, where, at the same time, the transmembrane segment of the protein is cleaved off. A large number of plasma membrane proteins are modified in this way and they are attached to the exterior of the plasma membrane by their GPI anchors. In principle, they can be released from cells in soluble form in response to signals that activate a specific phospholipase in the plasma membrane.

4.2.2 Lipid bilayer assembly in the ER

The ER membrane synthesizes nearly all of the major classes of lipids, including both phospholipids and cholesterol. There are two major exceptions: sphingomyelin and glycolipid synthesis begin in the ER and are completed in Golgi; some unique lipids in mitochondrial and chloroplast membranes are synthesized by enzymes residing in those organelles. The major phospholipid made in ER is phosphatidylcholine, which can be formed from choline, two fatty acids, and glycerol phosphate. Three steps are catalyzed by enzymes in the ER membrane that have their active sites facing the cytosol, where all of the required metabolites are found. Therefore, phospholipid synthesis occurs exclusively in the cytosolic leaflet of the ER membrane. Some of these lipid molecules are later flipped into the opposite leaflet through the action of flippases.

The ER also produces cholesterol and ceramide. Ceramide is made by condensing the amino acid serine with a fatty acid to form the amino alcohol sphingosine, a second fatty acid is then added to form ceramide. The ceramide is exported to the Golgi apparatus, where it serves as a precursor for the synthesis of two types of lipids: oligosaccharide chains are added to form glycosphingolipids, and phosphocholine head groups are transferred from phosphatidylcholine to other ceramide molecules to form sphingomyelin. Thus, both glycolipids and sphingomyelin are produced relatively late in the process of membrane synthesis. Because they are produced by enzymes exposed to the Golgi lumen, they are found exclusively in the noncytosolic leaflet of the lipid bilayers.

The two phospholipid leaflets in the cellular membranes are asymmetric. This asymmetry is established initially in the endoplasmic reticulum and maintained as membrane carriers bud from one compartment and fuse to the next. As a result, domains located at the cytosolic surface of the ER membrane can be identified on the cytosolic surface of transport vesicles, the cytosolic surface of Golgi cisternae, and the cytoplasmic surface of the plasma membrane. Similarly, domains situated at the luminal surface of the ER membrane are found at the exoplasmic surface of the plasma membrane. In fact, the lumen of the ER and other compartments of the secretory pathway is a lot like the extracellular space. They all have high calcium concentration and abundance of proteins with disulfide bonds and carbohydrate chains.

The membranes of different organelles have markedly different lipid compositions, which indicates that changes take place as the membrane flows through the cell. Several factors may contribute to such changes. First, most membranous organelles contain different enzymes that modify lipids already present within a membrane. Second, when vesicles bud from a compartment, some types of lipids may be preferentially included within the membrane of the forming vesicle, while others may be excluded. Third, cells contain phospholipid-transfer proteins that can bind and transport phospholipids through the aqueous cytosol from one membrane compartment to another. These enzymes facilitate the movement of specific phospholipids from the ER to other organelles. This is especially important in delivering lipid to mitochondria and chloroplasts, which are not part of the normal flow of membrane along the biosynthetic pathway.

4.2.3 Other functions in the ER

ER is the start site for biosynthesis in the cell. Some parts of RER are devoid of ribosomes and function as the exit sites where the first transport vesicles in the biosynthesis pathway are formed. The first step in vesicular transport begins from ER to Golgi complex as COP Ⅱ coated vesicles, which will be discussed later in this chapter.

The SER is extensively developed in cells from skeletal muscle, kidney tubules, and steroid-producing endocrine glands. SER functions as the synthesis site of steroid hormones in the endocrine cells of the gonad and adrenal cortex. In the liver, detoxification of a wide variety of organic compounds, including barbiturates and ethanol, is carried out by a system of oxygentransferring enzymes (oxygenases), including the cytochrome P450 family. There is no substrate specificity for these enzymes which are able to oxidize thousands of different hydrophobic compounds and convert them into more hydrophilic, more readily excreted derivatives. Cytochrome P450 metabolize many prescribed medications, and genetic variation in these enzymes among humans may explain differences from one person to the next in the effectiveness and side effects of many drugs.

Another function of ER is sequestering calcium ions from the cytoplasm of cells. SER is abundant in skeletal and cardiac muscle cells (known as the sarcoplasmic reticulum in muscle cells). A Ca^{2+}-ATPase resides on the ER membrane and pumps Ca^{2+} from cytosol into the ER lumen. A high concentration of Ca^{2+}-binding proteins in the ER facilitaes Ca^{2+} storage. The release of Ca^{2+} from ER triggers contraction of muscle cells.

4.3 Golgi apparatus

Golgi apparatus is constituted of a series of flattened membrane vesicles or sacs, called **cisternea**, surrounded by many spherical membrane vesicles. The Golgi cisternae can be divided into cis, medial, and trans cisternae. Each Golgi stack has two distinct faces: an entry (*cis*) face and an exit (*trans*) face (Figure 4-14). The *cis* face is adjacent to the ER, while the *trans* face points toward the plasma membrane. The *cis* face is composed of an interconnected network of tubules referred to as the *cis* Golgi network (CGN), which is thought to function primarily as a sorting station that distinguishes between proteins to be shipped back to the ER and those that are allowed to proceed to the next Golgi station. The *trans*-most face of Golgi contains a distinct network of tubules and vesicles called the *trans* Golgi network (TGN). The TGN is a sorting station where proteins are segregated into different types of vesicles heading either to the plasma membrane or to various intracellular destinations.

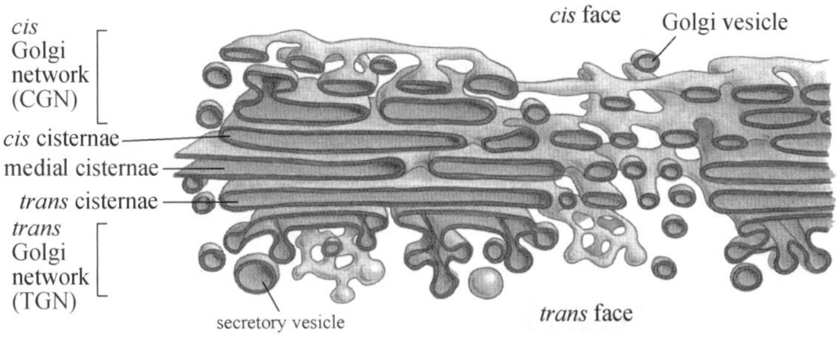

Figure 4-14 Structure of Golgi apparatus (From: Alberts B, et al. Molecular Biology of the Cell. 2008)

Golgi apparatus is a cell structure to further process the proteins synthesized in the ER. It receives proteins and lipids from the ER, modifies them, and then dispatches them to other destinations in the cell. Golgi is a major site of carbohydrate synthesis. Cells make many polysaccharides in the Golgi apparatus, including the pectin and hemicellulose of the cell wall in plants and most of the glycosaminoglycans of the extracellular matrix in animals. A large proportion of the carbohydrates are attached as oligosaccharide side chains to the proteins and lipids as they are transported from ER to Golgi. Most proteins and lipids, once receiving their appropriate oligosaccharides in the Golgi apparatus, are recognized for targeting into the transport vesicles to their destinations.

(1) Glycosylation in Golgi apparatus Many of the *N*-linked oligosaccharides that are added to proteins in the ER undergo further modifications in the Golgi apparatus. In contrast to the glycosylation events that occur in the ER, which assemble a single-core oligosaccharide, the glycosylation steps in the Golgi complex can produce carbohydrate domains of remarkable sequence diversity. As newly synthesized glycoproteins from ER pass through the *cis* and medial cisternae of the Golgi stack, most of the mannose (Man) residues are removed from the core oligosaccharides, while a variety of sugars, such as N-acetylglucosamine (GlcNAC), galactose (Gal), fucose, and sialic acid (NANA), are added to the oligosaccharide by specific glycosyltransferases to yield a finished *N*-linked oligosaccharide. There is a clear correlation between the position of an enzyme in the chain of

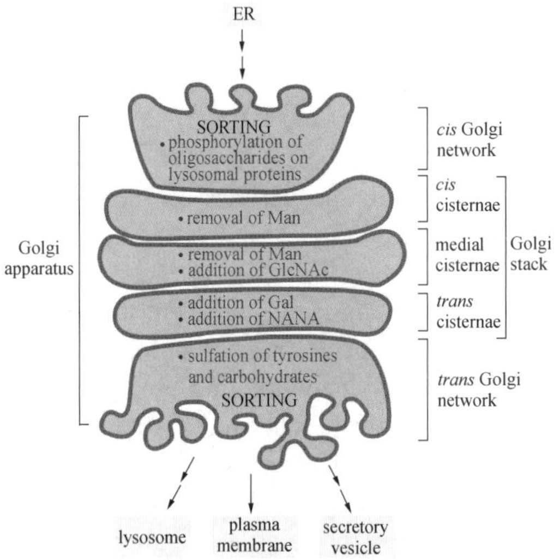

Figure 4-15 Oligosaccharide processing in Golgi apparatus (From: Alberts B, et al. Molecular Biology of the Cell. 2008)

processing events and its localization in the Golgi stack: enzymes that act early are found in cisternae close to the *cis* face, while enzymes that act late are found in cisternae near the *trans* face (Figure 4-15).

Except that many proteins are modified by *N*-linked oligosaccharides, some proteins are glycosylated by *O*-linked oligosaccharides in Golgi apparatus. *O*-linked oligosaccharides are bound to the hydroxyl groups in certain serine, threonine, or hydroxyl lysine residues. They are generally short, often containing only one to four sugar residues. *O*-linked oligosaccharides are formed by the sequential addition of sugars in the Golgi. Which is catalyzed by various glycosyltransferases that are specific for the donor sugar nucleotide and acceptor molecule. The Golgi apparatus confers the heaviest *O*-linked glycosylation on proteoglycan core proteins to produce proteoglycans. This process involves the polymerization of one or more glycosaminoglycan chains via a xylose link onto serines on a core protein. Many proteoglycans are secreted and become components of the extracellular matrix, while others remain anchored to the extracellular face of the plasma membrane. Some proteoglycans form a major component of slimy materials, such as the mucus that is secreted to form a protective coating on the surface of many epithelia. The sugars incorporated into glycosaminoglycans are heavily sulfated in the Golgi apparatus immediately after these polymers are made, thus adding a significant portion of their characteristically large negative charge. Some tyrosines in proteins also become sulfated shortly before they exit from the Golgi apparatus.

The vast abundance of glycoproteins and the complicated pathways to synthesize them suggest that the oligosaccharides on glycoproteins and glycosphingolipids have very important functions. First, it has a direct role in making folding intermediates more soluble, thereby preventing their aggregation. Second, the sequential modifications of the *N*-linked oligosaccharide establish a "glyco-code" that marks the progression of protein folding and mediates the binding of the protein to chaperones (discussed in the ER part) in guiding ER-to-Golgi transport. Third, the presence of oligosaccharides tends to make a glycoprotein more resistant to digestion by proteolytic enzymes. Glycosylation can also have important regulatory roles in cell signaling. The recognition of sugar chains by lectins in the extracellular space is important in many developmental processes and cell-cell recognition.

(2) Material transport through the Golgi apparatus How materials move through the various compartments of the Golgi complex has long been investigated. Up until the mid-1980s, it was generally accepted that Golgi cisternae were transient structures and Golgi cisternae formed at the *cis* face of the stack by fusion of membranous carriers from the ER and ERGIC (endoplasmic reticulum Golgi intermediate compartment) and that each cisterna physically moved from the *cis* to the trans end of the stack, changing in composition as it progressed. This is known as the **cisternal maturation model** (Figure 4-16A). In such a model, each cisterna "matures" into the next cisterna along the stack. From the mid-1980s to the mid-1990s, an alternate model called the **vesicular transport model** arose (Figure 4-16B). In the vesicular transport model, it was proposed that the cisternae of a Golgi stack remain in place as stable compartments. Cargo (i. e., secretory, lysosomal, and membrane proteins) is shuttled through the Golgi stack, from the CGN to the TGN, in vesicles that bud from one membrane compartment and fuse with a neighboring compartment farther along the stack. The vesicular transport and the cisternal maturation models are not mutually exclusive. Indeed, evidence suggests that transport may occur by a combination of the two mechanisms, in which some cargo is moved forward (anterograde) or backward (retrograde) rapidly in transport vesicles, whereas other cargo is moved forward more slowly as the Golgi apparatus constantly renews itself through cisternal maturation.

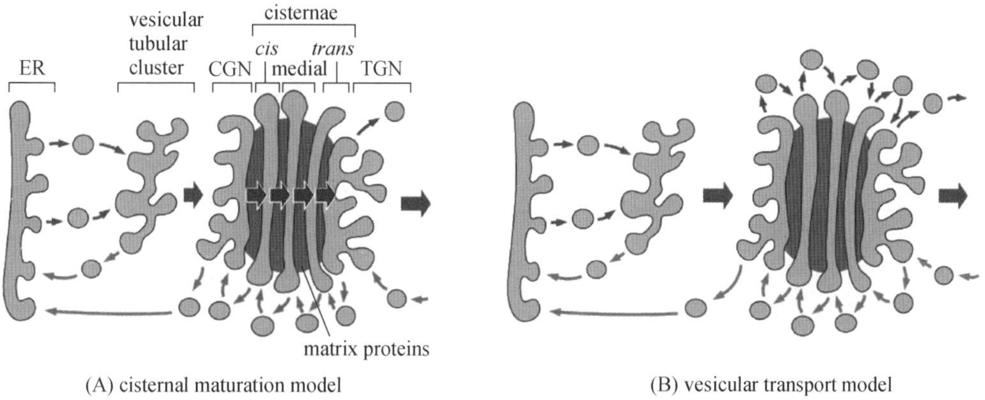

Figure 4-16 The dynamic transport through Golgi apparatus (From: Alberts B, et al. Molecular Biology of the Cell. 2008)

4.4 Lysosomes and peroxisomes

4.4.1 Lysosomes

Lysosomes are found in all eukaryotic cells as tiny organelles lined by a single layer of membrane. The appearance of lysosomes in electron micrographs is quit divergent and varies in size, shape and electron density. Lysosomes are formed off the membrane of the *trans* Golgi and contain digestive enzymes that degrade worn-out organelles, as well as macromolecules and particles taken into the cell by endocytosis.

The enzymes of a lysosome share an important property: all have their optimal activity at an acid pH and thus are acid hydrolases. The pH optimum of these enzymes is suited to the low pH, approximately 4.6, in the lysosomal compartment. The high internal proton concentration is maintained by a proton pump (an H^+-ATPase) present in the membrane. Lysosomal membranes contain a variety of highly glycosylated integral proteins whose carbohydrate chains are thought to form a protective layer that shields the membrane from attack by the enclosed enzymes. The lysosomal membrane contains transport proteins that allow the final products of the digestion of macromolecules, such as amino acids, sugars, and nucleotides, to be transported to the cytosol, from where they can be either excreted or utilized by the cell.

The specialized digestive enzymes and membrane proteins of the lysosome are synthesized in the ER and transported through the Golgi apparatus to the *trans* Golgi network. *N*-linked oligosaccharides are added to lysosomal enzymes cluring their transport to lysosomes and to prevent their secretion. The addition and initial processing of *N*-linked oligosaccharide precursors in the rough ER are the same for lysosomal enzymes as for membrane and secretory proteins. In the *cis* Golgi, one or more mannose residues in the resulting *N*-linked oligosaccharides become phosphorylated to obtain a mannose 6-phosphate (M6P) residue (Figure 4-17). The M6P residues are then bound to mannose 6-phosphate receptors, which are found primarily in the *trans* Golgi network. Vesicles containing the M6P receptors and bound lysosomal enzymes bud from the *trans* Golgi network and then fuse with a sorting vesicle, an organelle often termed the **late endosome**. The bound lysosomal enzymes are released within late endosomes. and, a phosphatase within late endosomes generally removes the phosphate from lysosomal enzymes, preventing their rebinding to the M6P receptor. Two types of vesicles bud from late endosomes. One type containing lysosomal enzymes but not the M6P receptor, delivering the lysosomal enzyme to their final destination. The other type of vesicle recycles the M6P receptor back to the *trans* Golgi network or, on occasion, to the cell surface.

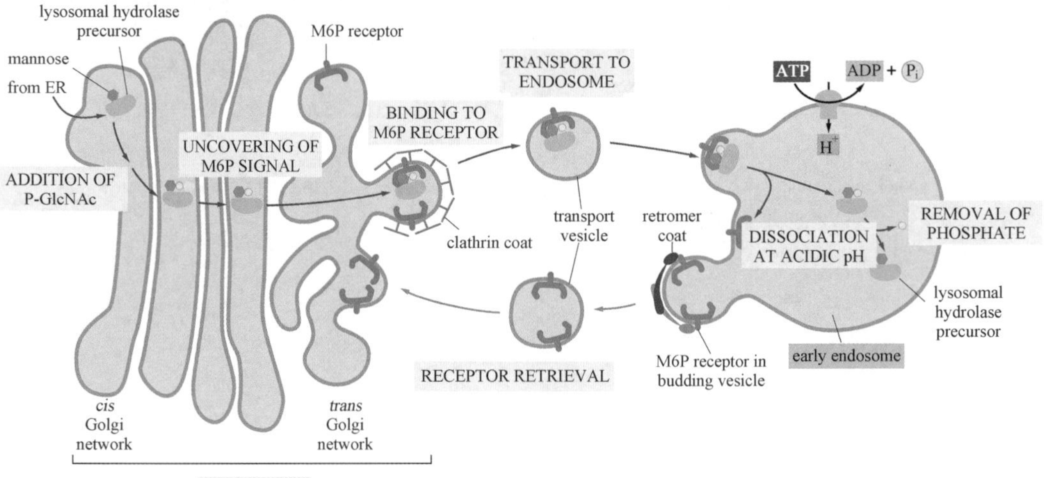

Figure 4-17 The mannose 6-phosphate (M6P) pathway for targeting lysosomal enzymes to lysosomes
(From: Alberts B, et al, Molecular biology of the cell, 2008)

Materials follow different paths to enter lysosomes depending on their source. The extracellular particles are taken up into **phagosomes**, which then fuse with lysosomes, and the extracellular fluid and macromolecules are taken up into smaller endocytic vesicles, which deliver their contents to lysosomes via endosomes. But cells have an additional pathway for supplying materials to lysosomes, this pathway is used for degrading obsolete parts of the cell itself. In electron micrographs of liver cells, for example, one often sees lysosomes digesting mitochondria, as well as other organelles. The process seems to begin with the enclosure of the organelle by a double membrane, creating an **autophagosome**, which then fuses with lysosomes.

The dysfunction of lysosomes might lead to severe diseases. The deficiency of a single lysosomal enzyme and the corresponding accumulation of undegraded substrate in lysosomes leads to so called **lysosomal storage disorders**. Such as Pompe disease caused by the absence of a lysosomal enzyme, α-glucosidase, could lead to undigested glycogen accumulating in lysosomes, swelling of the organelles and irreversible damage to the cells and tissues. In I-cell disease, deficiency of an enzyme (N-acetylglucosamine phosphotransferase) required for mannose phosphorylation of lysosomal enzymes in the Golgi complex leads to the loss of an "address" for delivery of these proteins to lysosomes. Many cells in these patients contain lysosomes that are bloated with undegraded materials. Gaucher's disease, a deficiency of the lysosomal enzyme glucocerebrosidase, is characterized by accumulating large quantities of glucocerebroside lipids in the lysosomes of patient macrophages, causing spleen enlargement and anemia.

Most plant cells and fungal cells (including yeasts) contain one or several very large fluid-filled vesicles called **vacuole**, which belongs to lysosomes. Plant vacuoles carry out a wide spectrum of essential functions. It functions as a storage for both nutrients and waste products. Vacuoles may also store a host of toxic compounds, such as cyanide-containing glycosides and glucosinolates, which are part of an arsenal of chemical weapons that are released when the cell is injured by an herbivore or fungus. The membrane that bounds the vacuole contains many active transport systems that pump ions into the vacuolar compartment. Such active transport leads to high ion concentration in vacuoles so that water enters the vacuole by osmosis. Hydrostatic pressure exerted by the vacuole not only provides mechanical support for the soft tissues of a plant, but also stretches the cell wall during cell growth.

4.4.2 Peroxisomes

Peroxisomes are small organelles enclosed by a single membrane. They contain enzymes used in a variety of oxidative reactions that break down lipids and destroy toxic molecules. Unlike mitochondria and chloroplasts, peroxisomes lack DNA and ribosomes and are lined by a single layer membrane. Thus, all peroxisomal proteins are encoded by nuclear genes, synthesized on ribosomes free in the cytosol, and then incorporated into pre-existing peroxisomes. Most peroxisomal membrane and matrix proteins are incorporated into the organelle post-translationally, although some of them enter the peroxisome membrane via the ER. The oxidative enzymes, such as catalase and urate oxidase, are folded in the cytosol and incorporated as a folded protein. As peroxisomes are enlarged by addition of protein (and lipid), they eventually divide, forming new ones (Figure 4-18).

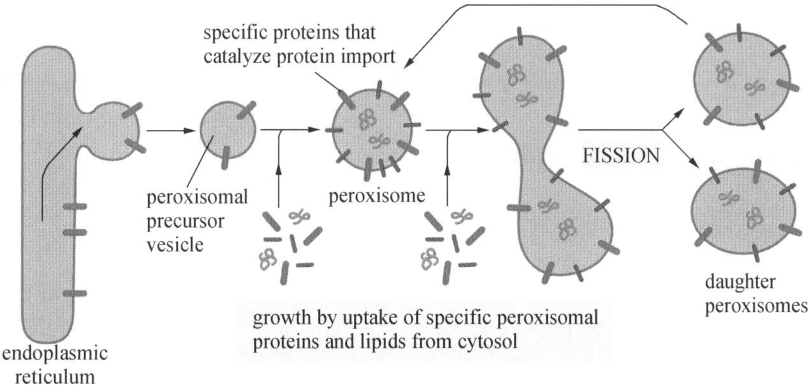

Figure 4-18 Model for peroxisome generation and proliferation (From: Alberts B, et al. Molecular Biology of the Cell. 2008)

Many peroxisomal matrix proteins, such as catalase and fatty acyl CoA oxidase, utilize a C-terminal Ser-Lys-Leu (**SKL**) **targeting sequence** that is not cleaved after import. Such proteins bind to the cytosolic receptor PTS1R, which escorts the imported protein to transporter proteins on the peroxisomal membrane. After transported across the membrane, the imported protein dissociates with PTSIR in the lumen, and PTS1R returns to cytosol to pick up another peroxisome-destined protein. Other peroxisomal matrix proteins, such as thiolase, are synthesized as precursors with an N-terminal uptake-targeting sequence. Proteins with this type of signal sequence bind to a cytosolic receptor protein named PTS2R and are escorted to the peroxisomal membrane. Following the import of such proteins, the N-terminal targeting sequence is cleaved.

Like mitochondria, peroxisomes are major sites of oxygen utilization. However, oxidative reactions performed in peroxisomes are not coupled to ATP formation. In peroxisomes, molecular oxygen is used to remove hydrogen atoms from specific organic substrates (designated as R) in an oxidation reaction that produces hydrogen peroxide (H_2O_2): $RH_2 + O_2 \rightarrow R + H_2O_2$. Catalase uses the H_2O_2 to oxidize a variety of other substrates — including phenols, formic acid, formaldehyde, and alcohol — by the "peroxidation" reaction: $H_2O_2 + R'H_2 \rightarrow R' + 2H_2O$. When excess H_2O_2 accumulates in the cell, catalase converts it to H_2O through the reaction: $2H_2O_2 \rightarrow 2H_2O + O_2$.

This type of oxidation reaction is particularly important in liver and kidney cells, where the peroxisomes detoxify various toxic molecules that enter the bloodstream. About 25% of the ethanol we drink is oxidized to acetaldehyde in this way. Another major function of the oxidation reactions performed in peroxisomes is the breakdown of fatty acid molecules through β oxidation. Thus the fatty acids are converted to acetyl CoA in peroxisomes and then exported to the cytosol for reuse in biosynthetic reactions. In mammalian cells, β oxidation occurs in both mitochondria and peroxisomes; in yeast and plant cells, however, this essential reaction occurs exclusively in peroxisomes. An essential biosynthetic function of animal peroxisomes is to catalyze the first reactions in the formation of plasmalogens, which are the most abundant class of phospholipids in myelin. Plasmalogen deficiencies cause profound abnormalities in the myelination of nerve cell axons, which is why many peroxisomal disorders lead to neurological disease.

Peroxisomes are also important in plants. Plant peroxisomes present in leaves participate in photorespiration. The other type of peroxisome is present in germinating seeds, where it converts the fatty acids stored in seed lipids into the sugars needed for the growth of the young plant. In a process called glyoxylate cycle, two molecules of acetyl CoA produced by fatty acid breakdown in the peroxisome are used to make succinic acid, which then leaves the peroxisome and is converted into glucose in the cytosol. The glyoxylate cycle does not occur in animal cells, and animals are therefore unable to convert the fatty acids in fats into carbohydrates.

4.5 Molecular mechanisms of vesicular transport

As to eukaryotic cells, the traffic of molecules between different membrane-enclosed compartments within a cell, or between the interior of the cell and its surroundings is highly organized by **vesicular transport**. In the outward **biosynthetic-secretory pathway** (**exocytosis**), molecules are packaged into **transport vesicles**, leading from ER, through Golgi apparatus, plasma membrane and to the cell surface. At the Golgi apparatus, a side branch budding of vesicles transports molecules to endosomes and then lysosomes. The major inward **endocytic pathway** (**endocytosis**) transports materials uptake at the cell plasma membrane, through endosomes, to lysosomes. (Figure 4-19).

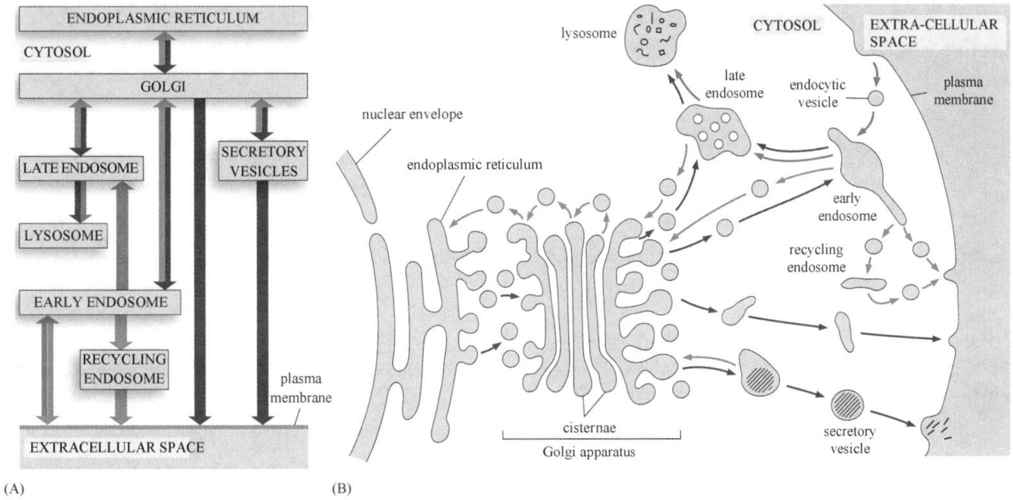

Figure 4-19 Exocytosis and endocytosis pathways (From: Alberts B, et al. Molecular Biology of the Cell. 2008)

4.5.1 Three types of coated transport vesicles

Transport vesicles have a distinctive protein coat (**coated vesicle**) on their cytosolic surface when budding from membranes. Based on the coat protein, transport vesicles are classified into three major types: **Clathrin**, **COP I** and **COP II** coated vesicles (Figure 4-20). COP I -coated vesicles are involved in transporting molecules from Golgi apparatus to the ER and between the Golgi cisternae. Vesicles with a COP II coat transport proteins from the ER to the *cis* Golgi apparatus. Clathrin-coated vesicles bud from the Golgi apparatus on the outward secretory pathway and from the plasma membrane on the inward endocytic pathway.

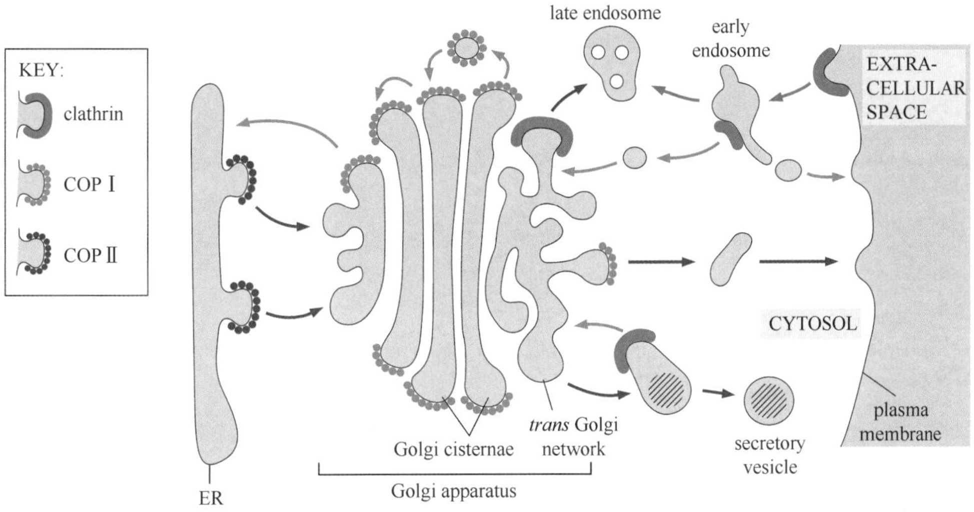

Figure 4-20 Different coats used for different steps in vesicle traffic (From: Alberts B, et al, Molecular biology of the cell. 2008)

Different coat proteins assembly requires a family of small monomeric GTP-binding proteins, so-called **coat-recruitment GTPases**, such as Arf proteins and Sarl protein (Table 4-3). These GTPases function to regulate coat assembly and disassembly by switching between an inactive GDP-bound and an active GTP-bound state.

Table 4-3 Some types of coated vesicles

type	GTPase	coat protein	origin	destination
clathrin	Arf	clathrin, adaptin2	plasma membrane	endosomes
clathrin	Arf	clathrin, adaptin1	*trans* Golgi	endosomes
clathrin	Arf	clathrin, adaptin3	Golgi	lysosomes, melanosome, or platelet vesicles
COP Ⅰ	Arf	COP proteins	*cis* Golgi	ER
COP Ⅰ	Arf	COP proteins	later Golgi cisterna	earlier Golgi cisterna
COP Ⅱ	Sarl	Sec23/Sec24 and Sec13/Sec31 complexes, Sec16, Sec12	ER	*cis* Golgi

In order to load correct cargo proteins in the correct transport vesicles, discrimination among potential membrane and soluble cargo proteins is required. The vesicle coat functions in selecting specific proteins as cargo by directly binding to specific sequences, or sorting signals (Table 4-4). The polymerized coat thus acts as an affinity matrix to cluster selected membrane cargo proteins into forming vesicle buds. Soluble proteins within the lumen of parent organelles can in turn be selected by binding to the luminal domains of certain membrane cargo proteins, which act as receptors for luminal cargo proteins.

Table 4-4 Known sorting signals that direct proteins to specific transport vesicles

signal sequence*	proteins with signal	signal receptor	vesicles
Lys-Asp-Glu-Leu (KDEL)	ER-resident luminal proteins	KDEL receptor in *cis* Golgi membrane	COP Ⅰ
Lys-Lys-X-X (KKXX)	ER-resident membrane proteins (cytosolic domain)	COP Ⅰ α and β subunits	COP Ⅰ
Di-acidic (e. g., Asp-X-Glu)	cargo membrane proteins in ER (cytosolic domain)	COP Ⅱ Sec24 subunit	COP Ⅱ
mannose 6-phosphate (M6P)	soluble lysosomal enzymes after processing in *cis* Golgi	M6P receptor in *trans* Golgi membrane	clathrin/AP1
	secreted lysosomal enzymes	M6P receptor in plasma membrane	clathrin/AP2
Asn-Pro-X-Tyr (NPXY)	LDL receptor in the plasma membrane (cytosolic domain)	AP2 complex	clathrin/AP2
Tyr-X-X-Φ (YXXΦ)	membrane proteins in *trans* Golgi (cytosolic domain)	AP1 (μ_1 subunit)	clathrin/AP1
	plasma membrane proteins (cytosolic domain)	AP2 (μ_2 subunit)	clathrin/AP2
Leu-Leu (LL)	plasma membrane proteins (cytosolic domain)	AP2 complexes	clathrin/AP2

* X: any amino acid. Φ: hydrophobic amino acid. Single-letter amino acid abbreviations are in parentheses.

(1) Formation of clathrin-coated vesicles Clathrin-coated vesicles are the best-studied vesicles. The formation of this transport vesicle requires various adapter proteins (**adaptin**) as well as **clathrin**, and their final budding off requires a small GTP-binding protein, called **dynamin**.

Each clathrin consists of three large and three small polypeptide chains that together form a three-legged structure called a triskelion. During vesicle formation, the cargo proteins carrying specific transport signals are recognized by cargo receptors in the compartment membrane. Adaptins bind to the cargo receptors via a four amino acid motif on the receptors, Tyr-X-X-Φ (X for any amino acid, and Φ for hydrophobic amino acids such as Phe, Leu or Met), to help selecting cargo molecules. There are several types of adaptins that are specific for a different set of cargo receptors to form distinct clathrin coated vesicles (Figure 4-21). On the other hand, adaptins bind and secure the clathrin coat to the vesicle membrane. The clathrins assembles into a basketlike network on the cytosolic surface of the membrane to shape it into a **clathrin-coated pit**. Dynamin assembles as a ring around the neck of each invaginated coated pit and contracts by consuming the energy generated from GTP hydrolysis until the vesicle pinches off. After budding from its parent organelle, the vesicle sheds its coat, allowing its membrane to interact directly with the membrane to which it will fuse.

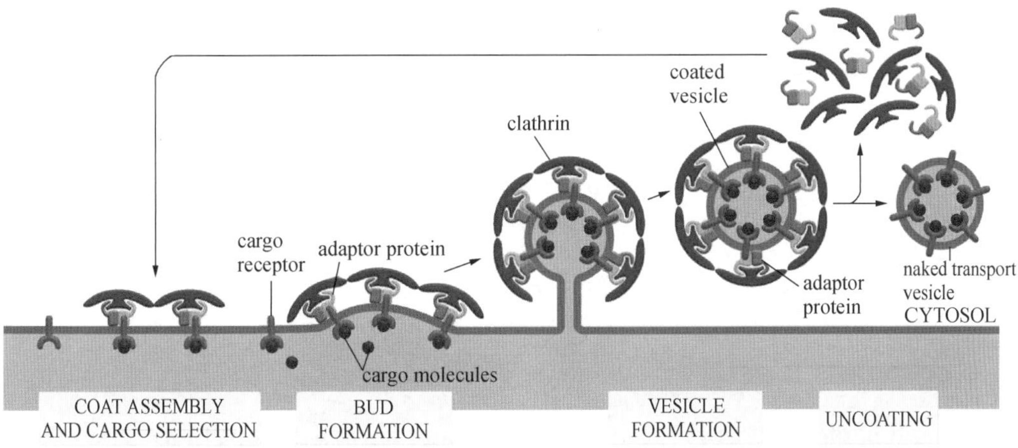

Figure 4-21 Formation of clathrin coated vesicles (From: Alberts B, et al. Molecular Biology of the Cell. 2008)

The inositol phospholipids of the vesicle membrane also play important role in vesicle formation (Figure 4-22). The adaptor proteins found in clathrin coats also bind to phosphoinositides (PIPs), which are used as molecular markers of compartment identity and direct when and where coats assemble in the cell. Various types of PIPs can be produced through rapid cycles of phosphorylation and dephosphorylation at the 3′, 4′, and 5′ positions of their inositol sugar head groups by the distinct sets of PI and PIP kinases and PIP phosphatases located on different organelles in the endocytic and biosynthetic-secretory pathways. As a consequence, the distribution of PIPs varies from organelle to organelle, and even within a continuous membrane from one region to another, thereby defining specialized membrane domains (Figure 4-22G). Adaptor proteins contain domains that bind with high specificity to the head groups of particular PIPs to help regulate vesicle formation and other steps in membrane transport.

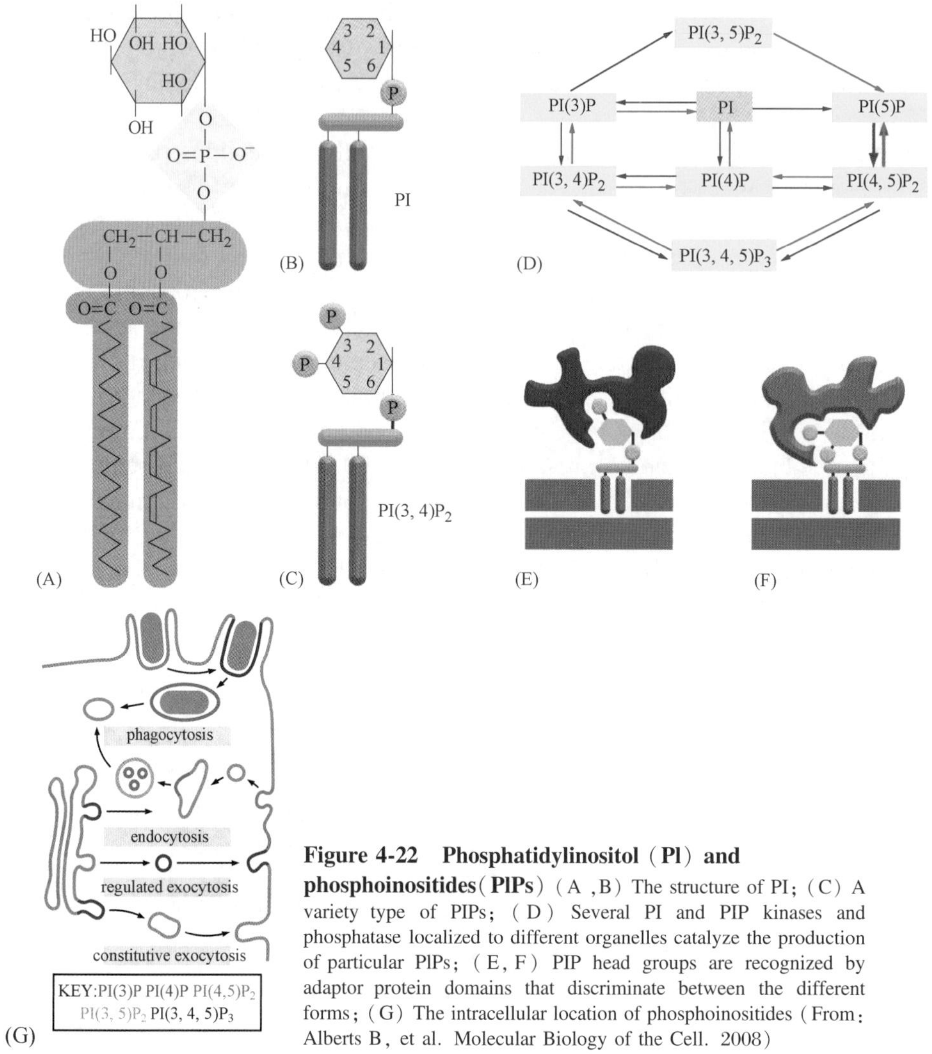

Figure 4-22 Phosphatidylinositol (PI) and phosphoinositides (PIPs) (A, B) The structure of PI; (C) A variety type of PIPs; (D) Several PI and PIP kinases and phosphatase localized to different organelles catalyze the production of particular PIPs; (E, F) PIP head groups are recognized by adaptor protein domains that discriminate between the different forms; (G) The intracellular location of phosphoinositides (From: Alberts B, et al. Molecular Biology of the Cell. 2008)

(2) Formation of COP Ⅱ vesicles The first step in the secretory pathway, from the ER to Golgi, is mediated by COP Ⅱ vesicles. The formation of COP Ⅱ vesicles is triggered by a guanine nucleotide — exchange factor, Sec12-mediated GDP/GTP exchange on Sar1 GTPase. The GTP-bound Sar1 is recruited to the ER membrane followed by binding of a complex of Sec23 and Sec24 proteins. A second complex comprising Sec13 and Sec31 proteins then binds to the previous complexes to complete the coating structure. Sec16, a large fibrous protein, is bound to the cytosolic surface of the ER, interacts with the Sec13/31 and Sec23/24 complexes, and acts to organize the other coat proteins, increasing the efficiency of coat polymerization. Certain integral ER membrane proteins are specifically recruited into COP Ⅱ vesicles for transport to the Golgi. Many of these proteins contain a di-acidic sorting signal (such as Asp-X-Glu) in the cytosolic segments. This sorting signal binds to the Sec24 subunit of the COP Ⅱ and is essential for the selective export of certain membrane proteins from the ER. These vesicles bud from specialized regions of the ER, called ER exit sites, whose membrane lacks bound ribosomes.

(3) Formation of COP Ⅰ vesicles The start of formation of COPⅠ vesicles requires the membrane recruitment of small GTP-binding protein Arf1. COP Ⅰ vesicles mediate retrograde transport within the Golgi and from the Golgi to the ER and transport escaped proteins back to the ER. Proteins that normally reside in the ER contain retrieval signals at their C-termini which can be captured by specific receptors in COP Ⅰ -coated vesicles when the proteins escape from ER vesicles to the Golgi apparatus. Soluble resident proteins of the ER lumen (such as protein disulfide isomerase and the molecular chaperones that facilitate folding) typically possess the retrieval signal "lys-asp-glu-leu" (KDEL). These proteins are recognized and returned to the ER by the KDEL receptor, an integral membrane protein that shuttles between the *cis* Golgi and the ER compartments. Membrane proteins that reside in the ER also have a retrieval signal with two closely linked basic residues, most commonly "lys-lys-X-X" (KKXX, where X is any residue), that binds to the COP Ⅰ coat, facilitating their return to the ER.

Another aspect of ER quality control is the retention of unassembled or misfolded proteins. Proteins that fold up incorrectly, and dimeric or multimeric proteins that fail to assemble properly, are actively retained in the ER by binding to chaperone proteins, such as Bip and Calnexin, that reside there. Interaction with chaperones holds the proteins in the ER until proper folding occurs; otherwise, the proteins are dragged to the cytosol and ultimately degraded. Antibody molecules, for example, are composed of four polypeptide chains that assemble into the complete antibody molecule in the ER. Partially assembled antibodies are retained in the ER until all four polypeptide chains have assembled; any antibody molecule that fails to assemble properly is ultimately degraded.

4.5.2 Targeting vesicles to a particular compartment

After a transport vesicle buds from a membrane, it is actively transported by motor proteins that move along cytoskeleton fibers to its destination. Once a transport vesicle has reached its target, it must recognize and dock with the organelle. The vesicle membrane fuses with the target membrane and unloads the vesicle's cargo. Targeting vesicles to a particular compartment requires specific interactions between different membranes. Two types of proteins are crucial in these processes: Rab proteins direct the vesicle to specific spots on the correct target membrane and then SNARE proteins mediate the fusion of the lipid membrane.

(1) Rabs Like coat-recruitment GTPases, Rabs belong to a family of small G proteins and associate with membranes by a lipid anchor when bound GTP. Over 60 different Rab genes were identified in humans, which constitute the most diverse group of proteins involved in membrane trafficking. Different Rabs associated with different membrane compartments give each compartment a unique surface identity. The GTP-bound Rabs play a key role in vesicle targeting by recruiting specific cytosolic tethering proteins to specific membrane surfaces. Each Rab appears to bind to a specific Rab effector, a typically long coiled-coil protein, associated with the target membrane.

The Rab effectors vary greatly in structure. Some are motor proteins that propel vesicles along the cell skeleton to their target membrane. Others are tethering proteins containing long threadlike domains that link two membranes. Rab effectors can also interact with SNAREs coupling membrane tethering to fusion (Figure 4-23). Rab5, for instance, is anchored to the membrane of endosomes upon GTP binding to mediate the capture of clathrin-coated vesicles budding from plasma membrane. Simultaneously, Rab5 activates PI3-kinase to convert PI to PI(3)P which in turn binds some Rab effectors and helps to establish functionally distinct membrane domains within a continuous membrane. Different

Rab proteins help to create multiple specialized membrane domains. Rab5 receives incoming vesicles from plasma membranes, Rab11 and Rab4 domains in the same membrane are thought to organize the budding of recycling vesicles that return proteins from endosomes to the plasma membrane.

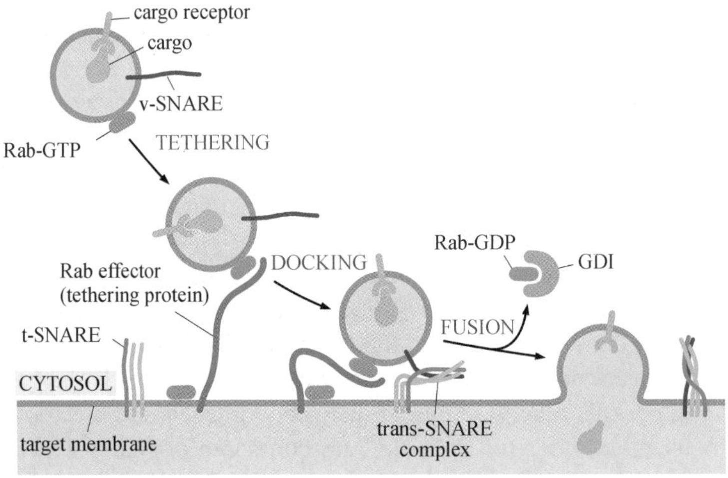

Figure 4-23 Tethering of a vesicle to a target membrane by Rabs and SNAREs (From: Alberts B, et al. Molecular Biology of the Cell. 2008)

(2) SNAREs SNAREs are transmembrane proteins responsible for the fusion of transport vesicles to the target membrane. Uncoating of transport vesicles exposes specific v-SNARE proteins on the surface of each type of vesicle. Each v-SNARE recognizes and binds to the complementary t-SNARE proteins on the membrane of target vesicles or organelles. Each organelle or each type of transport vesicle is believed to carry a unique SNARE, and the interactions between complementary SNAREs help ensuring that transport vesicles fuse only with the correct membrane.

Once a transport vesicle has recognized its target membrane and docked there, the vesicle has to fuse with the membrane to deliver its cargo. Fusion not only delivers the contents of the vesicle into the interior of the target organelle, but also adds the vesicle membrane to the membrane of the organelle. Membrane fusion does not always follow immediately after docking, however, it often awaits a specific molecular signal. Although the SNARE proteins themselves are thought to play a central role in the fusion process, some other proteins, probably Rabs, are also involved in membrane fusion.

4.6 Secretory pathways

Newly made proteins, lipids, and carbohydrates are delivered from the ER, via the Golgi apparatus, to the cell surface by **transport vesicles** that fuse with the plasma membrane in a process called **exocytosis** (Figure 4-24). Each molecule that travels along this route passes through a fixed sequence of membrane-enclosed compartments and is often modified along this route. During this process, polypeptides in the membrane and lumen of the ER undergo five principal modifications before they reach their final destinations: formation of disulfide bonds; proper folding; addition and processing of carbohydrates; specific proteolytic cleavages; and assembly into multimeric proteins.

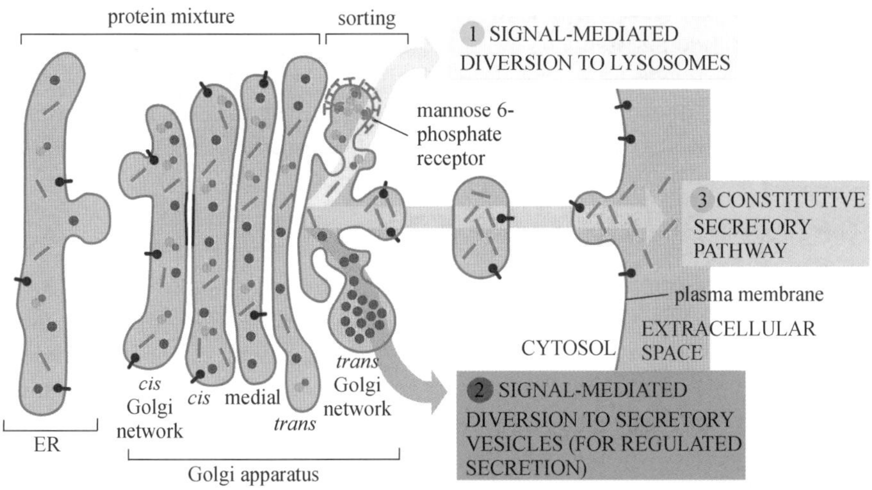

Figure 4-24 The regulated and constitutive pathways of exocytosis in secretory cells
(From: Alberts B, et al. Molecular Biology of the Cell. 2008)

4.6.1 Constitutive exocytosis

There are two exocytosis pathways that proteins and lipids from the *trans* Golgi network are transported to plasma membrane and outside cells. All eukaryotic cells have the **constitutive exocytosis** pathway that a steady stream of vesicles bud from the *trans* Golgi network and fuse with the plasma membrane. This constitutive exocytosis pathway operates continually and supplies newly made lipids and proteins to the plasma membrane. The constitutive pathway also carries proteins to the cell surface to be released to the outside, a process called **secretion**. Some of the released proteins adhere to the cell surface, where they become peripheral proteins of the plasma membrane, some are incorporated into the extracellular matrix or diffuse into the extracellular fluid to nourish or to signal other cells. Because entry into this nonselective pathway does not require a particular signal sequence, it is sometimes referred to as the **default pathway**. In an unpolarized cell such as a white blood cell, it seems any protein in the lumen of the Golgi apparatus is automatically carried by the constitutive pathway to the cell surface unless it is specifically returned to the ER, retained as a resident protein in Golgi or selected for the pathways that lead to regulated secretion or lysosomes.

4.6.2 Regulated exocytosis

In addition to the constitutive exocytosis pathway, which operates continually in all eukaryotic cells, there is a **regulated exocytosis pathway**, which operates only in cells that are specialized for secretion. Specialized secretory cells produce large quantities of particular products, such as hormones, mucus, or digestive enzymes, which are stored in **secretory vesicles** for later release. These vesicles bud off from the *trans* Golgi network and accumulate near the plasma membrane. There they wait for the extracellular signal that will stimulate them to fuse with the plasma membrane and release their contents to the cell exterior. An increase in blood glucose, for example, signals cells in the pancreas to secrete the hormone insulin. The signal is often a chemical messenger that binds to receptors on the cell surface. The resulting activation of the receptors generates intracellular signals, often including a transient increase in the concentration of free Ca^{2+} in the cytosol. In nerve terminals, the initial signal for exocytosis is usually an electrical excitation triggered by a chemical transmitter binding to receptors on the same cell surface. When the action potential reaches the nerve terminals, it causes an influx of Ca^{2+} through voltage-gated Ca^{2+} channels. The binding of Ca^{2+} ions to specific sensors then triggers the secretory vesicles (called synaptic vesicles) to fuse with the plasma membrane and release their contents to the extracellular space.

4.6.3 Secretory vesicles

Secretory vesicles bud from the *trans* Golgi network. Proteins destined for secretory vesicles are sorted and packaged in the *trans* Golgi network. Proteins that travel by this pathway have special surface properties that cause them to aggregate with one another under the ionic conditions (acidic pH and high Ca^{2+}) that prevail in the *trans* Golgi network. The aggregated proteins are packaged into secretory vesicles, which pinch off from the network. Proteins secreted by the constitutive pathway do not aggregate and are therefore carried automatically to the plasma membrane by the transport vesicles of the constitutive pathway. The selective aggregation has another function: it allows secretory proteins to be packaged into secretory vesicles at concentrations much higher than the concentration of the unaggregated protein in the Golgi lumen. This increase in concentration can reach up to 200 fold, enabling secretory cells to release large amounts of the protein promptly when triggered to do so.

Secretory and membrane proteins become concentrated as they move from the ER through the Golgi apparatus because of an extensive retrograde retrieval process mediated by COP I -coated transport vesicles. Membrane recycling is important for returning Golgi components to the Golgi apparatus, as well as for concentrating the contents of secretory vesicles. The vesicles that mediate this retrieval originate as clathrin-coated buds on the surface of immature secretory vesicles.

Besides the concentration of secretory proteins during secretory vesicles mature, proteins are often proteolytically processed. Many polypeptide hormones and neuropeptides, as well as many secreted hydrolytic enzymes, are synthesized as inactive protein precursors which require proteolysis to liberate the active molecules from these precursors. The cleavages begin in the *trans* Golgi network, and continue in the secretory vesicles and sometimes in the extracellular fluid after secretion has occurred.

When a secretory vesicle fuses with the plasma membrane, its contents are discharged from the cell by exocytosis, and its membrane becomes part of the plasma membrane. Very rapidly, membrane components are removed from the surface by endocytosis almost as fast as they are added by exocytosis. After their removal from the plasma membrane, the proteins of the secretory vesicle membrane are either recycled or shuttled to lysosomes for degradation. Control of membrane traffic thus has a major role in maintaining the composition of the various membranes of the cell. To maintain each membrane-enclosed compartment in the secretory and endocytic pathways at a constant size, the balance between the outward and inward flows of membrane needs to be precisely regulated.

Most cells in tissues are polarized and have two or more distinct plasma membrane domains that are the targets of different types of vesicles. Epithelial cells, for example, frequently secrete one set of products such as digestive enzymes or mucus in cells lining the gut at their apical surface and another set of products such as components of the basal lamina at their basolateral surface. By examining polarized epithelial cells in culture, it has been found that proteins from the ER destined for different domains travel together until they reach the TGN. There they are separated and dispatched in secretory or transport vesicles to the appropriate plasma membrane domain. The apical plasma membrane of most epithelial cells is greatly enriched in glycosphingolipids. Similarly, plasma membrane proteins that are linked to the lipid bilayer by GPI anchor are found predominantly in the apical plasma membrane. GPI-anchored proteins are thought to be directed to the apical membrane because they associate with glycosphingolipids in lipid rafts that form in the membrane of the TGN. Thus, similar to the selective partitioning of some membrane proteins into the specialized lipid domains in caveolae at the plasma membrane, lipid domains may also participate in protein sorting in the TGN. Membrane proteins destined for delivery to the basolateral membrane contain sorting signals in their cytosolic tails. Such signals are recognized by coat proteins that package them into appropriate transport vesicles in the TGN. The same basolateral signals that are recognized in the TGN also function in endosomes to redirect the proteins back to the basolateral plasma membrane after they have been endocytosed.

4.7 Endocytic pathways

Eukaryotic cells are continually taking up fluid, as well as large and small molecules, by the process of **endocytosis**. Specialized cells are also able to internalize large particles and even other cells. The material to be ingested is progressively enclosed by a small portion of the plasma membrane, which first buds inward and then pinches off to form an intracellular endocytic vesicle, called **endosomes**. Endosomes transport their contents in a series of steps to a lysosome, which subsequently digests the materials. The mebabolites generated by digestion are transferred directly out of the lysosome into the cytosol, where they can be used by the cell. In other instances, however, endosomes are used by the cell to transport various substances between different portions of the external cell membrane.

Two main types of endocytosis are distinguished based on the size of the endocytic vesicles formed. **Pinocytosis** (cellular drinking) involves the ingestion of fluid and molecules via small vesicles (< 150 nm in diameter). **Phagocytosis** (cellular eating) involves the ingestion of large particles, such as microorganisms and cell debris, via large vesicles called phagosomes (generally > 250 nm in diameter). All eukaryotic cells are continually ingesting fluid and molecules by pinocytosis, whereas large particles are ingested mainly by specialized phagocytic cells. In this final section, we trace the endocytic pathway from the plasma membrane to lysosomes.

4.7.1 Phagocytosis and pinocytosis

The most dramatic form of endocytosis, phagocytosis, was first observed more than a hundred years ago. In protozoa, phagocytosis is a form of feeding: microorganisms ingest large particles, such as bacteria, by taking them up into phagosomes; these phagosomes fuse with lysosomes, where the food particles are digested. Few cells in multicellular organisms can ingest large particles efficiently. In the animal gut, for example, large particles of food have to be broken down into individual molecules by extracellular enzymes before they can be taken up by the absorptive cells lining the gut. Nevertheless, phagocytosis is important in most animals for purposes other than nutrition. Phagocytic cells including macrophages defend us against infection by ingesting invading microorganisms. To be taken up by a macrophage or other white blood cell, bacteria must first bind to the phagocytic cell surface, and activates one of a variety of surface receptors, and induce the phagocytic cell to extend sheetlike projections of the plasma membrane, called pseudopods, which engulf the bacterium and fuse at their tips to form a phagosome. This compartment then fuses with a lysosome and the microbe is digested. Phagocytic cells also play an important part in scavenging dead and damaged cells and cellular debris.

Eukaryotic cells continually ingest bits of their plasma membrane in the form of small pinocytic vesicles that are later returned to the cell surface. The rate at which plasma membrane is internalized by pinocytosis varies from cell type to cell type, but it is usually surprisingly large. Pinocytosis is mainly carried out by the clathrin-coated pits and vesicles that we discussed earlier. After they pinch off from the plasma membrane, clathrin-coated vesicles rapidly shed their coat and fuse with an endosome. Extracellular fluid is trapped in the coated pits as it invaginates to form a coated vesicle, and so substances dissolved in the extracellular fluid are internalized and delivered to endosomes. This fluid intake is generally balanced by fluid loss during exocytosis.

4.7.2 Receptor-mediated endocytosis

The endocytic vesicles simply trap any molecules that happen to be present in the extracellular fluid and carry them into the cell. In most animal cells, however, pinocytosis via clathrin-coated vesicles also provides an efficient pathway for taking up specific macromolecules from the extracellular fluid. The macromolecules bind to complementary receptors on the cell surface and enter the cell as receptor-macromolecule complexes in clathrin-coated vesicles. This process, called **receptor-mediated endocytosis**, provides a selective concentrating mechanism that increases the efficiency of internalization of particular macromolecules more than 1,000 fold compared with ordinary pinocytosis, so that even minor components of the extracellular fluid can be taken up in large amounts without taking in a correspondingly large volume of extracellular fluid. An important example of receptor-mediated endocytosis is the ability of animal cells to take up the cholesterol they need to make new membrane.

Cholesterol is extremely insoluble and is transported in the bloodstream bound to protein in the form of particles called low-density lipoproteins (LDL). The LDL binds to receptors located on cell surfaces, and the receptor-LDL complexes are ingested by receptor-mediated endocytosis and delivered to endosomes. The interior of endosomes is more acid than the surrounding cytosol or the extracellular fluid, and in this acidic environment the LDL dissociates from its receptor: the receptors are returned in transport vesicles to the plasma membrane for reuse, while the LDL is delivered to lysosomes. In the lysosomes, the LDL is broken down by hydrolytic enzymes; the cholesterol is released and escapes into the cytosol, where it is available for new membrane synthesis. The LDL receptors on the cell surface are continually internalized and recycled (Figure 4-25).

Receptor-mediated endocytosis is also used to take up many other essential metabolites, such as vitamin B_{12} and iron, which cells cannot take up by the processes of membrane transport. Vitamin B_{12} and iron are both required, for example, for the synthesis of hemoglobin, the major protein in red blood cells; these metabolites enter immature red blood cells as a complex with protein. Many cell-surface receptors that bind extracellular signaling molecules are also ingested by this pathway: some are recycled to the plasma membrane for reuse, whereas others are degraded in lysosomes. Unfortunately, receptor-mediated endocytosis can also be exploited by viruses: the influenza virus and HIV, which

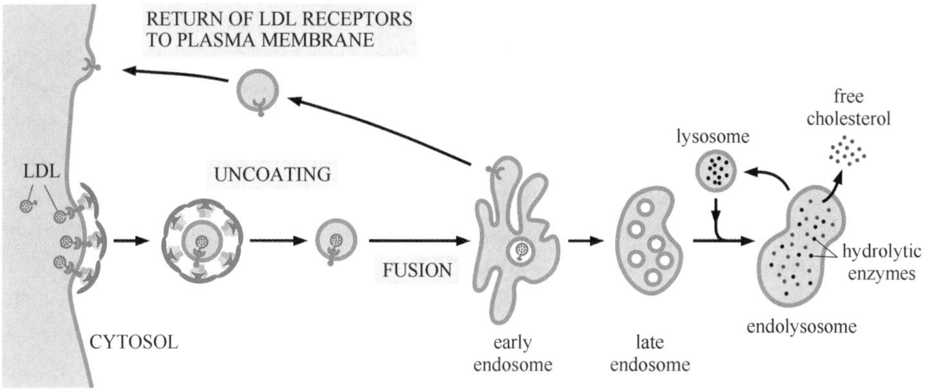

Figure 4-25 The LDL receptor-mediated endocytosis (From: Alberts B, et al. Molecular Biology of the Cell. 2008)

causes AIDS, gain entry into cells in this way.

4.7.3 Endocytosed macromolecules are sorted in endosomes

Extracellular material taken up by pinocytosis is rapidly transferred to endosomes. By an electron microscope, the endosomal compartment can be visualized to be a complex set of connected membrane tubes and larger vesicles when living cells are incubated in fluid containing an electron-dense marker. The marker molecules appear first in **early endosomes**, just beneath the plasma membrane; 5 ~ 15 minutes later they show up in **late endosomes**, near the nucleus. Early endosomes mature gradually into late endosomes as the vesicles within them fuse, either with one another or with a pre-existing late endosome. The interior of the endosome compartment is kept acidic by an ATP-driven H^+(proton) pump in the endosomal membrane that pumps H^+ into the endosome lumen from the cytosol.

The endosomal compartment acts as the main sorting station in the inward endocytic pathway, just as the *trans* Golgi network serves this function in the outward secretory pathway. The acidic environment of the endosome plays a crucial part in the sorting process by causing many receptors to release their bound cargo. The routes taken by receptors once they have entered an endosome differ according to the type of receptor: ①most are returned to the same plasma membrane domain from which they came, as is the case for the LDL receptor; ② some travel to lysosomes, where they are degraded; ③ some proceed to a different domain of the plasma membrane, thereby transferring their bound cargo molecules from one extracellular space to another, a process called **transcytosis**.

Cargo molecules that remain bound to their receptors share the fate of their receptors. Those that dissociate from their receptors in the endosome are doomed to destruction in lysosomes along with most of the contents of the endosome lumen. It remains uncertain how molecules move from endosomes to lysosomes. One possibility is that they are carried in transport vesicles; another is that endosomes gradually convert into lysosomes.

Summary

Eukaryotic cells are highly compartmented. These membrane-enclosed compartments or organelles perform different functions. Proteins synthesized at the ribosomes are selectively delivered to various compartments. The directing of newly made proteins to their correct organelles is determined by sorting signal sequence(s), a continuous stretch of amino acid sequence on proteins. Proteins destined to mitochondrion, chloroplast, peroxisome, and the interior of the nucleus are delivered directly from the cytosol. For others, including the Golgi apparatus, lysosomes, endosomes, and the nuclear membranes, proteins and lipids are delivered through transport vesicles via the ER, which is itself a major site of lipid and protein synthesis. The transport pathways mediated by transport vesicles extend outward from the ER to the plasma membrane, and inward from the plasma membrane to lysosomes, and thus provide routes of communication between the interior of the cell and its surroundings. As proteins and lipids are transported outward along these pathways, many of them undergo various types of chemical modification, such as the formation of disulfide bonds, and *N*-linked or *O*-linked glycosylation. Cells ingest fluid, molecules, and sometimes even particles, by endocytosis. Receptor-mediated endocytosis also provides a selective concentrating mechanism to take up macromolecules. Much of the endocytotic material is delivered to endosomes and then to lysosomes. Most of the components of the endocytic vesicle membrane, however, are recycled in transport vesicles back to the plasma membrane for reuse.

Questions

1. What are the main functions of membrane-enclosed organelles in a typical eukaryotic cell?
2. How proteins synthesized at the ribosomes are transported to various organelles?
3. How do the transport vesicles specifically select cargo molecules?
4. What is the difference between the *N*-linked and *O*-linked glycosylation?
5. How is the cholesterol taken by cells?

（张　儒）

Chapter 5

Mitochondria and chloroplasts

The basic energy source of living things comes from the energy of sunlight, but organisms cannot use sunlight energy directly unless it was transformed into chemical energy. This type of energy transformation takes place in chloroplasts of plant cells and on the photosynthetic lamellae of blue-green algae. **Chloroplasts** via photosynthesis change sunlight energy into chemical energy that is stored in organic substances. Animals and other organisms without chloroplasts acquire energy from breaking down organic nutrition from plants. **Mitochondrion** is this type of organelle that can decompose organic substances and turn chemical energy into ATP that cells can use directly. Evidently, mitochondria and chloroplasts are energy-producing organelles in cells.

5.1 Mitochondria and oxidative phosphorylation

Mitochondria are present in nearly all eukaryotic cells — in plants, animals, and most eukaryotic microorganisms — and it is in these organelles that most of a cell's ATP is produced. Without them, eukaryotes would be dependent on the relatively inefficient process of glycolysis for all of their ATP production. Therefore mitochondria are generally called "powerhouse" in a cell.

5.1.1 Structure of mitochondria

Mitochondria occupy a substantial portion of the cytoplasmic volume of the eukaryotic cells, and they have been essential for all kinds of eukaryotes. Mitochondria are usually depicted as stiff, elongated cylinders with a diameter of 0.5~1 μm, which may be fused to one another, although these attributes can vary depending on the cell type. In a living cell, mitochondria have the characteristics of polymorphism, variability, mobility and adaptivity. Their number varies dramatically in different cell types, and can change with the energy needs of the cell. In skeletal muscle cell, for example, the number of mitochondria may increase five-to ten-fold due to mitochondrial growth and division that occurs if the muscle has been repeatedly stimulated to contract. Their locations inside the cell are not fixed. As they move about in the cytoplasm, they often seem to be associated with microtubules, which can determine the unique orientation and distribution of mitochondria in different types of cells.

Each mitochondrion is bounded by two highly specialized membranes — one wrapped around the other — that play a vital part in its activities. The outer and the inner mitochondrial membranes create two mitochondrial compartments: a large internal space called the matrix (or inner chamber) and the much narrower intermembrane space (Figure 5-1).

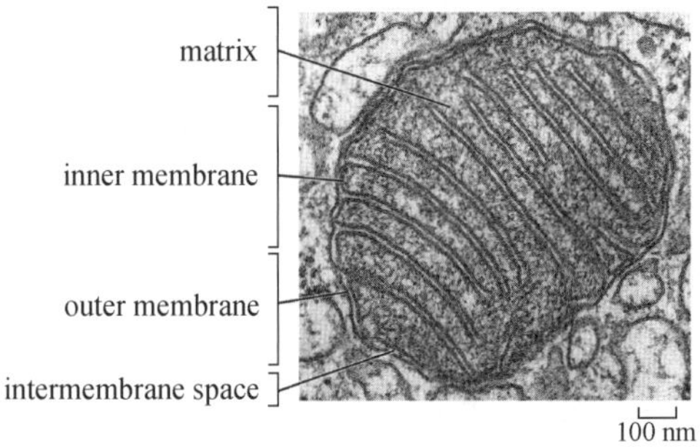

Figure 5-1 The structure of a mitochondrion

The outer membrane contains many molecules of transport protein called porin molecules, which form wide aqueous channels through the lipid bilayer. As a result, the outer membrane is like a sieve that is permeable to all molecules of 5,000 Daltons or less, including small proteins. This makes the intermembrane space chemically equivalent to the cytosol with respect to the small molecules it contains.

The inner membrane is usually highly convoluted, forming a series of infoldings, known as cristae, that project into the matrix space to increase the surface area of the inner membrane(Figure 5-1). These folds provide a large surface on which ATP synthesis can take place. For example, the number of cristae is three times greater in a mitochondrion of a cardiac muscle cell than in a mitochondrion of a liver cell, presumably because of the greater demand for ATP in heart cells.

5.1.2 Biochemical compositions of mitochondria

Main chemical components of mitochondria are proteins and lipids. Proteins can be classified into two types: soluble proteins and insoluble proteins. Soluble proteins are mainly enzymes and proteins located on the outside of the membrane — peripheral proteins; while insoluble proteins are proteins embedded inside the membrane — integral proteins and structural proteins. The lipids are largely phospholipids. The chemical components of the outer and inner membranes are different in their compositions of protein and lipid. The outer membrane is composed of about half protein and half lipid; the inner membrane is about 80 percent protein and 20 percent lipid — a higher proportion of protein than occurs in other cellular membranes. In the matrix there are enzymes of the citric acid cycle and other metabolic pathways.

The inner membrane is highly specialized, its lipid bilayer contains a high proportion of the "double" phospholipid cardioolipin, which has four fatty acids rather than two and make it especially impermeable to the passage of ions and most small molecules, except where a path is provided by membrane transport proteins. The mitochondrial matrix therefore contains only molecules that can be selectively transported into the matrix across the inner membrane, and its contents are highly specialized.

The inner mitochondrial membrane is the site of electron transport and proton pumping, and it contains the ATP synthase. Most of the proteins embedded in the inner mitochondrial membrane are components of the electron-transport chains required for oxidative phosphorylation. This membrane has a distinctive lipid composition and also contains a variety of transport proteins that allow the entry of selected small molecules, such as pyruvate and fatty acids, into the **matrix**.

5.1.3 Electron transport and oxidative phosphorylation

The metabolism of food molecules is completed in the mitochondria. Mitochondria can use both pyruvate and fatty acids as fuel. Pyruvate comes mainly from glucose and other sugars, and fatty acids come from fats. Both of these fuel molecules are transported across the inner mitochondrial membrane and then converted to the vital metabolic intermediate acetyl CoA by enzymes located in the mitochondrial matrix. The acetyl groups in acetyl

CoA are then oxidized in the matrix via the citric acid cycle. The cycle converts the carbon atoms in acetyl CoA to CO_2, which is released from the cell as a waste product. Most importantly, the cycle generates high-energy electrons, carried by the activated carrier molecules NADH and $FADH_2$. These high-energy electrons are then transferred to the inner mitochondrial membrane, where they enter the **electron-transport chain**; the loss of electrons regenerates the NAD^+ and FAD that are needed for continued oxidative metabolism. Electron transport along the chain now begins. The entire sequence of reactions is outlined in figure 5-2.

(1) Electron-transport chain (respiratory chain) The electron-transport chain that carries out **oxidative phosphorylation** is present in many copies in the inner mitochondrial membrane. Most of the proteins involved in the mitochondrial electron-transport chain are grouped into four large respiratory enzyme complexes, each containing multiple individual proteins. Each complex includes transmembrane proteins that hold the entire protein complex firmly in the inner mitochondrial membrane.

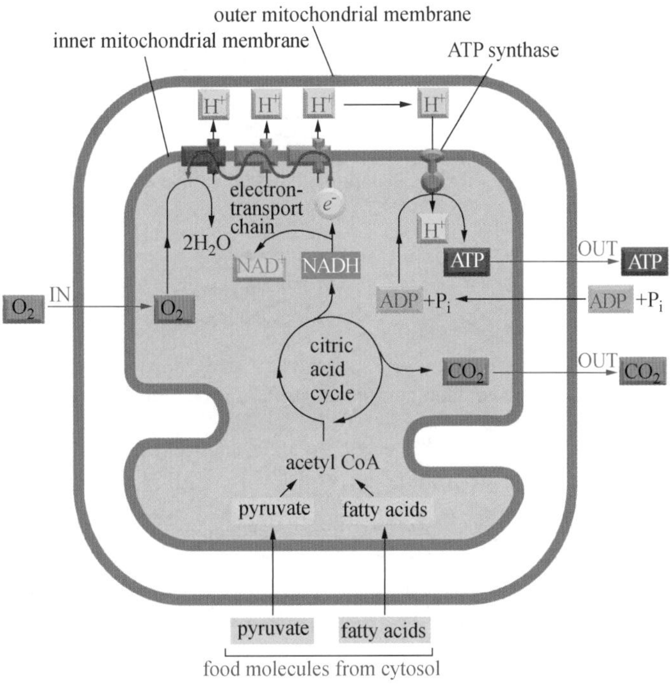

Figure 5-2 A summary of energy-generating metabolism in mitochondria

The four respiratory enzyme complexes are: ①NADH dehydrogenase complex is the largest of the respiratory enzyme complexes, containing more than 40 polypeptide chains. ②Succinate-CoQ reductase complex (complex Ⅱ). ③Cytochrome b-c_1 complex (complex Ⅲ) contains at least 11 different polypeptide chains and functions as a dimmer. ④Cytochrome oxidase complex (complex Ⅳ) also functions as a dimmer, each monomer contains 13 different polypeptide chains, including two cytochromes and two copper atoms. Each contains metal ions and other chemical groups that form a pathway for the passage of electrons through the complex. Complexes Ⅰ, Ⅲ, Ⅳ are the sites of proton pumping, and each can be thought of as a protein machine that pumps protons across the membrane as electrons are transferred through it.

The process of electron transport begins when the hydride ion (H^-) is removed from NADH (to generate NAD^+) and is converted into a proton and two high-energy electrons: $H^- \rightarrow H^+ + 2e^-$ (Figure 5-3). This reaction is catalyzed by the first of the respiratory enzyme complexes, the NADH dehydrogenase, which accepts the electrons. The electrons are then passed along the chain to each of the other enzyme complexes in turn. Within each of the three respiratory enzyme complexes, the electrons move mainly between metal atoms that are tightly bound to the proteins, travelling by skipping from one metal ion to the next. In contrast, electrons are carried between the different respiratory complexes by molecules that diffuse along the lipid bilayer, picking up electrons from one complex and delivering them to another in an orderly sequence. Quinone (also called ubiquinone, or coenzyme Q), a small hydrophobic molecule that dissolves in the lipid bilayer, is the only carrier that is not part of a protein. Ubiquinone picks up electrons from the NADH dehydrogenase complex and delivers them to the cytochrome b-c_1 complex (Figure 5-3).

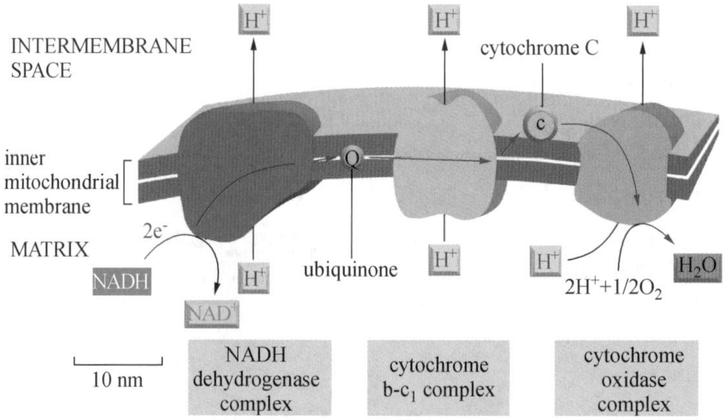

Figure 5-3 Electrons are transferred through three respiratory enzyme complexes in the inner mitochondrial membrane

All the rest of the electron carriers in the electron-transport chain are small molecules that are tightly bound to proteins. To get from NADH to ubiquinone, the electrons are passed inside the NADH dehydrogenase complex between a flavin group bound to the protein complex and a set of iron-sulfur centers.

In the pathway from ubiquinone to O_2, iron atoms in heme groups that are tightly bound to cytochrome proteins are commonly used as electron carriers, as in the cytochrome b-c_1 and cytochrome oxidase complexes. The cytochromes constitute a family of colored proteins; each contains one or more heme groups whose iron atom changes from the ferric (Fe^{3+}) to the ferrous (Fe^{2+}) state whenever it accepts an electron. Cytochrome C shuttles electrons between the cytochrome b-c_1 complex and the cytochrome oxidase complex.

At the very end of the respiratory chain, just before oxygen, the electron carriers are either iron atoms in heme groups or copper atoms that are tightly bound to the complex in the cytochrome oxidase complex. For cytochrome oxidase, the energetically favorable reaction is the addition of electrons to O_2 at the enzyme's active site. Cytochrome oxidase receives electrons from cytochrome c, thus oxidizing it, and donates these electrons to oxygen. It is here that nearly all of the oxygen we breathe is used, serving as the final repository for the electrons that NADH donated at the start of the electron-transport chain.

(2) The redox potential determines the electron transfer along the chain In biochemical reactions, any electrons removed from one molecule are always passed to another, so that whenever one molecule is oxidized, another is reduced. Like any other chemical reaction, the tendency of such oxidation-reduction reactions, or redox reactions, to proceed spontaneously depends on the relative affinities of the two molecules for electrons. The tendency to transfer electrons from any redox pair depends on the redox potential — the voltage difference between two redox pairs. Electrons will move spontaneously from a redox pair like NADH/NAD^+ with a low redox potential (a low affinity for electrons) to a redox pair like O_2/H_2O with a high redox potential (a high affinity for electrons). Thus NADH is a good molecule to donate electrons to the respiratory chain, while O_2 is well suited to act as the "sink" for electrons at the end of the pathway. The difference in redox potential is a direct measure of the standard free-energy change for the transfer of an electron from one molecule to another (Figure 5-4).

(3) Protons are moved with the transfer of electrons Whenever a molecule is reduced by acquiring an electron, the electron (e^-) brings with it a negative charge. In many cases, this charge is rapidly neutralized by the addition of a proton (H^+) from water, so that the net effect of the reduction is to transfer an entire hydrogen atom, $H^+ + e^-$. Similarly, when a molecule is oxidized, the hydrogen atom can be readily dissociated into its constituent electron and proton, allowing the electron to be transferred separately to a molecule that accepts electrons, while the proton is passed to the water. Therefore, in a membrane in which electrons are being passed along an electron-transport chain, it is a relatively simple matter, in principle, to pump protons from one side of the membrane to another. All that is required is that the electron carrier would be arranged in the membrane in a way that causes it to pick up a proton from one side of the membrane when it accepts an electron, while releasing the proton on the other side of the membrane as the electron is passed on to the next carrier molecule in the chain.

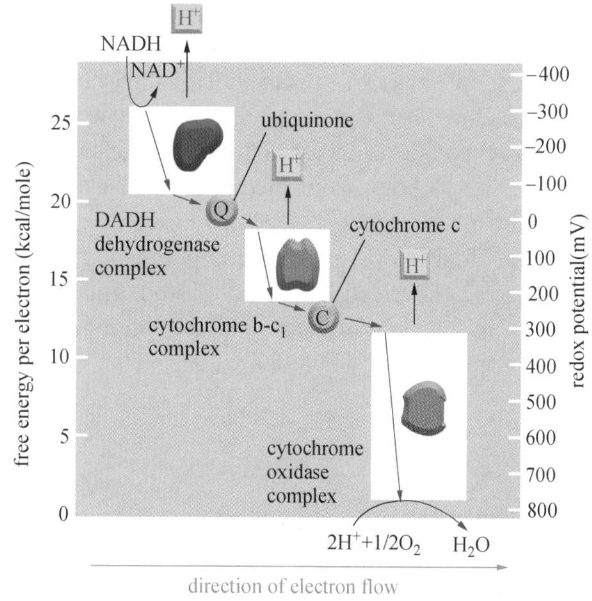

Figure 5-4 Redox potential increases along the mitochondrial electron-transport chain

(4) The mechanism of H^+ pumping Some respiratory enzyme complexes pump one H^+ per electron across the inner mitochondrial membrane, whereas others pump two. The detailed mechanism by which electron transport is coupled to H^+ pumping is different for the three different respiratory enzyme complexes. In the cytochrome b-c_1 complex, the quinones clearly have a role. As mentioned previously, a quinone picks up a H^+ from the aqueous medium along with each electron it carries and liberates it when it releases the electron. Since ubiquinone is freely

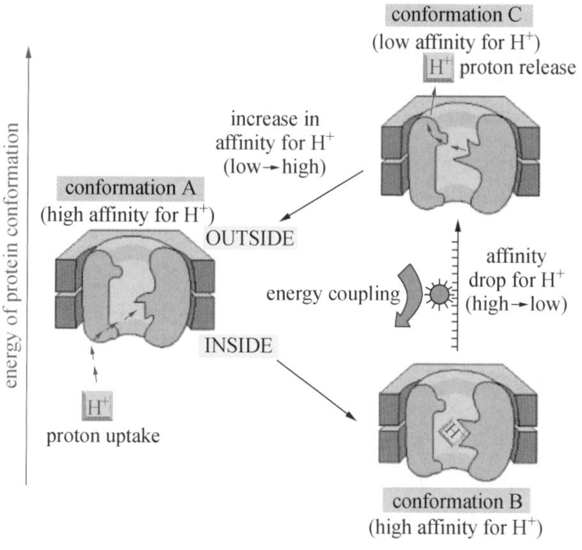

Figure 5-5 A general model for H^+ pumping

mobile in the lipid bilayer, it can accept electrons near the inside surface of the membrane and donate them to the cytochrome b-c_1 complex near the outside surface, thereby transferring one H^+ across the bilayer for every electron transported. However, two protons are pumped per electron in the cytochrome b-c_1 complex.

For both the NADH dehydrogenase complex and the cytochrome oxidase complex, it seems likely that electron transport drives sequential allosteric changes in protein conformation by altering the redox state of the components, which in turn cause the protein to pump H^+ across the mitochondrial inner membrane. This type of H^+ pumping requires at least three distinct conformations for the pump protein, a general mechanism is presented in Figure 5-5.

A schematic view of the reaction catalyzed by cytochrome oxidase is presented in Figure 5-6. In brief, four electrons from cytochrome c and four protons from the aqueous environment are added to each O_2 molecule in the reaction $4e^- + 4H^+ + O_2 \rightarrow 2H_2O$. In addition, four more protons are pumped across the membrane during electron transfer, building up the electrochemical proton gradient. Proton pumping is caused by allosteric changes in the conformation of the protein, which are driven by energy derived from electron transport.

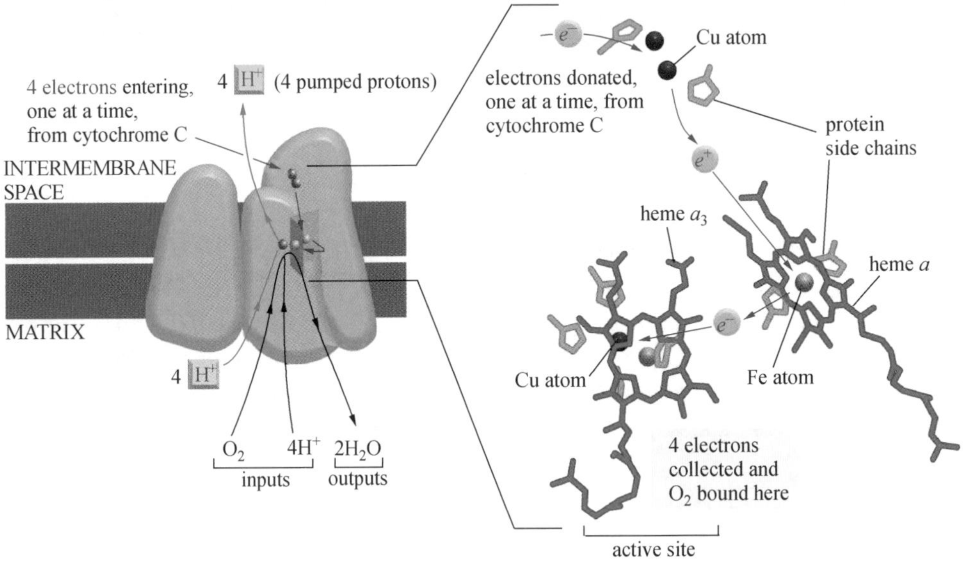

Figure 5-6 Cytochrome oxidase consumes nearly all the oxygen we breathe

(5) Electron transport generates a proton gradient across the inner membrane Each of the respiratory enzyme complexes couples the energy to an uptake of protons from water ($H_2O \rightarrow H^+ + OH^-$) in the mitochondrial matrix, accompanied by the release of protons on the other side of the inner membrane into the intermembrane space. As a result, the energetically favorable flow of electrons along the electron-transport chain pumps protons across the inner membrane out of the matrix, creating an electrochemical proton gradient across the inner mitochondrial membrane (Figure 5-7).

The active pumping of protons thus has two major consequences: ①It generates a pH gradient across the inner mitochondrial membrane, with the pH higher in the matrix than in the intermembrane space, where the pH is generally close to 7. ②It generates a voltage gradient (membrane potential) across the inner mitochondrial membrane, with the inside negative and the outside positive as a result of a net outflow of positive ions-protons.

Because protons are positively charged, they will move more readily across a membrane if the membrane has an excess of negative electrical charges on the other side. In the case of the inner mitochondrial membrane, the pH gradient and membrane potential work together to create a steep electrochemical proton gradient that makes it energetically very favorable for H^+ to flow back into the mitochondrial matrix. The membrane potential adds to the driving force pulling H^+ back across the membrane; hence this potential increases the amount of energy stored in the proton gradient (Figure 5-7).

In addition to the three proton pumps (complex Ⅰ, Ⅱ, Ⅳ), one of the enzymes in citric acid cycle, succinate dehydrogenase (complex Ⅱ), is embedded in crista membrane. Complex Ⅱ can capture electrons in the form of a tightly bound FADH2 molecule and pass them to a molecule of ubiquinone. The reduced ubiquinone then passes its two electrons to complex Ⅲ. Complex Ⅱ is not a proton pump, and it does not contribute directly to the electrochemical potential for ATP production in mitochondria.

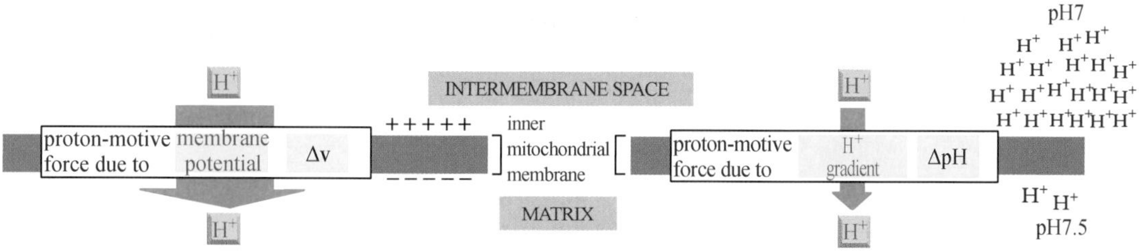

Figure 5-7 The two components of the electrochemical proton gradient. The total proton-motive force across the inner mitochondrial membrane consists of large force due to the membrane potential (ΔV) and a smaller force due to the H^+ gradient (ΔpH). Both forces combine to drive H^+ into the matrix.

(6) The proton gradient drives ATP synthesis The electrochemical proton gradient across the inner mitochondrial membrane can be used to drive ATP synthesis in the process of **oxidative phosphorylation** — a process involves both the consumption of O_2 and the synthesis of ATP through the addition of a phosphate group to ADP (Figure 5-8). This is made possible by a large membrane-bound enzyme called **ATP synthase**. This enzyme creates a hydrophilic pathway across the inner mitochondrial membrane that allows protons to flow down their electrochemical gradient. As these ions thread their way through the ATP synthase, they are used to drive the energetically unfavorable reaction between ADP and Pi that makes ATP. The ATP synthase is of ancient origin; the same enzyme occurs in the mitochondria of animal cells, the chloroplasts of plants and algae, and in the plasma membrane of bacteria and archaea.

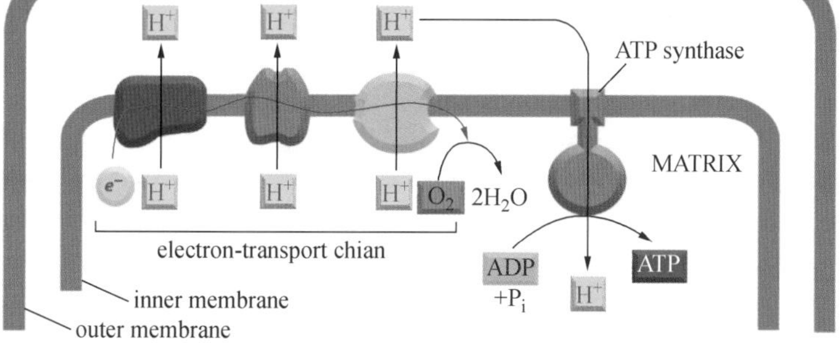

Figure 5-8 The general mechanism of oxidative phosphorylation

The structure of ATP synthase is shown in Figure 5-9. It is a large, multisubunit protein. A large enzymatic portion, shaped like a lollipop head and composed of 6 subunits, projects on the matrix side of the inner mitochondrial membrane and is attached through a thinner multi-subunit "stalk" to a transmembrane proteins that form a "stator". The stator is in contact with a "rotor" composed of a ring of 10 to 14 identical trans-membrane protein subunits. As protons pass through a narrow channel formed at the stator-rotor contact, their movement causes the rotor ring to spin rapidly within the head, inducing the head to make ATP. The synthase essentially acts as an energy-generating molecular motor, converting the energy of proton flow down a gradient into the mechanical energy of two sets of proteins rubbing against one another — rotating stalk proteins pushing against stationary head proteins. The changes in protein conformation driven by the rotating stalk subsequently convert this mechanical energy into the chemical bond energy needed to generate ATP. This marvelous device is capable of producing more than 100 molecules of ATP per second, calling for about three protons pass through the synthase to make each molecule of ATP. Through the above processes proton gradients across the inner mitochondrial membrane produce most of the ATP in cells.

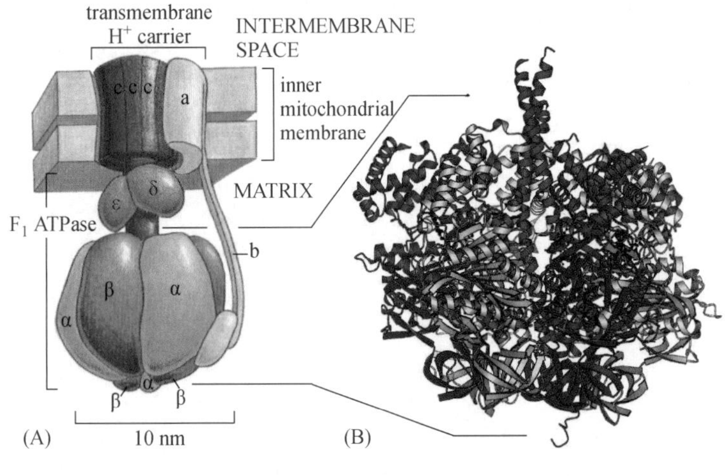

Figure 5-9 ATP synthase

(7) The proton gradient drives coupled transport across the inner membrane The synthesis of ATP is not the only process driven by the electrochemical proton gradient. In mitochondria, many charged small molecules, such as pyruvate, ADP, and Pi, are pumped into the matrix from the cytosol, while others, such as ATP, must be moved in the opposite direction. Transporters that bind these molecules can couple their transport to the energetically favorable flow of H^+ into the mitochondrial matrix. Thus, for example, pyruvate and inorganic phosphate (Pi) are co-transported inward with H^+ as the H^+ moves into the matrix.

In contrast, ADP is co-transported with ATP in opposite directions by a single carrier protein. As an ATP molecule has one more negative charge than ADP, each nucleotide exchange results in the moving of a total of one negative charge out of the mitochondrion. The ADP-ATP co-transport is therefore driven by the voltage difference across the membrane (Figure 5-10). All in all, a typical ATP molecule in the human body shuttles out of a mitochondrion and back into it (as ADP) for recharging more than once per minute, keeping the concentration of ATP in the cell about 10 times higher than that of ADP.

(8) The efficiency of respiration During the biological oxidations — the total amount of energy generated and stored in the phosphate bonds of ATP, the efficiency with which oxidation energy is converted into ATP bond energy is often greater than 40%. This is considerably better than the efficiency of most nonbiological energy-conversion devices. If cells worked only with the efficiency of an electric motor or a gasoline engine (10%~20%), an organism would have to eat voraciously in order to maintain itself.

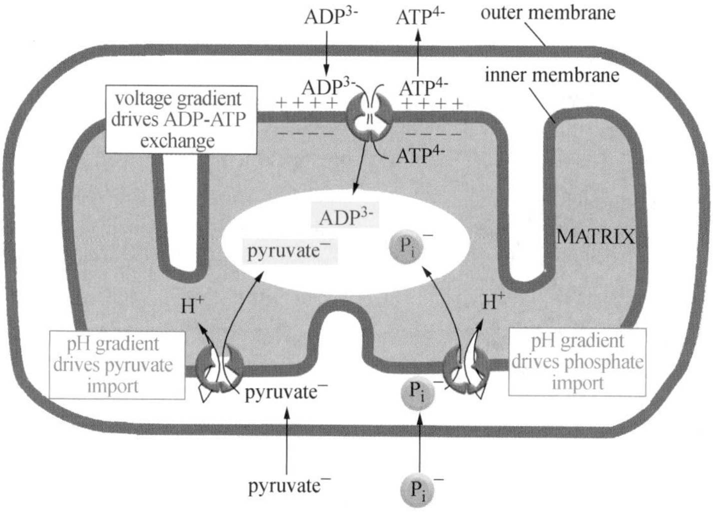

Figure 5-10 Some of the active transport processes driven by the electrochemical proton gradient across the inner mitochondrial membrane

Oxidation produces huge amounts of free energy, which can be utilized efficiently only in small bits. Biological oxidative pathways involve many intermediates, each differing only slightly from its predecessor. The energy released is thereby parceled out into small packets that can be efficiently converted to high-energy bonds in useful molecules, such as ATP and NADH, by means of coupled reactions.

5.2 Chloroplasts and photosynthesis

The living organisms on Earth rely on solar energy to survival, and **photosynthesis** is the only biological way to harvest this energy. Photosynthesis is a process used by plants and other organisms to capture the solar energy to the photolysis of water. Nearly all of the organic material and energy required for surviving is from photosynthesis. Therefore, photosynthesis has a very important significance for the human beings, and is regarded as the most important chemical reaction on Earth.

In plants, photosynthesis is carried out in the **chloroplast**. Chloroplasts perform photosynthesis during the daylight hours and thereby produce ATP and NADPH, which are used to convert CO_2 into sugars inside the chloroplast. The sugars produced are exported to the surrounding cytosol, where they are used as fuel to make ATP and as starting materials for many of the other organic molecules that the plant cell needs. Sugar is also exported to all those cells in the plant that lack chloroplasts. As is the case for animal cells, most of the ATP present in the cytosol of plant cells is made by the oxidation of sugars and fats in mitochondria.

5.2.1 Morphology and structure of chloroplasts

Chloroplasts are specific organelles in plant cells, and they are the most prominent members of the plastid family of organelles in plant cells. Plastids form a distinct group of organelles in plants and are one of the characteristics by which plants are different to animals. Chloroplasts are the most prominent form of plastid occurring in all green plant tissues and perform photosynthesis and other cellular metabolism. The shape, size and number of chloroplast are dependent on the plant species, cell types, ecological environment and the physiological state. Most chloroplasts in leaf mesophyll cells are typically ellipsoidal in shape but with defined poles, a feature that is crucial to their division. Chloroplasts are observable as flat discs usually 2 to 10 μm in diameter and 1 μm thick.

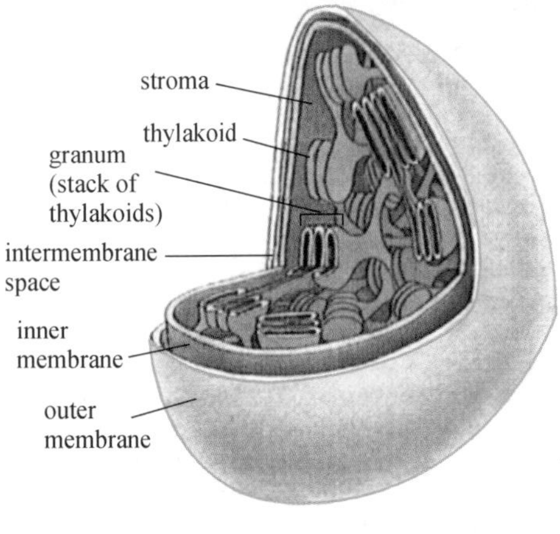

Figure 5-11 Structure of a chloroplast

Like all plastids, they are bounded by a double envelope membrane, that is, a highly permeable outer membrane and a much less permeable inter membrane — in which membrane transport proteins are embedded, and a narrow intermembrane space in between (Figure 5-11). The inner membrane surrounds a large space called the **stroma**, which, analogous to the mitochondrial matrix, contains various metabolic enzymes. Besides the inner and outer membranes of the envelope, chloroplasts have a third internal membrane system, called the **thylakoid** membrane. The thylakoid membrane forms a network of flattened discs called thylakoids, which are frequently arranged in stacks called **granum** (Figure 5-11). The space inside each thylakoid is thought to be connected with that of other thylakoids, thereby defining a continuous third internal compartment that is separated from the stroma by the thylakoid membrane (Figure 5-12). Need to point out that the light-capturing systems, the electron-transport chains, and ATP synthases are all incorporated in the thylakoid membrane, a third membrane that forms a set of flattened disclike sacs, the thylakoids (Figure 5-12).

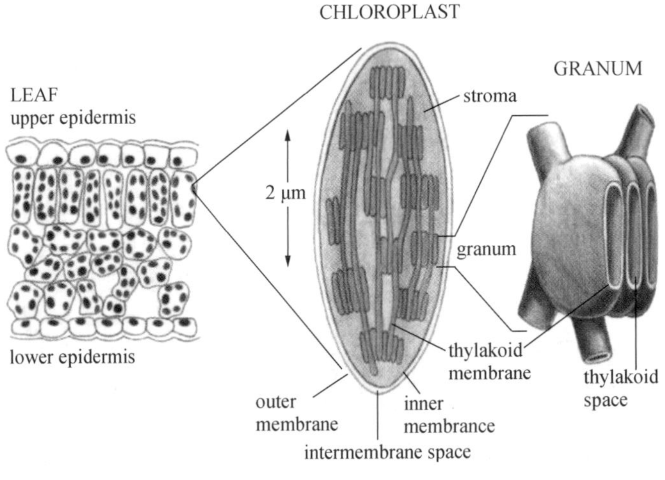

Figure 5-12 A chloroplast contains a third internal compartment

5.2.2 Photosynthesis of chloroplast

Photosynthesis is a process that plants, alage, and prokaryotes directly use light energy to synthesize organic compounds. The process of photosynthesis encompasses a complex series of chemical reactions that involve light absorption, energy conversion, electron transfer, carbon dioxide fixation and water photolysis. Most photosynthetic organisms convert light energy into stable chemical products, according to the following reaction equation:

$$2H_2A + CO_2 \rightarrow (CH_2O) + 2A + H_2O$$

As shown in the equation, photosynthesis is a biological oxidation-reduction process. CO_2 is the electron acceptor, and H_2A (for example, H_2O or H_2S) is any reduced compound that can serve as the electron donor. CH_2O represents the carbohydrate generated by the reduction, and A represents the product formed by oxidation of H2A (for example, O_2 or S). The photosynthetic process involved two stages: **light reactions and dark reaction or carbon dioxide fixation**. Photosynthesis is a very complex reaction process, and based on current information photosynthesis can be divided into two stages: the light-dependent process and the light-independent process. The first process, the light dependent process (also called the "light reactions"), requires the direct energy of light to make energy carrier molecules — ATP and NADPH that are used in the second process. Water is split in this process, releasing oxygen gas. The second process, the light independent process (also called the " dark reactions"), occurs when the products of the light reaction — the ATP and the NADPH — serve as the source of energy and reducing power respectively, to drive the conversion of CO_2 to carbohydrate (Figure 5-13).

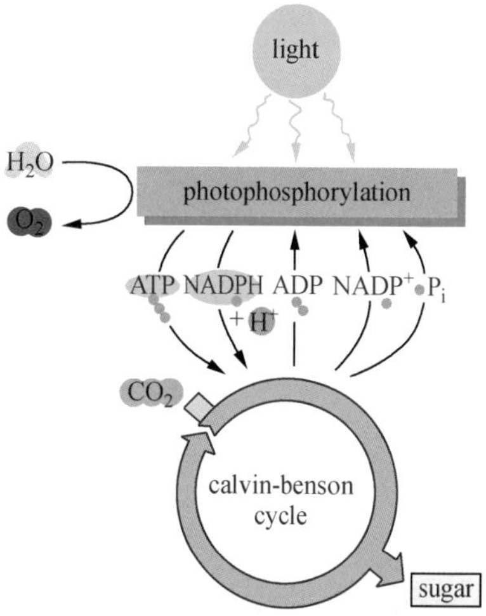

Figure 5-13 Overview of the two stages in the photosynthesis process

(1) Light reactions (photosynthetic electron-transfer reactions) White light is separated into the different colors of light when it passes through a prism. The order of colors is determined by the wavelength of light. Thus, not all colors of light have equal energy. The particle properties are demonstrated by the photoelectric effect. When we consider events at the level of a single molecule — such as the absorption of light by a molecule of chlorophyll, we have to picture light as being composed of discrete packets of energy called photons. Light of different colors is distinguished by photons of different energy, with longer wavelengths corresponding to lower energies. Thus photons of red light have a lower energy than photons of green light.

1) **Chlorophyll and accessory pigments**: For light energy to be used by any system, the light must first be absorbed. Almost all photosynthetic organisms contain chlorophyll or a related pigment. Most photosynthetic organisms contain chlorophyll, one form of the light-absorbing pigment. **Chlorophyll** is a complex molecule and it contains tetrapyrrole ring-like structure, which is structurally similar to and produced through the same metabolic pathway as other porphyrin pigments such as heme. At the center of the chlorin ring is a magnesium ion which coordinated to the four modified pyrrole rings. Additionally, chlorophyll also contains a long hydrocarbon tail that makes the molecules hydrophobic (Figure 5-14).

All photosynthetic organisms have chlorophylla. Among organisms that perform oxygenic photosynthesis, only land plants, green algae, and a few groups of cyanobacteria (prochlorophytes) possess chlorophyllb. Chlorophylla and chlorophyllb have different spectrophotometry characteristic, and in diethyl ether, chlorophylla has approximate absorbance maxima of 430 nm and 662 nm, while chlorophyllb has approximate maxima of 453 nm and 642 nm (Figure 5-15). **Carotenoids** are tetraterpenoid organic pigments derived

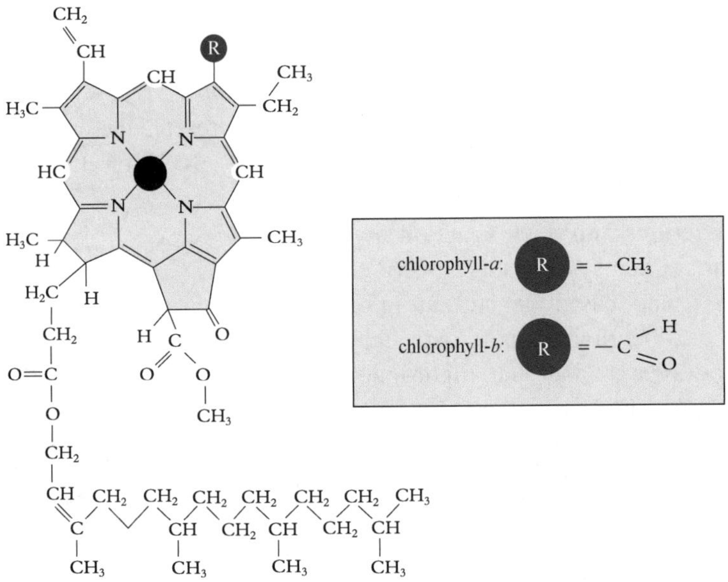

Figure 5-14 The molecular structure of chlorophylls

from eight isoprene units and they include the carotenes and xanthophylls. Carotenoids, which are responsible for the orange-yellow colors observed in the leaves of plants, absorb light which wavelength is between 400 and 500 nm (Figure 5-15), a range in which absorption by chlorophylla is relatively weak. Thus, carotenoids play a minor role as accessory light-harvesting pigments absorbing and transferring light energy to chlorophyll molecules, but they play important structural role in the assembly of light-harvesting complexes and have an essential function in protecting the photosynthetic apparatus from photooxidative damage. There is another group of photosynthetic pigments **phycobilins**.

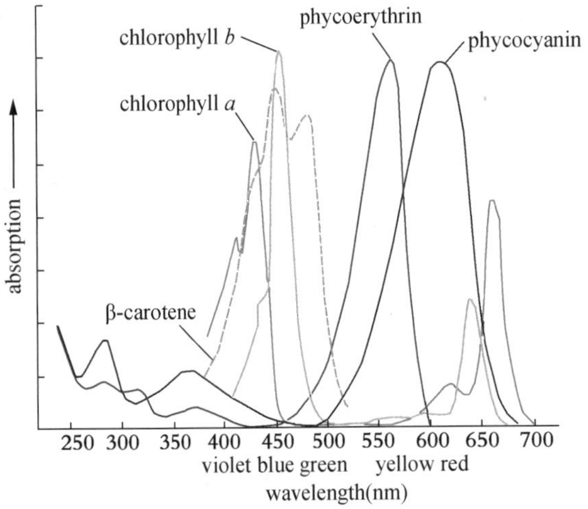

Figure 5-15 Absorption spectrum of several plant pigments

The action spectrum of photosynthesis is the relative effectiveness of different wavelengths of light at generating electrons. If a pigment absorbs light energy, one of three things will occur. Energy is dissipated as heat; the energy may be emitted immediately as a longer wavelength, a phenomenon known as fluorescence; energy may trigger a chemical reaction, as in photosynthesis. Chlorophyll only triggers a chemical reaction when it is associated with proteins embedded in a membrane (as in a chloroplast) or the membrane infoldings found in photosynthetic prokaryotes such as cyanobacteria and prochlorobacteria.

2) **Photosystems**: **Photosystems** are arrangements of light-absorbing chlorophyll and large multiprotein complexes in chloroplast thylakoid membranes of plant and in the membranes of photosynthetic bacteria. A photosystem consists of two closely linked components: an

antenna complex and a photochemical reaction center (Figure 5-16). The former consists of proteins bound to a large set of pigment molecules that capture light energy and feed it to the reaction centre, and the latter consists of a complex of proteins and chlorophyll molecules that enable light energy to be converted into chemical energy.

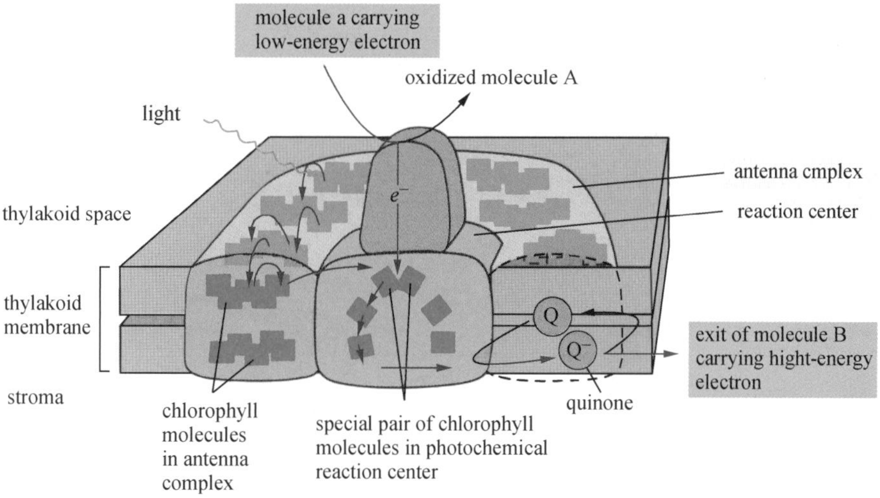

Figure 5-16 A photosystem contains a reaction center and an antenna

The antenna complex is important for capturing light energy, and it consists of light-harvesting complexes. A light-harvesting complex is a complex of proteins and photosynthetic pigments and surrounds a photosynthetic reaction center to focus energy, attained from photons absorbed by the pigment, toward the reaction center using Förster resonance energy transfer. The antenna complex in plants also contains accessory another pigments called carotenoids, which protect the chlorophylls from oxidation and can help the antenna complex. When a chlorophyll molecule in the antenna is excited by light, the excitation energy may also be transferred from one molecular to another molecule nearby by resonance energy transfer until it reaches a special pair of chlorophyll molecules in the reaction center.

The photochemical reaction center is a transmembrane complex of several proteins, pigments and other co-factors assembled together to perform the primary energy conversion reactions of photosynthesis. The special pair of chlorophyll molecules in the reaction center acts as an irreversible trap for an excited electron, as these chlorophylls are positioned to pass a high-energy electron to a precisely positioned neighboring molecule in the same protein complex, creating a charge separation by moving the energized electron rapidly away from the chlorophylls, which are transferred by reaction center is much more stable.

The chlorophyll molecule in the reaction center loses an electron, it becomes positively charged. As illustrated in figure 5-17A, this chlorophyll rapidly regains an electron from an adjacent electron donor (orange) to return to its unexcited, uncharged state. Then, in slower reactions, the electron donor has its missing electron replaced with an electron removed from water, and the high-energy electron that was generated by the excited chlorophyll is transferred to the electron transport chain(Figure 5-17B).

Green plants and algae have two different types of reaction centres that are part of larger supercomplexes known as photosystem Ⅰ(PS Ⅰ) and photosystem Ⅱ(PS Ⅱ). PS Ⅱ, called water-plastoquinone oxidoreductase, located in the thylakoid membrane of

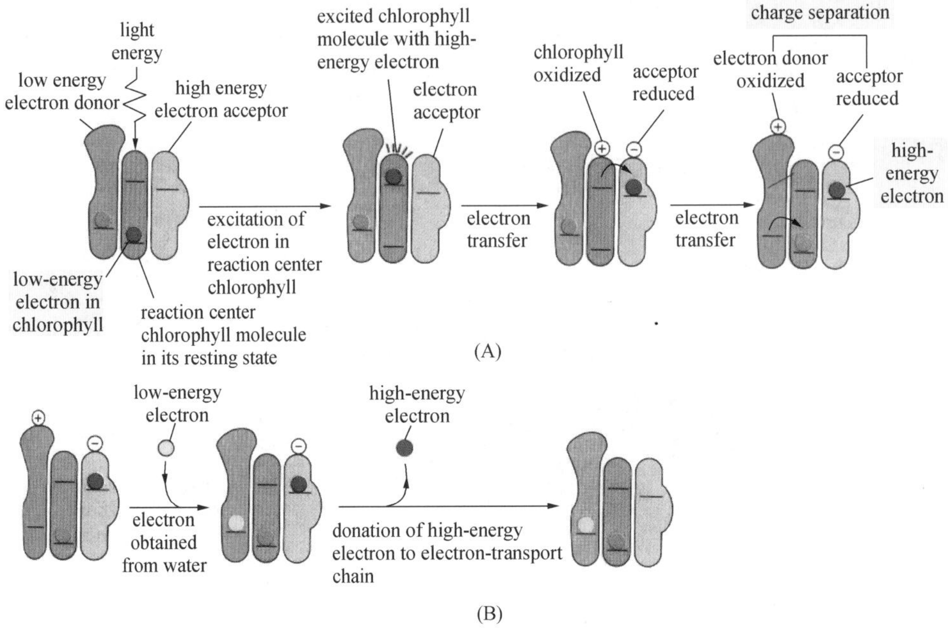

Figure 5-17 Light energy is harvested by a reaction center chlorophyll molecule

plants, algae, and cyanobacteria, is the first protein complex in the light-dependent reactions. In cyanobacteria and green plants, PS Ⅱ is composed of about 20 subunits as well as other accessory, light-harvesting proteins. PS Ⅱ captures light photons to energize electrons that are then transferred through a variety of coenzymes and cofactors to reduce plastoquinone to plastoquinol. The energized electrons are replaced by oxidizing water to form hydrogen ions and molecular oxygen. By obtaining these electrons from water, PS Ⅱ provides the electrons for all of photosynthesis to occur. PS Ⅰ is an integral membrane protein complex that was discovered in the 1950s and it is the second photosystem in the photosynthetic light reactions of algae, plants, and some bacteria. The PS Ⅰ system comprises more than 110 co-factors, significantly more than PS Ⅱ. These various components have a wide range of functions. PS Ⅰ uses light energy to mediate electron transfer from plastocyanin to ferredoxin.

3) **Electron-transport chain and photophosphorylation**: Photosynthetic **phosphorylation** is the process of phosphate group transfer into ADP to synthesize energy rich ATP molecule making use of light as external energy source (Figure 5-18). Photosynthesis in plants and cyanobacteria has at least two photophosphorylation ways that is noncyclic photophosphorylation and cyclic photophosphorylation. Noncyclic photophosphorylation produces both ATP and NADPH directly, while cyclic photophosphorylation only produces ATP but no reduced NADP (NADPH). In the non-cyclic reaction, light energy is captured in the light-harvesting antenna complexes of photosystem Ⅱ by chlorophyll and other accessory pigments. When a chlorophyll molecule at the core of the photosystemII reaction center obtains sufficient excitation energy from the adjacent antenna pigments, an electron is transferred to the primary electron-acceptor molecule, pheophytin, through a process called photo-induced charge separation. These electrons are shuttled through a Z-scheme electron transport chain, that initially functions to generate a chemiosmotic potential across the membrane. An ATP synthase enzyme uses the chemiosmotic potential to make ATP during photophosphorylation, whereas NADPH is a product of the terminal redox reaction in the Z-

scheme (Figure 5-19A). The electron enters a chlorophyll molecule in photosystem I. The electron is excited due to the light absorbed by the photosystem. A second electron carrier accepts the electron, which again is passed down lowering energies of electron acceptors. The energy created by the electron acceptors is used to move hydrogen ions across the thylakoid membrane into the lumen. The electron is used to reduce the co-enzyme NADP, which has functions in the light-independent reaction.

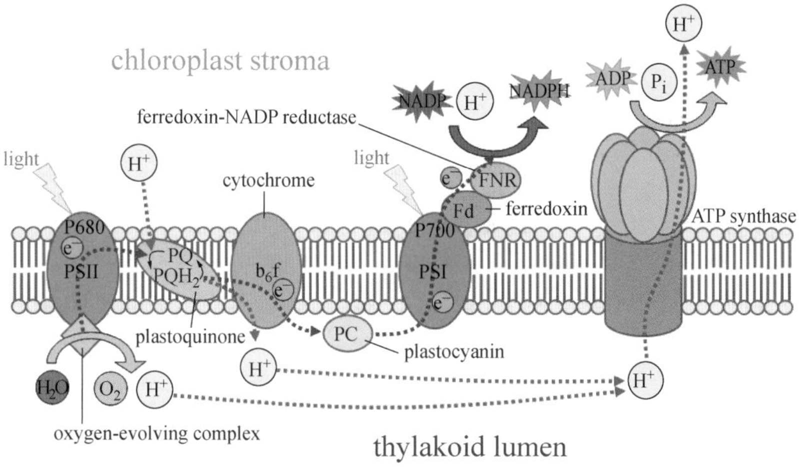

Figure 5-18 During photosynthesis electrons travel down an electron transport chain in the thylakoid membrane

In cyclic electron flow, the electron begins in a pigment complex called photosystemI, passes from the primary acceptor to ferredoxin, then to cytochrome b_6-f (a similar complex to that found in mitochondria), and then to plastocyanin before returning to chlorophyll. This transport chain produces a proton-motive force, pumping H^+ ions across the thylakoid membrane, which produces a concentration gradient that can be used to power ATP synthase during chemiosmosis. This cyclic photophosphorylation produces neither O_2 nor NADPH. Unlike non-cyclic photophosphorylation, $NADP^+$ does not accept the electrons; they are instead sent back to photosystem I(Figure 5-19B).

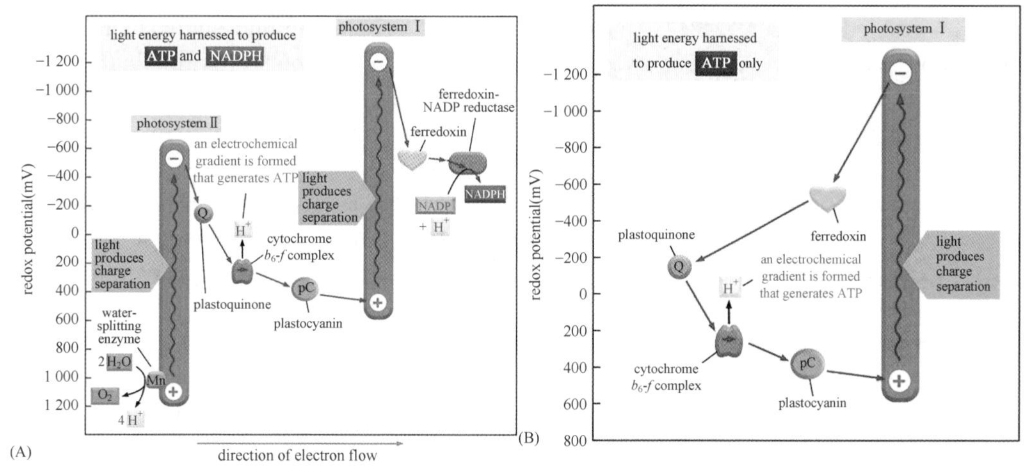

Figure 5-19 The components in the electron-transport chain have different redox potentials. (A) noncyclic photophosphorylation; (B) cyclic photophosphorylation.

(2) Dark reactions (carbon-fixation reactions)

1) The calvin cycle (C-3 pathway): The central reaction of photosynthetic carbon fixation, in which a molecule of CO_2 is converted to organic carbon, is illustrated in figure 5-20. CO_2 combines with the five-carbon sugar derivative ribulose 1,5-bisphosphate plus water to give two molecules of the three-carbon compound 3-phosphoglycerate. This reaction, which was discovered by Melvin Calvin in 1948, is catalyzed in the chloroplast stroma by a large enzyme called ribulose bisphosphate carboxylase (RuBisCO). The RuBisCO enzyme usually consists of two subunit, called the large chain (L) and the small chain (S). A total of eight large chains (= 4 dimers) and eight small chains assemble into a larger complex of about 540,000 Da. Since this enzyme works extremely sluggishly compared with most other enzymes (processing about three molecules of substrate per second compared with 1,000 molecules per second for a typical enzyme), many enzyme molecules are needed. RuBisCO often represents more than 50% of the total chloroplast protein, and it is widely claimed to be the most abundant protein on earth.

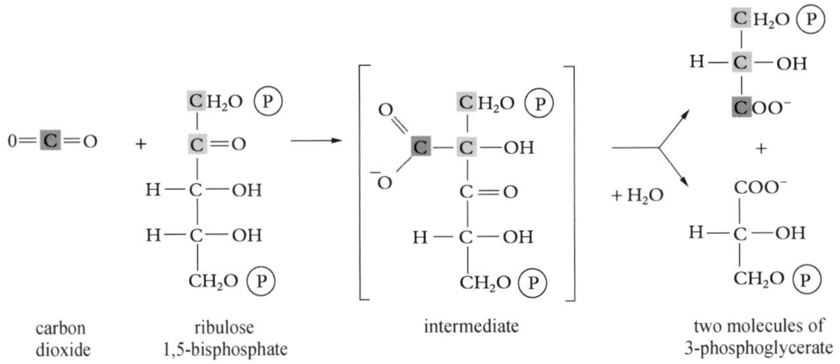

Figure 5-20 The initial reaction in carbon fixation. Carton dioxide is converted into organic carbon, is catalyzed in the chloroplast stroma by the abundant enzyme ribulose bisphosphate carboxylase.

This initial CO_2 fixing reaction is energetically favorable, but only when it receives a continuous supply of the energy-rich compound ribulose 1,5-bisphosphate, to which each molecule of CO_2 is added. The elaborate metabolic pathway by which this compound is regenerated requires both ATP and NADPH. This carbon-fixation cycle (also called Calvin cycle) is outlined in figure 5-21; it is a cyclic process, beginning and ending with ribulose 1,5-bisphosphate. However, for every three molecules of carbon dioxide that enter the cycle, one new molecule of glyceraldehyde 3-phosphate is produced, the three-carbon sugar that is the net product of the cycle. This sugar then provides the starting material for the synthesis of many other sugars and organic molecules.

In the carbon-fixation cycle, three molecules of ATP and two molecules of NADPH are consumed for each CO_2 molecule converted into carbohydrate. Thus both phosphate bond energy (as ATP) and reducing power (as NADPH) are required for the formation of sugar molecules from CO_2 and H_2O.

2) C-4 pathway and CAM pathway: In most plants, carbon fixation occurs when CO_2 reacts with a five-carbon compound called RuBisCO (ribulose 1, 5-bisphosphate) as described above. The product splits immediately to form a pair of three-carbon compounds, and therefore this pathway is called the C-3 pathway. Further reaction leads to the production of a sugar (glyceraldehyde-3-phosphate) and the regeneration of RuBisCO. However, RuBisCO is not very efficient at grabbing CO_2, and it has an even worse problem.

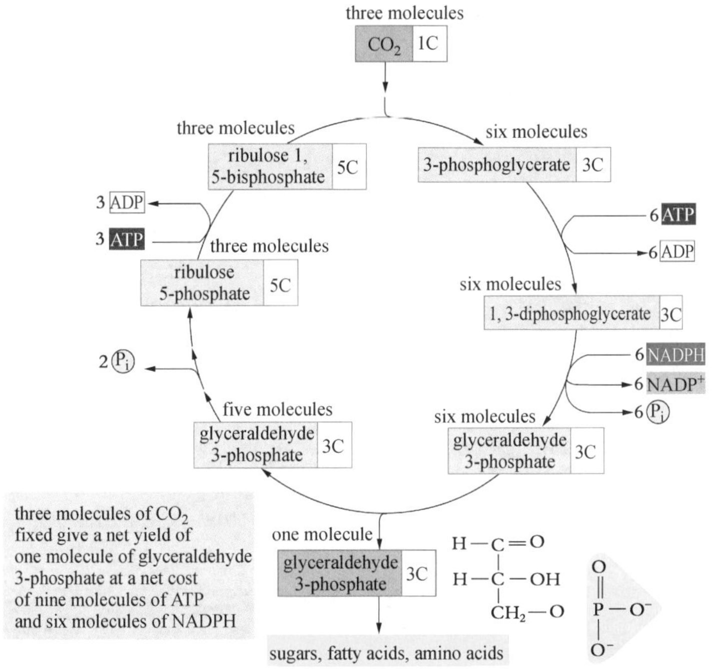

Figure 5-21 The carbon-fixation cycle forms organic molecules from CO_2 and H_2O

When the concentration of CO_2 in the air inside the leaf falls too low, RuBisCO starts combining oxygen instead. The ultimate consequence of this process, called photorespiration (Figure 5-22), is that sugar is burned up rather than being created. Photorespiration becomes a significant problem for plants during hot, dry days, when they must keep their stomates closed to prevent water loss.

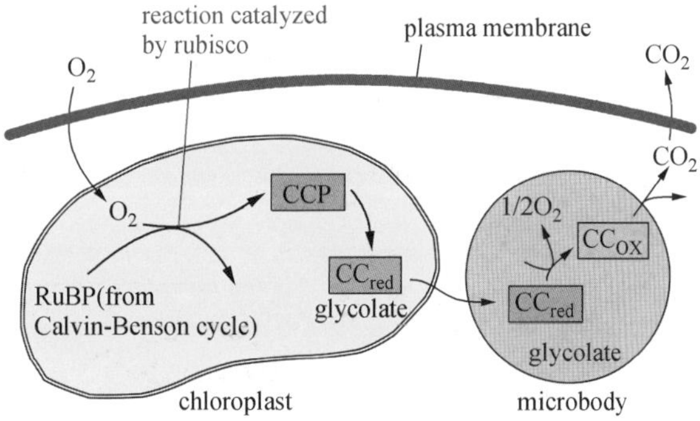

Figure 5-22 Photorespiration

Some plants have developed a preliminary step to the Calvin cycle (C-3 pathway), which is known as C-4. While most C-fixation begins with RuBP, C-4 begins with a new molecule, phosphoenolpyruvate (PEP), a 3-C chemical that is converted into oxaloacetic acid (OAA, a 4-C chemical) when carbon dioxide is combined with PEP. The OAA is converted to malic acid and then transported from the mesophyll cell into the bundle-sheath cell, where OAA is broken down into PEP plus carbon dioxide. The carbon dioxide then enters the Calvin cycle, with PEP returning to the mesophyll cell. The resulting sugars are now adjacent to the leaf veins and can readily be transported throughout the plant.

The capture of carbon dioxide by PEP is mediated by the enzyme PEP carboxylase, which has a stronger affinity for carbon dioxide than does RuBP carboxylase. When carbon dioxide levels decline below the threshold for RuBP carboxylase, RuBP is catalyzed with oxygen instead of carbon dioxide, which forms glycolic acid, a chemical that can be broken down by photorespiration, producing neither NADH nor ATP, in effect dismantling the Calvin cycle (Figure 5-23). C-4 plants, which often grow close together, have had to adjust to decreased levels of carbon dioxide by raising the carbon dioxide concentration in certain cells to prevent photorespiration. C-4 plants evolved in the tropics and are adapted to higher temperatures than are the C-3 plants found at higher latitudes. Common C-4 plants include crabgrass, corn, and sugar cane. By concentrating CO_2 in the bundle sheath cells, C-4 plants promote the efficient operation of the Calvin-Benson cycle and minimize photorespiration. C-4 plants include corn, sugar cane, and many other tropical grasses.

Crassulacean acid metabolism, also known as CAM photosynthesis, is another carbon fixation pathway that evolved in some plants as an adaptation to acid conditions. CAM plants initially grab CO_2 to PEP and form OAA (Figure 5-24). CAM plants fix carbon at night rather than during the day, and store the OAA in large vacuoles within the cell. This allows them to have their stomates open in the cool of the evening, avoiding water loss, and to use the CO_2 for the Calvin-Benson cycle during the day, when it can be driven by the sun's energy. CAM plants are more common than C-4 plants and include cacti and a wide variety of other succulent plants.

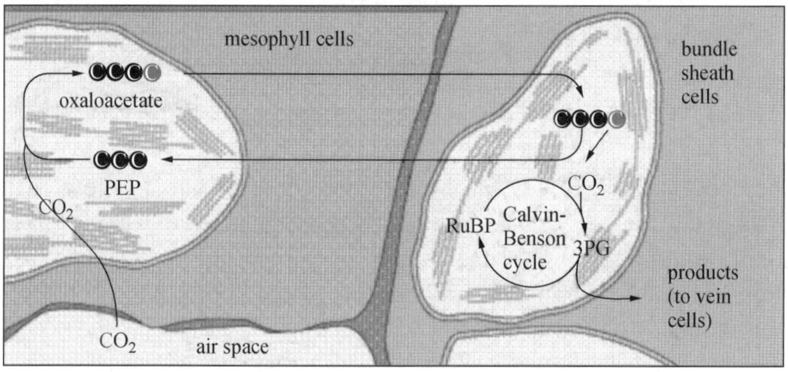

Figure 5-23 C-4 photosynthsis involves the separation of carbon fixation and carbohydrate systhesis in space and time

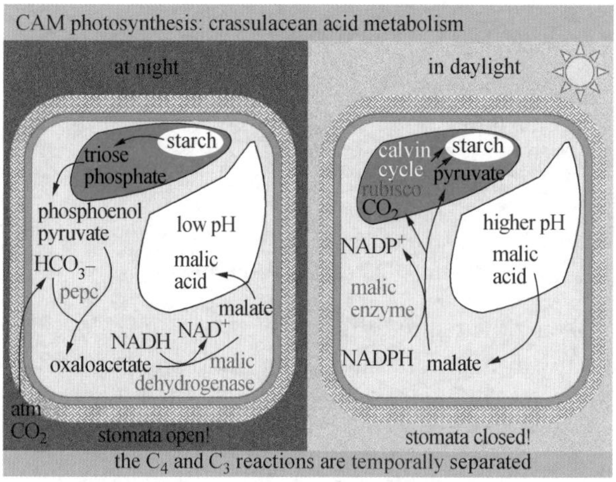

Figure 5-24 CAM photosynthsis (From: http://plantphys.info/plant_physiology/c4cam.shtml)

3) Carbon fixation in chloroplasts generates sucrose and starch: Much of the glyceraldehyde 3-phosphate produced in chloroplasts is moved out of the chloroplast into the cytosol. Some of it enters the glycolytic pathway, where it is converted to pyruvate that is then used to produce ATP by oxidative phosphorylation in plant cell mitochondria. The glyceraldehyde 3-phosphate is also converted into many other metabolites, including the disaccharide sucrose. Sucrose is the major form in which sugar is transported between plant cells: just as glucose is transported in the blood of animals, sucrose is exported from the leaves via the vascular bundle to provide carbohydrate to the rest of the plant.

The glyceraldehyde 3-phosphate that remains in the chloroplast is mainly converted to starch in the stroma. Like glycogen in animal cells, starch is a large polymer of glucose that serves as a carbohydrate reservoir. The production of starch is regulated so that it is synthesized and stored as large grains in the chloroplast stroma during periods of excess photosynthetic capacity. At night, starch is broken down to sugars to help support the metabolic needs of the plant. Starch forms an important part of the diet of all animals that eat plants.

5.3 The biogenesis of mitochondria and chloroplasts

5.3.1 Genome of mitochondria and chloroplasts

Both mitochondria and chloroplasts contain their complete genetic system, including the mitochondrial (chloroplast) genome (circle), numerous factors for maintaining, regulating and expressing the genome, and the mitochondrial (chloroplast) ribosomes that differ in size and composition from cytosolic ribosomes. Mitochondria and chloroplasts reproduce through the growth and division of pre-existing organelles (Figure 5-25). The fusion and fission of the organelles to change their morphology are in response to a multitude of signal.

The growth and proliferation of mitochondria and chloroplasts are complicated, as their component proteins are encoded by two separate genetic systems — one in the organelle and one in the cell nucleus. Animal mitochondria in fact contain a uniquely simple genetic system: the human mitochondrial genome, for example, contains only 16, 569 nucleotide pairs of DNA encoding 37 genes. The vast majority of mitochondrial proteins-including those needed to make the mitochondrion's RNA polymerase and ribosomal proteins, and all of the enzymes of its citric acid cycle — are produced from nuclear genes, and these proteins must therefore be imported into the mitochondria.

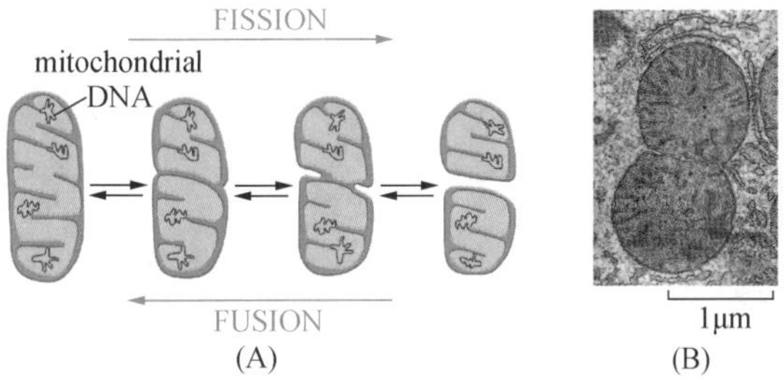

Figure 5-25 A mitochondrion divides

Like mitochondria, chloroplasts also contain their own genome. Most organism chloroplast genomes contain a pair inverted repeat (IRA and IRB) which are separated by long and short single copy regions (LSC and SSC). In a general, plastids possess a genome (plastome) of between 120 and 160 kb that encodes between 120 and 135 genes.

The mitochondrial matrix proteins encoded by the mitochondrial genome are typically inserted into the inner membrane in a co-translational mechanism by coupling translating ribosomes to the membrane integrated insertase, termed oxidase assembly (OXA).

5.3.2 Protein transfer from the cytosol to mitochondria and chloroplasts

The majority of mitochondrial proteins that are encoded in the nucleus are produced on cytosolic ribosomes, and then transport into mitochondrial in a post-translational manner. Post-translational import is a characteristic of protein import into mitochondria, chloroplasts, peroxisomes, and the nucleus. As soon as a nascent peptide emerges from the ribosomal exit tunnel, their unproductive association with other cellular components must be avoided to prevent protein misfolding or aggregation. Thereby, cytosolic chaperones (Hsp70, Hsp40, and Hsp90 families) bind and protect nascent proteins to assist their folding and to maintain precursor proteins in an import-competent conformation. Chaperones can also act as recognition factors that promote the delivery of synthesized substrates to the required subcellular location. Substrate proteins often harbor targeting signals that are maintained accessible to be read due to the activity of chaperones and that determine the final localization of the protein.

The translocase of the outer mitochondrial membrane (TOM complex) forms the entry gate for most preproteins. During targeting, preproteins must be maintained in an unfolded or partially folded form to allow passage through the TOM channel. After passing through the TOM channel, the preproteins are recognized by the other translocators in the mitochondrial membranes (TIM22, TIM23, SAM) in different ways that transfer the proteins to their functional destinations in the four mitochondrial subcompartments: outer and inner membranes, intermembrane space, and matrix. ATP hydrolysis and membrane potential drive protein import into the destinations. So far, five major protein transport pathways have been identified, each one characterized by a different type of targeting signal (Figure 5-26).

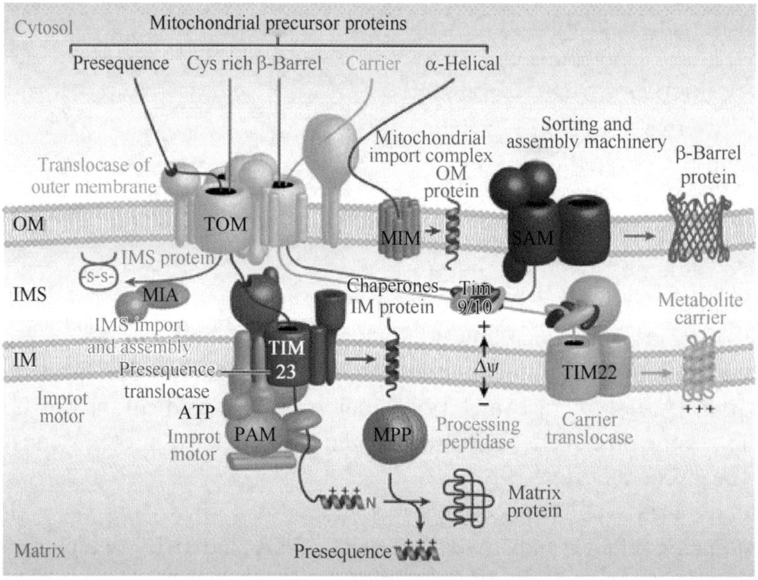

Figure 5-26 Overview of Major protein import pathways of mitochondia(From: Wiedemann N, Annu Rev Biochem. 2017)

The classical transport pathway is termed the presequence pathway. The vast majority of matrix proteins and many inner-membrane proteins are synthesized with N-terminal presequences that function as cleavable targeting signals. About 60% of mitochondrial preproteins are synthesized with cleavable targeting signals that are removed by the mitochondrial processing peptidase after import. Presequence-carrying precursors are imported by TOM and the presequence translocase of the inner membrane (TIM23). The hydrophilic proteins translocate into the matrix with the help of presequence translocase-associated motor (PAM), where the mitochondrial processing peptidase (MPP) cleaves off the presequences. Proteins with a hydrophobic sorting signal can be released into inner membrane(IM).

In the four other major protein import pathways, the mitochondrial proteins contain different types of non-cleavable internal targeting signals, such as transmembrane segments and their flanking regions in the case of membrane proteins. The TOM complex, however, functions as the main mitochondrial entry gate for both cleavable preproteins and most noncleavable precursors. The second protein transport pathway is the carrier pathway. The precursors of the multispanning hydrophobic carrier proteins of the IM are imported by TOM, the small TIM chaperones of the intermembrane space (Tim9/10), and the carrier translocase of the inner membrane (TIM22). The membrane potential ($\Delta\psi$) across the IM drives protein translocation by TIM22 and TIM23 complex.

The third protein transport pathway is the β-barrel pathway. The precursors of β-barrel proteins of the outer membrane use the TOM complex and Tim9/10, followed by insertion into the outer membrane by the sorting and assembly machinery (SAM).

The fourth protein transport pathway is used by cysteine-rich proteins located in intermembrane space. They are imported via TOM and the mitochondrial import and assembly machinery (MIA) of the intermembrane space.

The fifth protein transport pathway is about outer-membrane proteins with α-helical transmembrane segments. The mitochondrial import complex (MIM) promotes the efficient import of outer-membrane proteins with an N-terminal signal-anchor sequence as well as multispanning outer-membrane proteins.

These 5 pathways cannot summarize the transport of all mitochondrial proteins. There will be more and more complex pathways that need to be identified.

Protein import into chloroplasts resembles import into mitochondria. However, chloroplasts have extra membrane system and membrane-enclosed compartment, the thylakoid and its membrane, both of which contain many proteins. Like the precursors in some mitochondrial protein, the transport of these precursors from the cytosol to their destination occurs in two steps. First, they translocate through the double membrane at the special contact sites into stroma in chloroplast, then they pass either into the thylakoid membrane or into the thylakoid space in different pathways. The N-terminal thylakoid signal sequence will be removed from the precursors in thylakoid space (Figure 5-27).

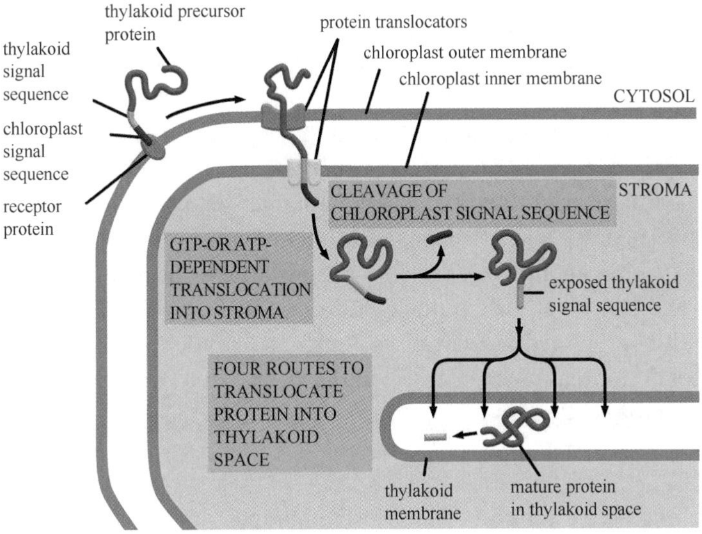

Figure 5-27 Translocation of chloroplast precursor proteins into thylakoid space (From: Molecular Biology of the Cell. 5/e, Garland Science, 2008)

5.3.3 Numerous functions of mitochondria

A large variety of functions have been revealed to mitochondrial proteins and protein complexes and are indicated in the figure 5-28: energy metabolism with respiration and synthesis of ATP; metabolism of amino acids, lipids and nucleotides; biosynthesis of iron-sulfur (Fe/S) clusters and cofactors; expression of the mitochondrial genome; quality control and degradation processes including mitophagy and apoptosis; signaling and redox processes; membrane architecture and dynamics; and the import and processing of precursor proteins that are synthesized on cytosolic ribosomes.

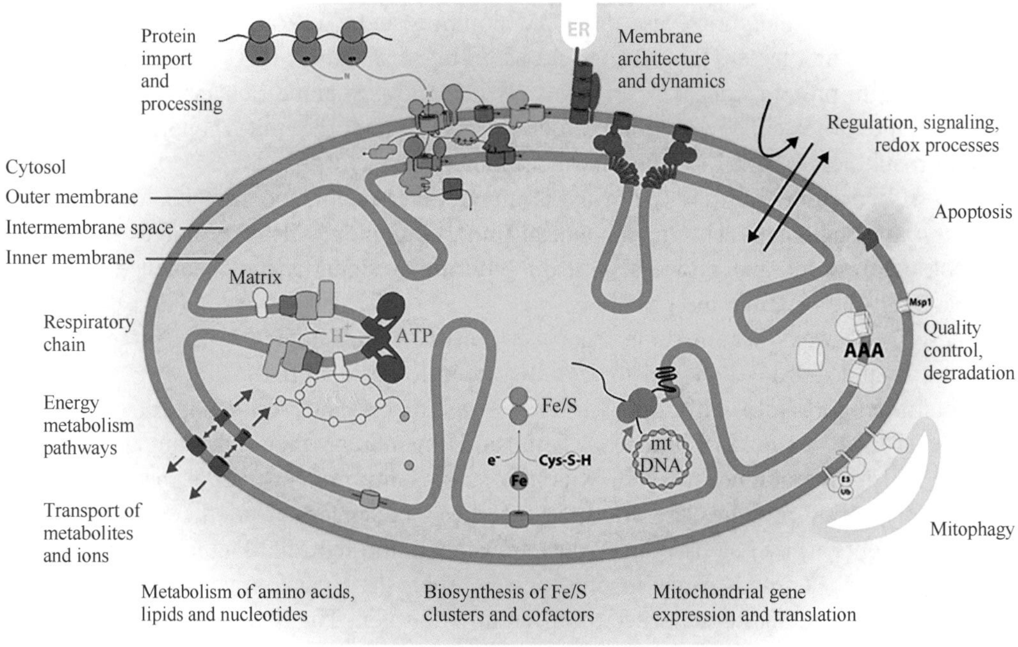

Figure 5-28 Overview of mitochondria and their functions. AAA, ATP-dependent proteases of the inner membrane; E3, ubiquitin ligase; Msp1, mitochondrial sorting of proteins, extracts mistargeted proteins; TCA, tricarboxylic acid cycle; Ub, ubiquitin(From: Pfanner N et al., Nat Rev Mol Cell Biol, 2019)

Hot research areas on mitochondria include mitochondrial signaling processes, regulation and quality control. The metabolic activity of the organelle serves as a measure for mitochondrial fitness and quality, and elaborate pathways for mitochondrial stress responses, selective degradation of damaged mitochondria by mitophagy (autophagy) and programmed cell death (apoptosis) via mitochondria have been identified. Mitochondrial quality control and turnover of mitochondrial proteins have three major pathways: autophagy, ubiquitin protein degradation system, and mitochondria internal proteases. In addition, mitochondria are a major site of cellular production of reactive oxygen species and contain numerous redox pathways.

Summary

Mitochondria are enclosed by two concentric membranes, the innermost of which encloses the mitochondrial matrix. The matrix space contains many enzymes, including those of the citric acid cycle. These enzymes produce large amounts of NADH and $FADH_2$ from the oxidation of acetyl CoA. In the inner mitochondrial membrane, high-energy electrons donated by NADH and $FADH_2$ pass along an electron-transport chain — the respiratory chain — eventually combining with molecular oxygen in an energetically favorable reaction.

Much of the energy released by electron transfers along the respiratory chain is harnessed to pump H^+ out of the matrix into the intermembrane space, thereby creating a transmembrane electrochemical proton (H^+) gradient. The proton pumping is carried out by three large respiratory enzyme complexes embedded in the inner membrane. The resulting electrochemical proton gradient across the inner mitochondrial membrane is harnessed to make ATP when H^+ ions flow back into the matrix through ATP synthase, an enzyme located in the inner mitochondrial membrane.

In photosynthesis in chloroplasts and photosynthetic bacteria, high energy electrons are generated when sunlight is absorbed by chlorophyll; this energy is captured by protein complexes known as photosystems, which are located in the thylakoid membranes of chloroplasts.

Electron-transport chains associated with photosystems transfer electrons from water to $NADP^+$ to form NADPH, with the concomitant production of an electrochemical proton gradient across the thylakoid membrane. Molecular oxygen is generated as a by-product. The proton gradient is used by an ATP synthase embedded in the membrane to generate ATP.

The ATP and the NADPH made by photosynthesis are used within the chloroplast to drive the carbon-fixation cycle in the chloroplast stroma, thereby producing carbohydrate from CO_2.

Mitochondria and chloroplasts are semi-autonomous organelles that contain their own genome. But the vast majority of proteins are encoded by the nuclear genome, which are synthesized in the cytoplasm and transported to the mitochondria and chloroplasts. Mitochondria perform crucial multi-functions in bioenergetics, metabolism and signaling.

Questions

1. Please explain the following terms: ATP synthase, mitochondrion, matrix, electron-transport chain, cytochrome, quinone, oxidative phosphorylation, chloroplast, stroma, photosynthesis, photosystem, reaction center, light reaction, carbon fixation.
2. Which of the following statements are correct? Explain your answers.
   A. Many, but not all, electron-transfer reactions involve metal ions.
   B. The electron-transport chain generates an electrical potential across the membrane because it moves electrons from the intermembrane space into the matrix.
   C. The electrochemical proton gradient consists of two components: a pH difference and an electrical potential.
   D. Ubiquinone and cytochrome c are both diffusible electron carriers.
   E. Plants have chloroplasts and therefore can live without mitochondria.
3. Please describe the structure of mitochondria.
4. Please describe the structure of chloroplasts.
5. Please describe the structural differences between leaves of C-3 plant and C-4 plant.
6. Explain the working mechanism of ATP synthase.
7. Please describe quality control of mitochondria.

（张　森　余庆波，李　瑶修改）

Chapter 6

Cytoskeleton

The cytoskeleton is an elaborate network of protein filaments that support and spatially organize the cytoplasm of a eukaryotic cell, allowing the cell to mechanically interact with its environment and make coordinated movements. The primary building blocks of the cytoskeleton are three types of protein filaments: **microtubules**, **actin filaments** and **intermediate filaments**(Figure 6-1).

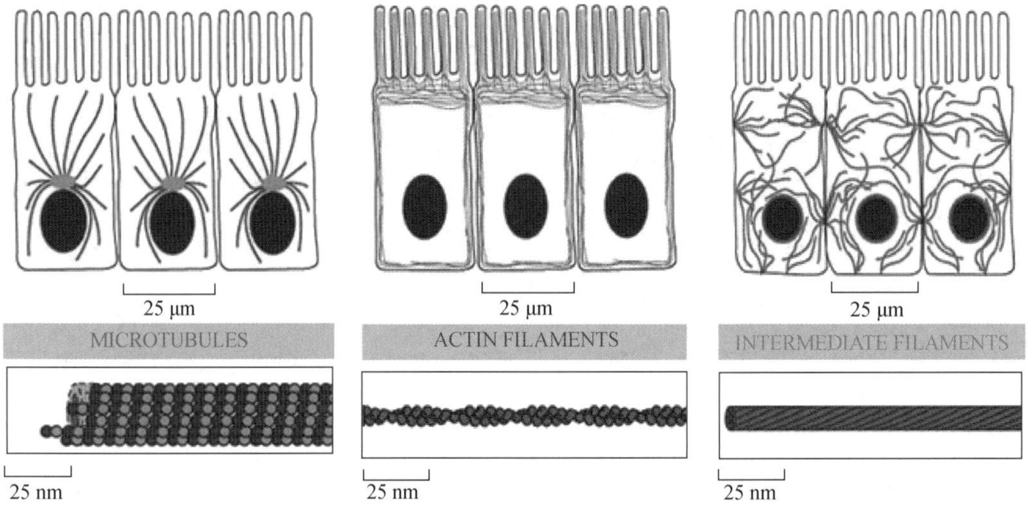

Figure 6-1 Three types of protein filaments form the cytoskeleton. The cells illustrated are epithelial cells lining the gut(From: Alberts B, et al. Essential Cell Biology. 2004)

6.1 Microtubules

6.1.1 Structure and functions of microtubules

Microtubules are long, straight and relatively stiff hollow cylindrical tubes that are ubiquitous in all eukaryotic cells. Microtubules have many functions in a cell. They are stiff enough to provide mechanical support to cells. Their cytoplasmic distribution helps to determine and maintain the cell shape. They are thought to help organizing internal cell contents. They provide a set of "tracks" for cell organelles and vesicles to move on. During mitosis, microtubules form the spindle fibers for separating chromosomes. When arranged in geometric patterns inside flagella and cilia, microtubules are used for locomotion.

A microtubule has an outer diameter of approximately 24 ~ 25 nm, a wall thickness of about 4 nm, and may extend across the length and breadth of a cell. In Figure 6-2A, the bumps at the surface of the microtubules are **microtubule-associated proteins (MAPs)**. Each MAP has its one domain attached to the side of a microtubule and another domain projects outward as a filament. The general functions of MAPs are to increase the stability of microtubules and to promote microtubule assembly. Some MAPs form cross-bridges connecting microtubules to each other to maintain their parallel alignment.

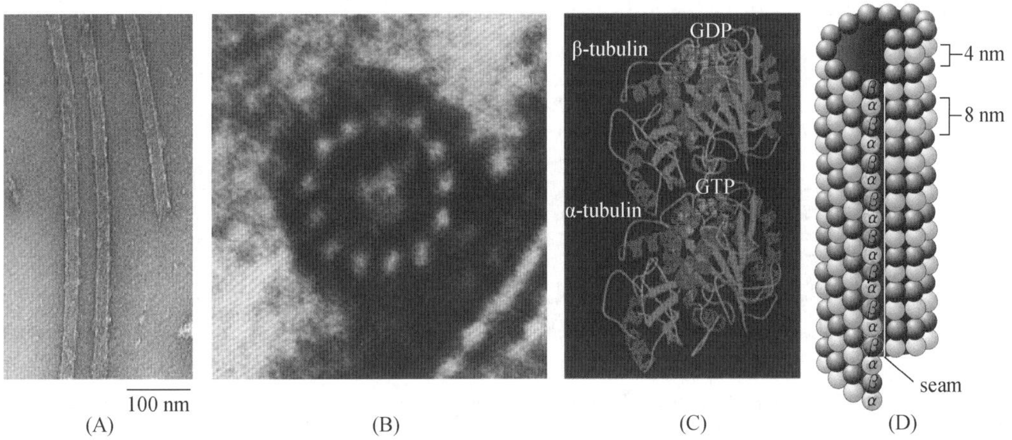

Figure 6-2 The structure of microtubules. (A) Electron micrograph of negative stained microtubules from brain showing the globular subunits that make up the protofilaments. The bumps at the surface of the microtubules are microtubule-associated proteins (MAPs). (B) Electron micrograph of a cross section through a microtubule of a *Juniperus* root tip cell revealing the 13 subunits arranged within the wall of the tubule. (C) A ribbon model showing the three-dimensional structure of the α/β-tubulin heterodimer. Note the complementary shapes of the subunits at their interacting surfaces. The α-tubulin subunit has a bound GTP, which is not hydrolyzed and is nonexchangeable. The β-tubulin subunit has a bound GDP, which is exchanged for a GTP prior to assembly into a polymer. The plus end of the dimer is at the top. (D) Diagram of a longitudinal section of a microtubule shown in the B-lattice, which is the structure thought to occur in the cell. The wall consists of 13 protofilaments composed of α/β-tubulin heterodimers stacked in a head-to-tail arrangement. Adjacent protofilaments are not aligned in register, but are staggered about 1 nm so that the tubulin molecules form a helical array around the circumference of the microtubule. The helix is interrupted at one site where α and β subunits make lateral contacts. They produce a "seam" that runs the length of the microtubule. (From: Karp G. Cell and Molecular Biology — concepts and experiments 17th edition. 2013)

The wall of a microtubule composed of globular proteins arranged in longitudinal rows, termed protofilaments, that are aligned parallel to the long axis of the tubule. The cross section of a microtubule consists of 13 protofilaments aligned side by side in a circular pattern within the wall (Figure 6-2B). Each protofilament is built from dimeric protein subunits. Each **tubulin dimer** is made of one **α-tubulin** and one **β-tubulin**. These two tubulins are very similar in 3-dimensional structure and they bind tightly together by noncovalent bonding. Tubulin dimers stack tightly together again by noncovalent bonding to form the wall of microtubules (Figure 6-2D). Each tubulin dimer binds two GTP molecules, with the α-tubulin binding to one GTP irreversibly and the β-tubulin binding to another GTP reversibly, and β-tubulin hydrolyzes GTP to GDP (Figure. 6-2C). As is discussed later in the chapter, the guanine bound to β-tubulin regulates the addition of tubulin subunits at the ends of a microtubule. Because each protofilament is made of a linear chain of heterodimers with alternating α and β tubulins, it has a structural polarity with an α-tubulin at one end and a β-tubulin at the other end. All protofilaments of one microtubule have the same polarity, therefore, the entire microtubule has polarity with one end made of all β-tubulin subunits called the plus end, and the other end made of all α-tubulin subunits called the minus end. This structural polarity is crucial in microtubule formation and function, as the plus end and the minus end differ in their rate of microtubule assembly.

Most microtubules in a cell consist of a simple tube, a *singlet* microtubule, built from 13 protofilaments. In addition to this simple singlet structure, *doublet* or *triplet* microtubules are found in specialized structures such as cilia and flagella (*doublet* microtubules) and centrioles and basal bodies (*triplet* microtubules) that we will explore later in the chapter. Each *doublet* or *triplet* contains one complete 13-protofilament microtubule (called the A tubule) and one or two additional tubules (B and C) consisting of 10 protofilaments each (Figure 6-3).

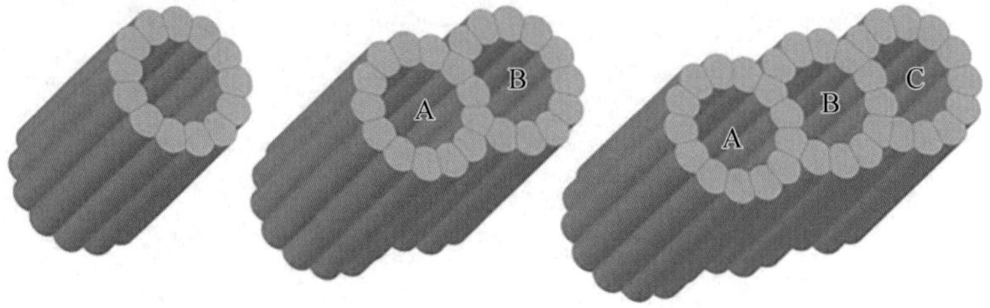

Figure 6-3 Singlet, doublet, and triplet microtubules. In cross section, a typical microtubule, a *singlet*, is a simple tube built from 13 protofilaments. In a *doublet* microtubule, an additional set of 10 protofilaments forms a second tubule (B) by fusing to the wall of a singlet (A) microtubule. Attachment of another 10 protofilaments to the (B) tubule of a doublet microtubule creates a (C) tubule and a *triplet* structure (From: Lodish H, et al. Molecular Cell Biology (8th edition). 2016)

6.1.2 Microtubule organizing center (MTOC)

Microtubules in cells grow from a variety of specialized organizing centers called **microtubule-organizing centers (MTOCs)**. The **centrosome** is the main MTOC in animal cells. In cilia and flagella, microtubules are assembled from an MTOC called a **basal body**. Plants do not have centrosomes and basal bodies, but use other mechanisms to nucleate the assembly of microtubules. The best-studied MTOC is the **centrosome** . It is composed of a pair of barrel-shaped centrioles surrounded by amorphous and electron-dense **pericentriolar material (PCM)**. **Centrioles** are cylindrical structures about 0.2 μm in diameter and 0.4 μm in length. Centrioles contain nine evenly spaced triplet, and only the A tubule connects to the center of the centriole by a radial spoke, thus giving the cross section of the centriole a pinwheel look (Figure 6-4).

All MTOCs share similar functions in all cells. They control the number of microtubules, microtubule polarity, the number of protofilaments that make up microtubule walls, as well as the time and location of microtubule assembly. All MTOCs have a common protein component called **γ-tubulin**. Approximately 80% of γ-tubulin in cells is part of a 25S complex named the γ-tubulin ring, a ring-shaped structure that serves as the starting point, or nucleation site, for the growth of one microtubule (Figure 6-5). The α/β-tubulin dimers add to the γ-tubulin ring in a specific direction such that the end result is that the minus end of each microtubule is embedded in the centrosomes while the plus end of each microtubule faces outward where further growth (polymerization) can occur (Figure 6-6).

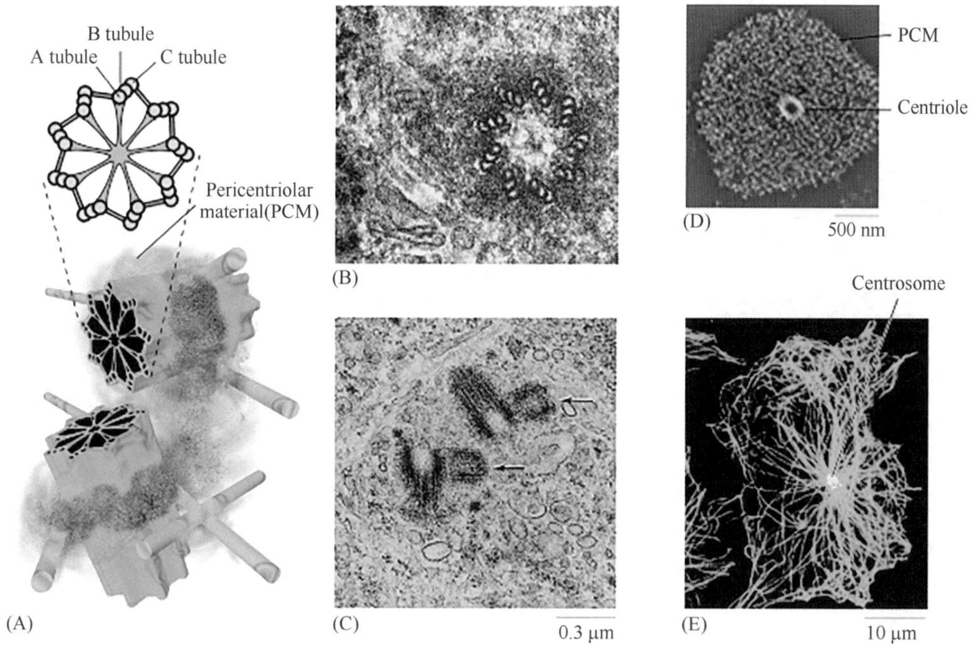

Figure 6-4 The centrosome. (A) Schematic diagram of a centrosome showing the paired centrioles; the surrounding pericentriolar material (PCM); and microtubules emanating from the PCM, where nucleation occurs. (B) Electron micrograph of a cross section of a centriole showing the pinwheel arrangement of the nine peripheral fibrils, each of which consists of one complete microtubule and two incomplete microtubules. (C) Electron micrograph showing two pairs of centrioles. Each pair consists of a longer parental centriole and a small daughter centriole (arrow), which is undergoing elongation in this phase of the cell cycle. (D) Electron micrographic reconstruction of a 1.0 mol/L potassium iodide — extracted centrosome, showing the PCM to contain a loosely organized fibrous lattice. (E) Fluorescence micrograph of a cultured mammalian cell showing the centrosome at the center of an extensive microtubular network. (From: Karp G, Cell and Molecular Biology-concepts and experiments (7th edition). 2013)

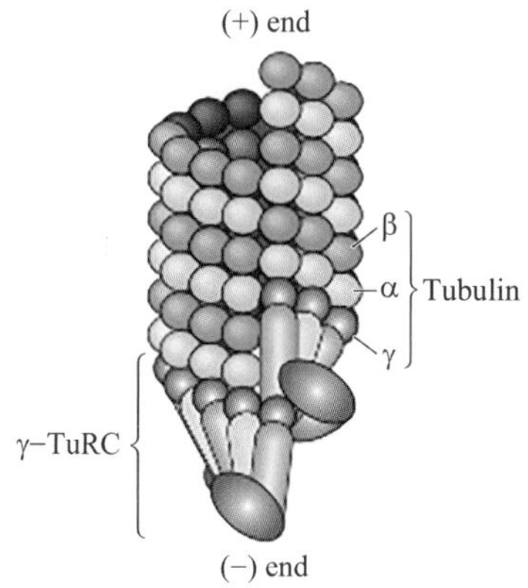

Figure 6-5 The γ-tubulin ring complex (γ-TuRC) that nucleates microtubule assembly. Model of how γ-TuRC nucleates assembly of a microtubule by forming a template corresponding to the (−) end (From: Lodish H, et al. Molecular Cell Biology (8th edition). 2016)

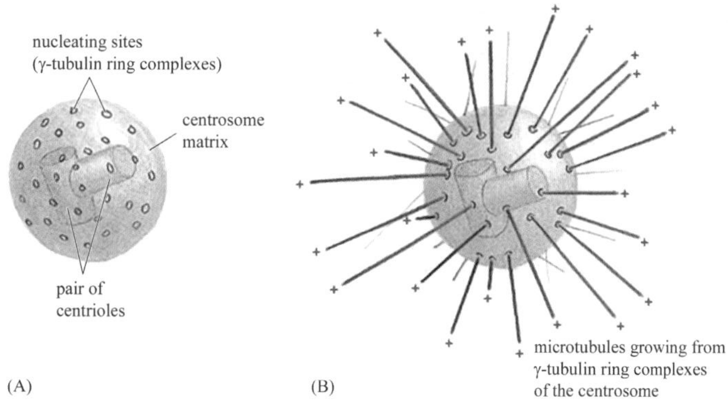

Figure 6-6 Tubulin polymerizes from nucleation sites on a centrosome. (A) Schematic drawing showing that a centrosome consists of an amorphous matrix of protein containing the γ-tubulin rings that nucleate microtubule growth. (B) A centrosome with attached microtubules. the minus end of each microtubule is embedded in the centrosome, having grown from a nucleating ring, whereas the plus end of each microtubule is free in the cytoplasm (From: Albert B, et al. Essential cell biology (3rd edition). 2009)

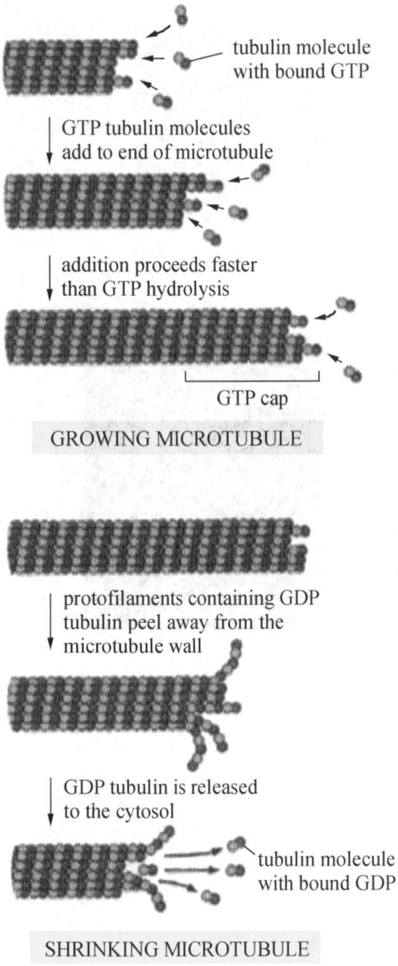

Figure 6-7 GTP hydrolysis controls the growth of microtubules. Tubulin dimers carrying GTP bind more tightly to one another than do tubulin dimers carrying GDP. Therefore, microtubules that have freshly added tubulin dimers at their end with GTP bound tend to keep growing. From time to time, however, especially when microtubule growth is slow, the subunit in this "GTP cap" will hydrolyze their GTP to GDP before fresh subunits loaded with GTP have time to bind. The GTP cap is thereby lost; the GDP-carrying subunits are less tightly bound in the polymer and are readily released from the free end, so that the microtubule begins to shrink continuously (From: Alberts B, et al. Essential Cell Biology (3rd edition). 2009)

Growing microtubules show dynamic instability: once a microtubule has been nucleated, its plus end grows outward by the addition of subunits for minutes then suddenly shrinks by losing subunits from its free end and subsequently start growing again. Thus a given microtubule oscillates unpredictably between assembly (growing/polymerization) and disassembly (shrinking/depolymerization) phases; growing and shrinking microtubules can coexist in the same region of a cell. This dynamic instability results mainly from GTP hydrolysis. As mentioned before, each tubulin dimer has one tightly bound GTP molecule that is hydrolyzed to GDP (still tightly bound) shortly after adding to a growing microtubule. After disassembly, the GDP bound to the released dimer is replaced by a new GTP. This nucleotide exchange allows the dimer to be once again ready to become a building block for polymerization.

The GTP-bound tubulin molecules pack more strongly to each other compared with GDP-bound tubulin molecules. When polymerization is progressing rapidly, tubulin molecules add to the end of microtubule faster than GTP hydrolysis rate, giving rise to a microtubule end full of GTP-tubulin subunits called GTP cap. However, occasionally, when the tubulin at the free end of the microtubule hydrolyzes its GTP to GDP before the next tubulin has been added, the end of microtubule consists of GDP bound subunits. Because GDP-tubulins bind less effectively to each other, this favors disassembly of the microtubule. Once disassembly starts, it tends to continue at a catastrophic rate. Therefore, microtubules shrink rapidly (Figure 6-7).

Drugs can prevent the polymerization or depolymerization of tubulin. The drug colchicine binds tightly to free tubulin and prevents its polymerization into microtubules. The drug taxol has the opposite action at the molecular level. It binds tightly to microtubules and prevents them from losing subunits.

In a normal cell, the MTOCs continuously explore randomly by shooting out new microtubules in all directions and retracting them. This process ceases when the plus end is somehow stabilized by attachment to another molecule or cell structure and a relatively stable link is established between that structure and the centrosome. The random exploration and selective stabilization enable MTOCs to set up a highly organized system of microtubules linking specific parts of the cell.

Cells can modify the dynamic instability of their microtubules for specific reasons. For example, when cells enter mitosis, microtubules become more dynamic, growing and shrinking more frequently than they normally do, which allows them to disassemble rapidly and reassemble into a mitotic spindle. However, differentiated cells often suppress the dynamic instability of their microtubules by having proteins bind to the ends or along the length of microtubules to stabilize them. Stabilized microtubules can maintain the organization of the cell.

6.1.3 Motor proteins

Polarized systems of microtubules determine a cell's polarity, help to position intracellular organelles in the cell and guide the intracellular movement. Movement of particles along microtubules is a lot faster and more efficient than free diffusion.

Directed intracellular movement of cell materials along microtubules is accomplished by motor proteins. This is also the case with the **actin filaments**, which will be discussed later in the chapter. Motor proteins derive their energy from repeated cycles of ATP hydrolysis. There are two major families of motor proteins that move along cytoplasmic microtubules: kinesins and dyneins. **Kinesins** usually move toward the plus end of a microtubule (outward, away from the centrosome) while **dyneins** move toward the minus end (inward, toward the centrosome). They both have two globular ATP-binding heads and a single tail (Figure 6-8A). The heads determine which microtubule to attach by interacting with microtubules of specific orientation while the tail binds stably to some specific cell materials for transport.

The heads of both kinesins and dyneins are enzymes with ATP-hydrolyzing (ATPase) activity. ATP hydrolysis provides energy for a cycle of conformation changes in the head that enables it to move along the microtubule by a cycle of binding, release, rebinding to the microtubule (Figure 6-8B).

Microtubules and their associated motor proteins play an important role in positioning intracellular organelles (such as membranes) within a cell. For example, the normal positioning of endoplasmic reticulum relies on its membrane receptors that bind to kinesin and subsequently are pulled outward along microtubules, stretched like a net; in the case of Golgi apparatus, it is dragged by dyneins along microtubules to reach its position at the center of a cell (Figure 6-9). In this way the cell internal membrane arrangement, on which the successful function of the cell relies, are established and maintained.

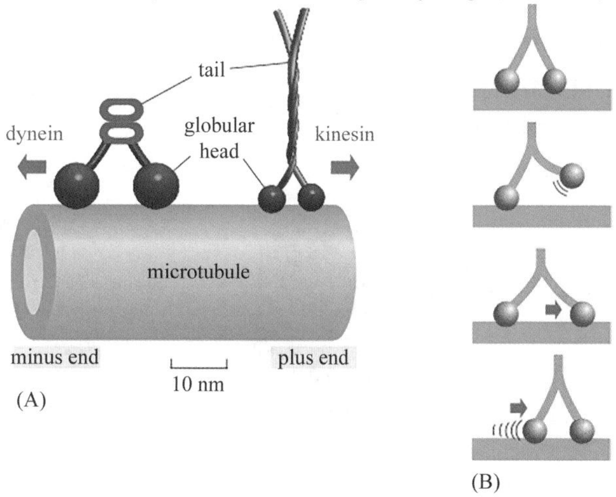

Figure 6-8 Motor proteins move along microtubules using their globular heads. (A) Kinesins and cytoplasmic dyneins are microtubule motor proteins that generally move in opposite directions along a microtubule. Each of these proteins (drawn here to scale) has two heavy chains and several smaller light chains. Each heavy chain forms a globular head that interacts with microtubules. (B) Diagram of a motor protein showing ATP-dependent "walking" along a filament(From: Alberts B, et al. Essential Cell Biology (3rd edition). 2009)

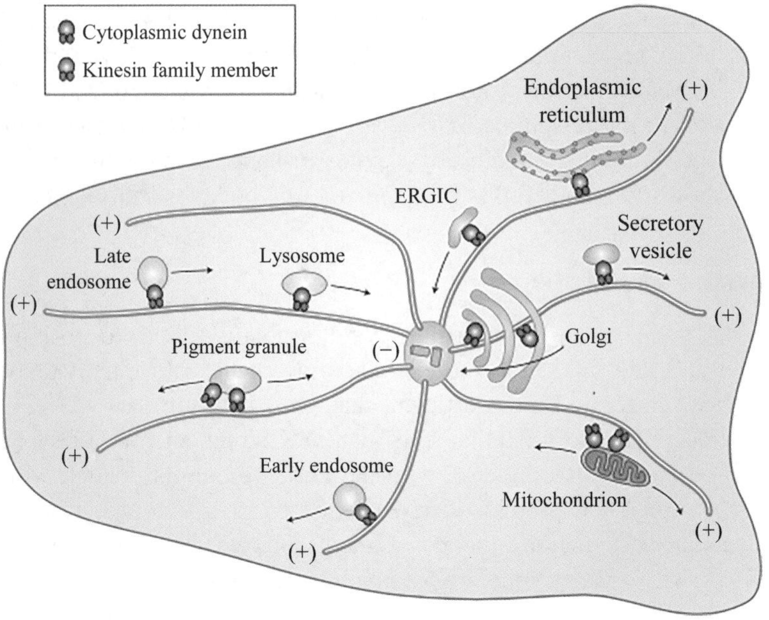

Figure 6-9 Organelle transport by microtubule motors. Cytoplasmic dyneins (red) mediate retrograde transport of organelles toward the (−) ends of microtubules (cell center); kinesins (purple) mediate anterograde transport toward the (+) ends (cell periphery). Most organelles have one or more microtubule-based motors associated with them. ERGIC = ER-to-Golgi intermediate compartment (From: Lodish H, et al. Molecular Cell Biology (8th edition). 2016)

6.1.4 The microtubules in cilia and flagella

Unlike cytoplasmic microtubules, the microtubules in **cilia** and **flagella** are arranged in a strikingly distinctive pattern revealed by electron microscopy(Figure 6-10).

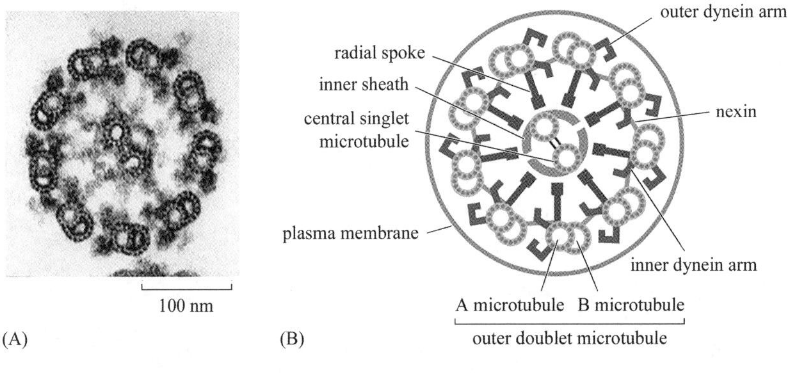

Figure 6-10 Microtubules in a cilium or in a flagellum are arranged in a "9+2" array. (A) Electron micrograph of a flagellum of *chlamydomonas* shown in cross section. Illustrating the distinctive "9+2" arrangement of microtubules. (B) Diagram of the flagellum across section. The nine outer microtubules (each a special paired structure) carry two rows of dynein molecules. The heads of these dyneins appear in this view like pairs of arms reaching toward the adjacent microtubule. In a living cilium, these dynein heads periodically make contact with the adjacent microtubule and move along it, thereby producing the force for ciliary beating. Various other links and projections shown are proteins that serve to hold the bundle of microtubules together and to convert the sliding motion produced by dyneins into bending(From: Alberts B, et al. Essential Cell Biology (3rd edition). 2009)

The cross section of a cilia or flagella shows a ring of nine doublet microtubules with a pair of single microtubule at the center. This "9+2 array" is seen in all forms of eukaryotic cilia and flagella, ranging from protozoa to human. This is a useful reminder that all living eukaryotes have evolved from a common ancestor. The movement of a cilium or a flagellum is initiated by the bending of its core as the microtubules slide against each other. The bending motion of the core is generated from ciliary dynein, a motor protein that resembles cytoplasmic dynein in both its structure and function. Ciliary dynein has its tail attached to one microtubule and its head to another microtubule to generate a sliding force between two filaments. Because the adjacent doublets are linked together, the parallel sliding movement between free microtubules is converted to a bending motion in the cilium (Figure 6-11).

In humans, hereditary defects in ciliary dynein cause Kartagener's syndrome. Men with this disorder are infertile because their sperm are non-motile, and all those affected have an increased susceptibility to bronchial infections because the cilia that line their respiratory tract are paralyzed and thus unable to clear bacteria and debris from the lungs.

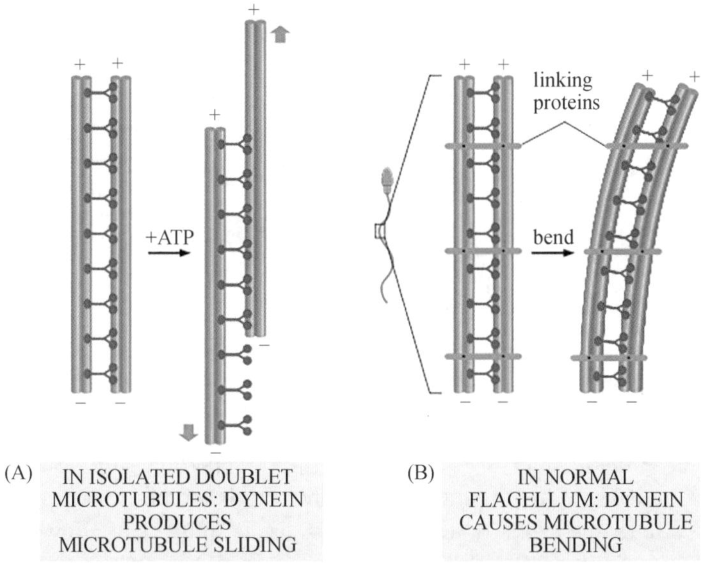

Figure 6-11 The movement of dynein causes the flagellum to bend. (A) If the outer doublet microtubules and their associated dynein molecules are freed from other components of a sperm flagellum and then exposed to ATP, the doublets slide against each other, telescope-fashion, due to the repetitive action of their associated dyneins. (B) In an intact flagellum, however, the doublets are tied to each other by flexible protein links so that the action of the system produces bending rather than sliding (From: Albert B, et al. Essential cell biology (3rd edition). 2009)

6.2 Actin filaments

Actin filaments (microfilaments) are thin and flexible filaments that are the most abundant intracellular proteins in a eukaryotic cell. They are essential for cell locomotion and intracellular motile processes. They also help to shape the cell.

6.2.1 Globular monomers and filamentous polymers

Actin exists either as G-actin, a globular monomer, or as F-actin, a filamentous polymer that is a linear chain of G-actin subunits. Small actin binding proteins such as thymosin and profilin binds to G-actin in the cytosol regulating the polymerization of G-actin subunits into filaments. Many other actin-binding proteins, however, bind to F-actin and control the behavior of the intact filaments to form different types of stable actin filament structures (Figure 6-12). Each actin molecule contains an Mg^{2+} ion complexed with either ATP or ADP. Thus, there are four states of actin: ATP-G-actin, ADP-G-actin, ATP-F-Actin and ADP-F-actin. The more prominent forms of actin in a cell are ATP-G-actin and ADP-F-actin. The interaction between the ATP and ADP forms of actin is important in the cytoskeleton assembly.

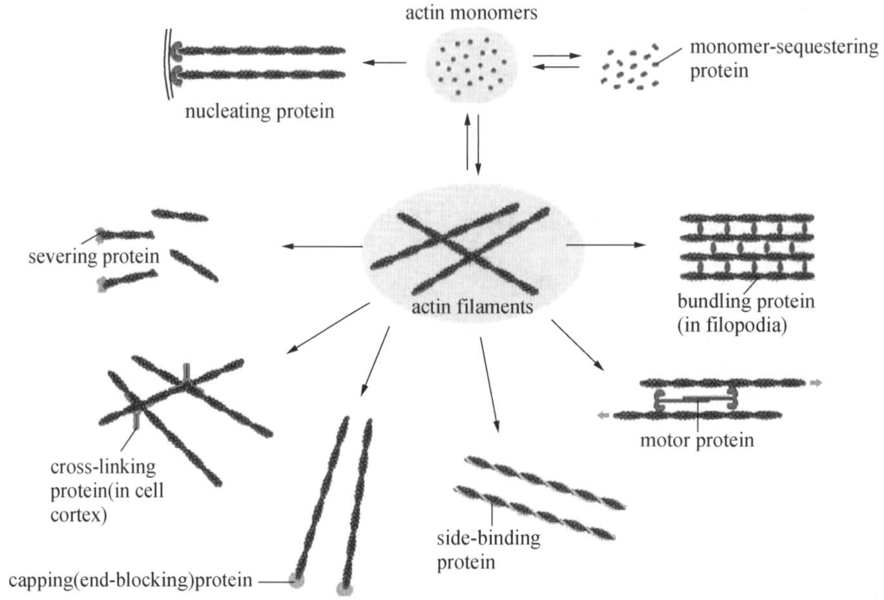

Figure 6-12 Actin-binding proteins control the behavior of actin filaments in vertebrate cells (From: Alberts B, et al. Essential Cell Biology (3rd edition). 2009)

X-ray crystallographic analysis shows G-actin is separated into two lobes by a deep cleft where ATP and Mg^{2+} are bound. Two lobes and a cleft make up an ATPase fold (Figure 6-13A) where the floor of the cleft acts as a hinge that allows the lobes to flex relative to each other and ATP holds together the two lobes of the actin monomer. The nucleotide affects the conformation of the actin molecule and stabilizes it. Without a bound nucleotide, G-actin denatures very quickly. When negatively stained by uranyl acetate for electron microscopy, F-actin appears as long twisted strings of beads whose diameter varies between 7 and 9 nm. (Figure 6-13B) From x-ray diffraction studies (Figure 6-13A), scientists have produced a model of an actin filament shown in figure 6-13C in which the subunits are organized into tightly wound helix. Each subunit is surrounded by four other subunits: one above, one below and two to one side. Each subunit corresponds to a bead seen in electron micrographs of F-actin filaments in Figure 6-13B.

6.2.2 The actin cytoskeleton is organized into bundles and networks of filaments

There are two most common organization styles of actin filaments in a cell: bundles and networks of filaments (Figure 6-14). Electron micrograph or immunofluorescence micrograph

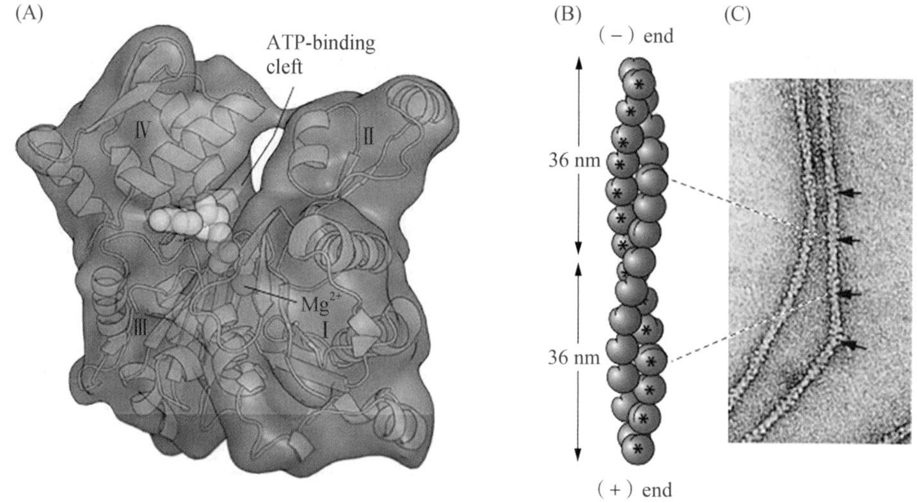

Figure 6-13 Structures of monomeric G-actin and F-actin filament. (A) Structure of the actin monomer (measuring 5.5×5.5×3.5 nm), which is divided by a central cleft into two approximately equal-sized lobes and four subdomains, numbered Ⅰ-Ⅳ. ATP (yellow) binds at the bottom of the cleft and contacts both lobes (the green ball represents Mg^{2+}). The N- and C-termini lie in subdomain I. (B) An actin filament appears as two strands of subunits. One repeating unit consists of 28 subunits (14 in each strand, indicated by * for one strand), covering a distance of 72 nm. The ATP-binding cleft of every actin subunit is oriented toward the same end of the filament. The end of a filament with an exposed binding cleft is the (−) end; the opposite end is the (+) end. (C) In the electron microscope, negatively stained actin filaments appear as long, flexible, and twisted strands of beaded subunits. Because of the twist, the filament appears alternately thinner (7-nm diameter) and thicker (9-nm diameter) (arrows) (From Lodish H, et al. Molecular Cell Biology (8th edition). 2016)

shows that bundles start from the protrusion of the cell surface membrane and continue to fan out to become a network of filaments. In bundles, the actin filaments are tightly packed in parallel arrays whereas in a network the actin filaments are loosely packed in a criss-cross pattern and often at right angles. There are two types of actin networks. One is planar or two-dimensional, like a net or a web, and is associated with the plasma membrane; the other is three-dimensional, present in the cell and provides the cytosol gel-like properties. Actin cross-linking proteins hold the filaments together in all bundles and networks. Each cross-linking protein has two actin-binding sites, with one site for each filament. Parallel actin filaments in bundles are held close together by short cross-linking protein while

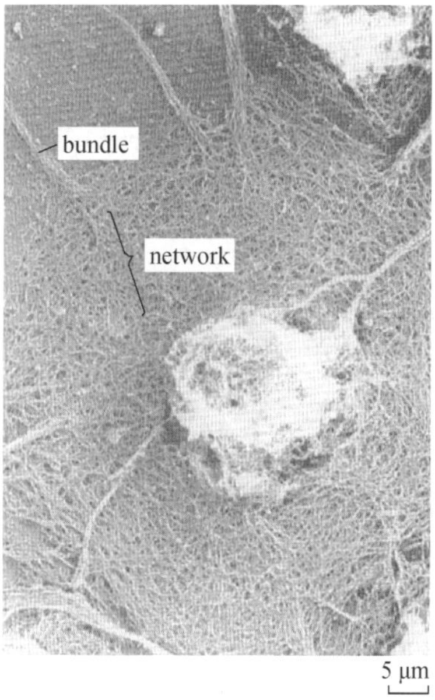

Figure 6-14 Micrograph revealing bundles and networks of actin filaments in the cytosol of a spreading platelet treated with detergent to remove the plasma membrane. Actin bundles project from the cell to form the spikelike filopodia. In the lamellar region of the cell, the actin filaments form a network that fills the cytosol. In contrast to the roughly parallel alignment of bundled filaments. The filaments in networks lie at various angles approaching 90° (From: Lodish H, et al. Molecular Cell Biology (4th edition). 2004)

orthogonally oriented actin filaments in networks are tethered by long, flexible cross-linking proteins. Functionally, both bundles and networks support the plasma membrane and help to determine a cell's shape.

6.2.3 Mechanism of actin polymerization

Actin and tubulin polymerize by similar mechanisms. *In vitro*, the addition of ions Mg^{2+}, K^{+}, or Na^{+} to a solution of G-actin induces the polymerization of G-actin into F-actin filaments, which are indistinguishable from microfilaments (actin) isolated from cells. This process is reversible in that once the ion concentration is reduced in the solution, the F-actin filaments depolymerize into G-actin. This reversible assembly of actin lies at the core of many cell movements. The assembly of G-actin into F-actin is accompanied by the hydrolysis of ATP to ADP and Pi, although ATP hydrolysis is not necessary for polymerization to occur.

There are three sequential phases in the polymerization of actin filaments (Figure 6-15). In phase one (lag state or nucleation state), G-actin keeps on aggregating into short, unstable oligomers until the oligomers reach a certain length (three or four subunits), then oligomers can act as a stable seed, or nucleus. In phase two (elongation state), the nucleus rapidly elongates into a filament due to the addition of actin monomers to both of its ends until the concentration of G-actin monomers decrease enough to reach its equilibrium state with the filament. In phase three (steady state), G-actin monomers only exchange with subunits at the filament ends but there is no net change in the total mass of the filaments. The equilibrium concentration of the pool of unassembled free subunits is called the **critical concentration** (**Cc**).

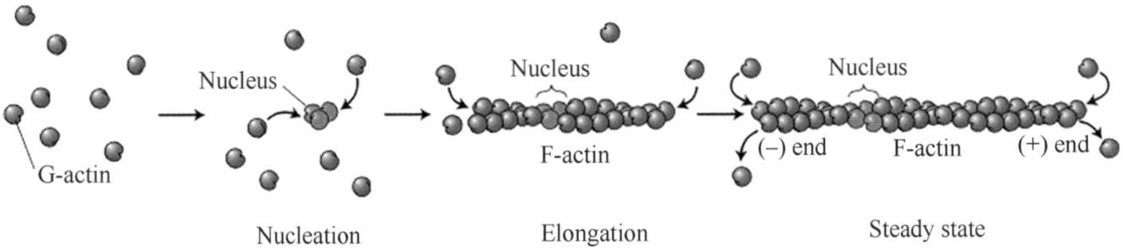

Figure 6-14 The three phases of G-actin polymerization in *vitro*. During the initial nucleation phase, ATP-G-actin monomers (red) slowly form stable complexes of actin (purple). These nuclei are rapidly elongated in the second phase by addition of subunits to both ends of the filament. In the third phase, the ends of actin filaments are in equilibrium with monomeric ATP-G-actin. After their incorporation into a filament, subunits slowly hydrolyze ATP and become stable ADP-F-actin. Note that the ATP-binding clefts (black triangles) of all the subunits are oriented in the same direction in F-actin (From: Lodish H, et al. Molecular Cell Biology (8th edition). 2016)

Toxins easily perturb the equilibrium between F-and G-actins. **Cytochalasin**, a fungal alkaloid, depolymerizes actin filaments by binding to the plus end of F-actin and blocks further addition of subunits. In contrast, **latrunculin**, a toxin secreted by sponges, binds to G-actin and inhibits it from adding to a filament end. Both toxins inhibit the polymerization; therefore, with the presence of these two toxins in the bath, the actin cytoskeleton disappears and cell movements are inhibited. A third toxin, **phalloidin**, has the opposite effect on actin. It binds at the interface between F-actins, locking adjacent subunits together, and thus preventing actin filaments from depolymerization, even when G-actin is diluted below its Cc.

A cell regulates actin polymerization through several actin-binding proteins that either promote or inhibit polymerization. For example, thymosin β4 (Tβ4, an abundant cytosolic protein, binds ATP-G-actin (but not F-actin) in a 1 : 1 complex. In the complex, G-actin cannot polymerize. Thus, thymosin β4 serves as a buffer for monomeric actin and inhibits actin assembly. Another cytosolic protein profilin, on the other hand, promotes actin assembly, although it also binds ATP-actin monomers in a stable 1 : 1 complex. Some actin binding proteins control the lengths of actin filaments by binding to actin filaments and breaking them into shorter fragments. Another group of proteins can cap the ends of actin filaments and stabilize actin filaments.

Table 6-1 shows a list of cytosolic proteins that regulates actin polymerization. The activity of actin-binding proteins that regulates the length, location, organization, and dynamic behavior of actin filaments are in turn, regulated by extracellular signals, allowing cells to rearrange its cytoskeleton to respond to its environment. In the case of actin filaments, a group of closely related monomeric GTP-binding proteins called Rho protein family (Cdc42, Rac, and Rho) act as signaling switch molecules (Figure 6-16). Activation of different Rho proteins causes different dramatic and complex structural changes in the cell.

Table 6-1 Some cytosolic proteins that control actin polymerization

Protein	MW	Activity
cofilin	15,000	severing
gCAP39	40,000	capping [(+) end]
severin	40,000	severing, capping
gelsolin	87,000	severing, capping [(+) end]
villin	92,000	cross-linking, severing, capping
capZ	36,000(α)32,000(β)	capping [(+) end]
tropomodulin	40,000	capping [(-) end]

(From: Lodish H, et al. Molecular Cell Biology. 2000)

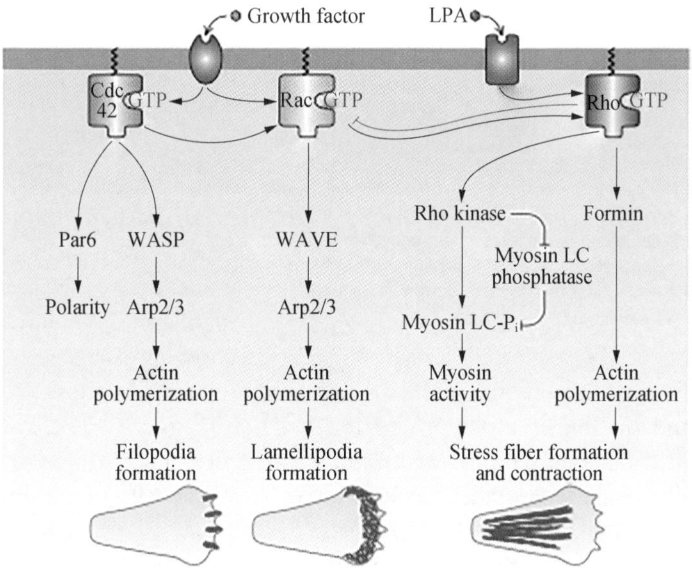

Figure 6-16 Summary of signal-induced changes in the actin cytoskeleton. Specific signals, such as growth factors and lysophosphatidic acid (LPA), are detected by cell-surface receptors. Detection leads to the activation of the small GTP-binding proteins, which then interact with effectors to bring about cytoskeletal changes as indicated. (From Lodish H, et al. Molecular Cell Biology (8th edition). 2016)

6.2.4 Cell movement by manipulating actin polymerization

Functionally, the cell derives its driving force for many types of movement by manipulating actin polymerization and depolymerization. For example, cell crawling, a major type of cell movement, depends on actin filaments. Cell crawling has three steps: ①the cell pushes out at its protrusion leading edge; ②new anchorage points created at the leading edge cause the adherence between protrusions and the cell crawling surface; ③the rest of the cell drags itself forward by traction on anchorage points (Figure 6-17).

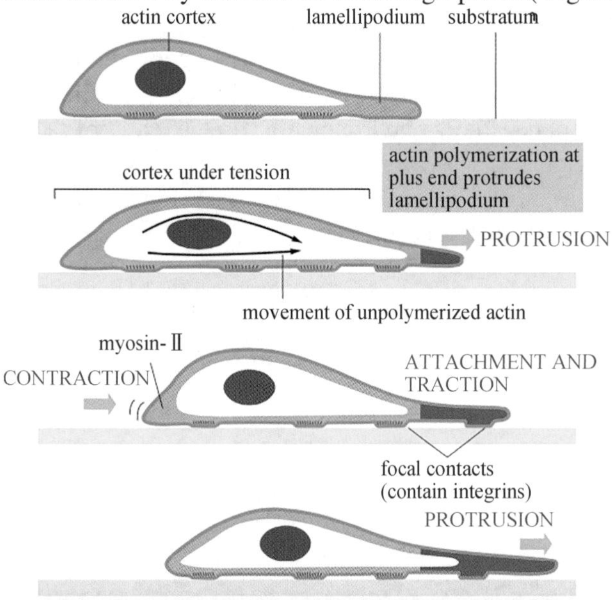

Figure 6-17 Forces generated in the actin-rich cortex move a cell forward. In this proposed mechanism for cell movement, actin polymerization at the leading edge of the cell *pushes* the plasma membrane forward (protrusion) and forms new regions of actin cortex, shown here in *red*. New points of anchorage are made between the actin filaments and the surface on which the cell is crawling (attachment). Contraction at the rear of the cell then draws the body of the cell forward (traction). New anchorage points are established at the front, and old ones are released at the back as the cell crawls forward. The same cycle is repeated over and over again, moving the cell forward in a stepwise fashion (From: Albert B, et al. Essential cell biology (3rd edition). 2009)

All three steps involve actin in different ways. The first step is driven by rapid polymerization of actin filaments, which are nucleated at the plasma membrane and push out the membrane without tearing it to form exploratory, motile structures such as lamellipodia or filopodia. **Actin-related proteins (ARPs)** form a complex that promotes the formation of a two-dimensional, tree like web of actin by binding to an existing actin filament and nucleating the creation of new side growths of filaments. With the help of other actin binding proteins, this web assembles at the front end and dissembles at the rear end, pushing the lamellipodium or filopodium forward (Figure 6-18).

In step two, the lamellipodia and filopodia stick to a favorable patch surface through integrins, a type of transmembrane proteins embedded in their plasma membrane. Integrins adhere to the crawling surface on the extracellular side and capture actin filaments intracellularly, creating anchorage sites. In step three, cells use anchorages and internal contractions to exert a pulling force. These are based on the interaction of actin filaments with motor proteins called myosins. Although how this pulling force is generated is not clear, the mechanism of how myosin motor proteins and actin filament interaction produce movement is well understood.

All actin-dependent motor proteins belong to the myosin family. They bind to and hydrolyze ATP, which provides the energy for their movement along actin filaments from the minus end of the filament toward the plus end (with one exception, the *myosin VI* working in the opposite direction). Yet different myosins perform very different types of functions. Many cell movements, ranging from the intracellular motion of cellular

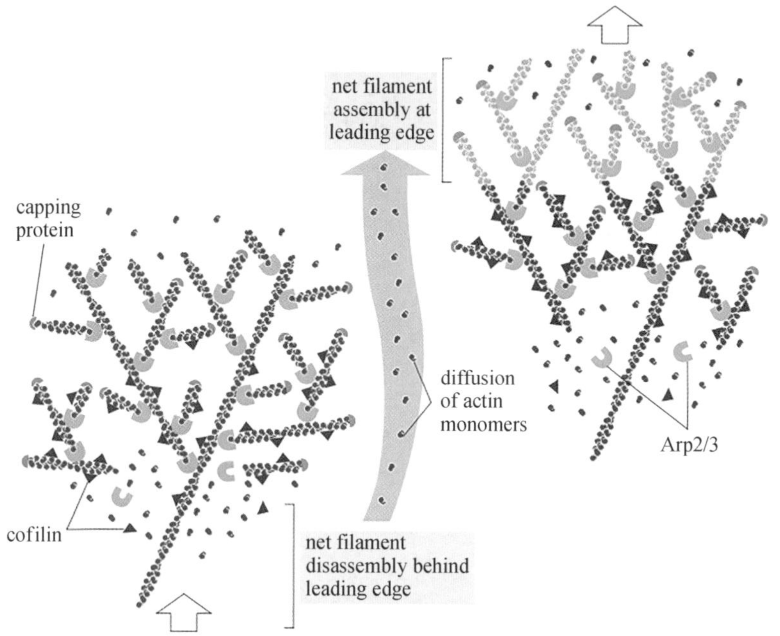

Figure 6-18 Assembly of an actin meshwork pushes forward the leading edge of a lamellipodium. Two time points during advance of the lamellipodium are illustrated, with newly assembled structures at the later time point shown in a lighter color. Nucleation of new actin filaments is mediated by ARP(actin-related proteins) complexes at the front of the web. Newly formed filaments are thereby attached to the sides of preexisting filaments. As these filaments elongate, they push the plasma membrane forward. The actin filament plus ends will become protected by capping proteins, preventing further assembly or disassembly from the old plus ends at the front of the array. After newly polymerized actin subunits hydrolyze their bound ATP in the filament lattice, the filaments become susceptible to depolymerization by cofilin. The spatial separation of assembly and disassmbly allows the network as a whole to move forward at a steady rate (From: Albert B, et al. Molecular biology of the cell (6th edition). 2014)

components to whole cell locomotion, depend on the interaction of actin and myosin. Among these, the best-studied contractile structures are muscles.

6.2.5 Muscle contraction

(1) Structure of muscle cell Vertebrates and many invertebrates have two major classes of muscle—skeletal muscle and smooth muscle—each serves different functions. A typical muscle cell, muscle fiber or a myofiber, is a cylindrical structure that is large (1~40 mm in length and 10~50 μm in width) and multinucleated (containing as many as 100 nuclei). A myofiber is filled with bundles of thinner, cylindrical filaments called myofibrils that extend the length of the cell. Each myofibril consists of a repeating linear array of sarcomeres, each about 2 μm long in resting muscle. Each sarcomere displays a characteristic dark and light banding pattern, which in turn gives the muscle cell a striated appearance. The dark band (A band) is bisected by a dark region (H zone) while the light band (I band) is bisected by a dark line (the Z disk or Z line). Each sarcomere is a segment from one Z disk to the next, consisting of two halves of an I band and an A band(Figure 6-19). Electron microscopy of stained muscle fibers reveals the banding pattern to be the result of the partial overlap of two distinct types of filaments: thin filaments composed of actin and thick filaments containing myosin Ⅱ.

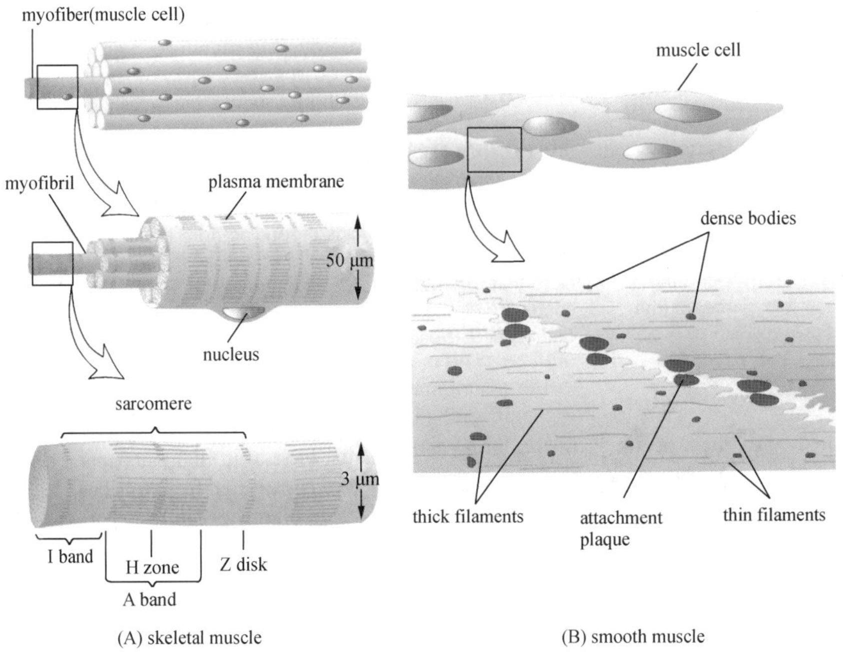

Figure 6-19 General structure of skeletal and smooth muscle. (A) Skeletal muscle tissue is composed of bundles of multinucleated muscle cell, or myofibers. Each muscle cell is packed with bundles of actin and myosin filaments, organized into myofibrils that extend the length of the cell. Packed end to end in a myofibril is a chain of sarcomeres, the functional units of contraction. The internal organization of the filaments gives skeletal muscle cells a striated appearance. (B) Smooth muscle is composed of loosely organized spindle-shaped cells that contain a single nucleus. Loose bundles of actin and myosin filaments pack the cytoplasm of smooth muscle cells. These bundles are connected to dense bodies in the cytosol and to the membrane at attachment plaques(From: Lodish H, et al. Molecular Cell Biology. 2000)

Myosin II is a dimer made of two identical myosin molecules held together by their tails. Each myosin II has two globular ATPase heads at one end and a single coiled-coil tail at the other. Myosin II molecules bind to one another through their tails and form a bipolar myosin filament in which the heads project from the sides(Figure 6-20).

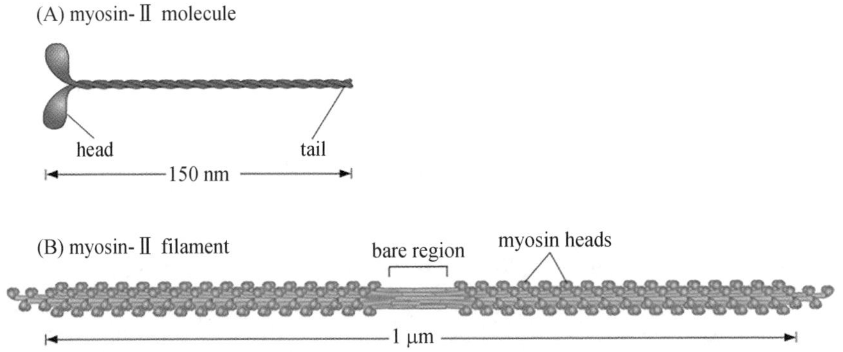

Figure 6-20 Myosin-II molecules can associate with one another to form myosin filaments. (A) A molecule of myosin-II has two globular heads and a coiled-coil tail. (B) The tails of myosin-II associate with one another to form a bipolar myosin filament in which the heads project outward from the middle in opposite direction. The bare region in the middle of the filament consists of tails only(From: Albert B. Essential Cell Biology (3rd edition). 2009)

(2) Actin filaments slide against myosin filament during muscle contraction A myosin filament has two sets of heads pointing in opposite directions from the middle. One set of heads binds to actin filaments in one direction and moves them one way and the other set of heads binds to actin filaments in the opposite direction and moves them accordingly (Figure 6-21). When actin and myosin are organized together in a bundle, the bundle can generate a contractile force, thus making sarcomere a contractile unit.

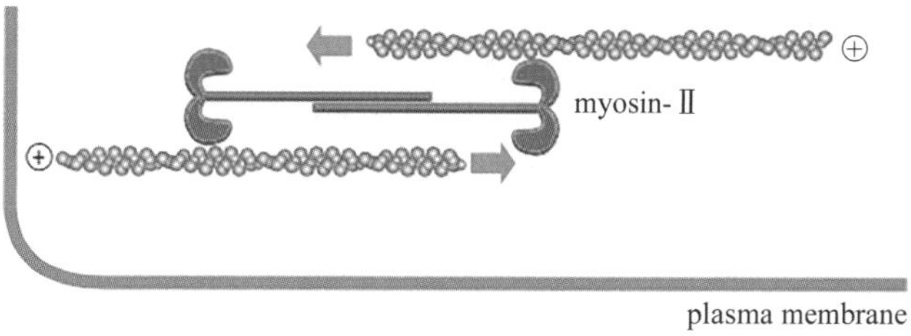

Figure 6-21 Even small bipolar filaments composed of myosin Ⅱ molecules can slide actin filaments over each other, thus mediating local *shortening* of an actin filament bundle. The head group of myosin Ⅱ walks toward the plus end of the actin filament it contacts (From: Alberts B, et al. Essential Cell Biology (3rd edition). 2009)

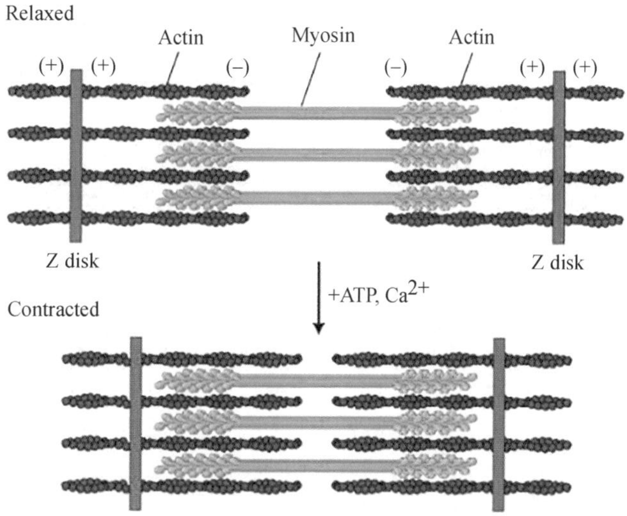

Figure 6-22 The sliding-filament model of contraction in striated muscle. The arrangement of thick myosin and thin actin filaments in the relaxed state is shown in the upper diagram. In the presence of ATP and Ca^{2+}, the myosin heads extending from the thick filaments walk toward the (+) ends of the thin filaments. Because the thin filaments are anchored at the Z disks (purple), movement of myosin pulls the actin filaments toward the center of the sarcomere, shortening its length in the contracted state, as shown in the lower diagram (From: Lodish. H, et al. Molecular Cell Biology (8rd edition). 2016)

The contraction of a muscle cell is caused by a simultaneous shortening of all the sarcomeres, which in turn is caused by the actin filaments sliding past the myosin filaments, with no change in the length of either type of filament (Figure 6-22). The sliding motion is generated by myosin heads that project from the sides of the myosin filament and interact with adjacent actin filaments. When a muscle is stimulated to contract, the myosin heads start to walk along the actin filament in repeated cycles of attachment and detachment. During each cycle, a myosin head binds and hydrolyzes one molecule of ATP. This causes a series of conformational changes in the myosin molecule that move the tip of the head by about 5 nm along the actin filament toward the plus end. This movement, repeated with each round of ATP hydrolysis, propels the myosin molecule unidirectionally along the actin filament (Figure 6-23). In so doing, the myosin heads pull against the actin filament, causing it to slide against the myosin filament. The concerted action of many myosin heads pulling the actin and myosin filaments past each other causes the sarcomere to contract. After a contraction is completed, the myosin heads lose contact with the actin filaments completely, and the muscle relaxes.

When a muscle receives signals from the nervous system, there is a sudden rise in cy-

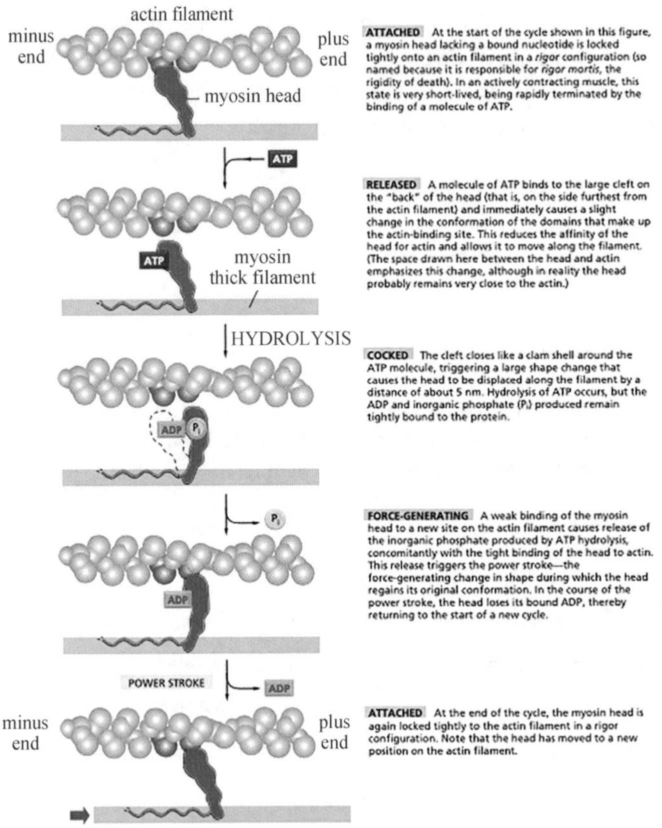

Figure 6-23 A myosin molecule walks along an actin filament through a cycle of structural changes (From: Albert B, et al. Essential cell biology (3rd edition). 2009)

tosolic Ca^{2+} in the muscle cell. The Ca^{2+} interacts with a molecular switch made of proteins such as tropomyosin and troponin that are specifically associated with actin filaments. Tropomyosin is a rigid, rod-shaped molecule that prevents the myosin heads from associating with the actin filament because it binds in the groove of the actin helix overlapping seven actin monomers. Troponin is a protein complex that includes a Ca^{2+}-sensitive protein troponin-C, which is associated with the end of a tropomyosin molecule. (Figure 6-24A) When the cytosolic Ca^{2+} rises, Ca^{2+} binds to troponin and induces a change. This in turn causes the tropomyosin molecules to shift their position slightly, allowing myosin heads to bind to the actin filament and initiating contraction (Figure 6-24B). As soon as the Ca^{2+} level returns to its resting level, troponin and tropomyosin molecules recover to their original positions where they block myosin binding and end contraction.

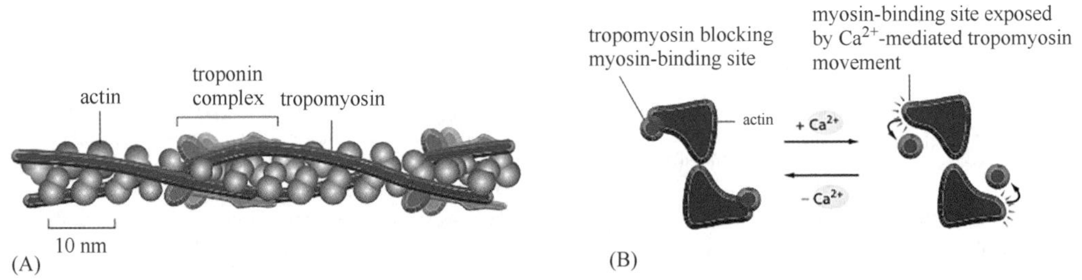

Figure 6-24 Skeletal muscle contraction is controlled by troponin. (A) A muscle thin filament showing the positions of tropomyosin and troponin along the actin filament. Every tropomyosin molecule has seven evenly spaced regions of homologous sequence, each of which is thought to bind to an actin subunit in the filament. (B) When Ca^{2+} binds to troponin, the troponin moves the tropomyosin that otherwise blocks the interaction of actin with the myosin heads (From: Alberts B, et al. Essential Cell Biology (3rd edition). 2009)

6.3 Intermediate filaments

The third sets of cytoskeletal fibers in eukaryotic cells are **intermediate filaments (IFs).** IFs typically form a network throughout the cytoplasm of most animal cells. They surround the nucleus and extend out to the cell periphery. They are often interconnected with other types of filaments by thin, wispy cross-bridges that are made of a huge, elongated accessory protein called plectin that has a binding site for an intermediate filament at one end and another binding site at the other end for another type of filament depending on the plectin isoform(Figure 6-25).

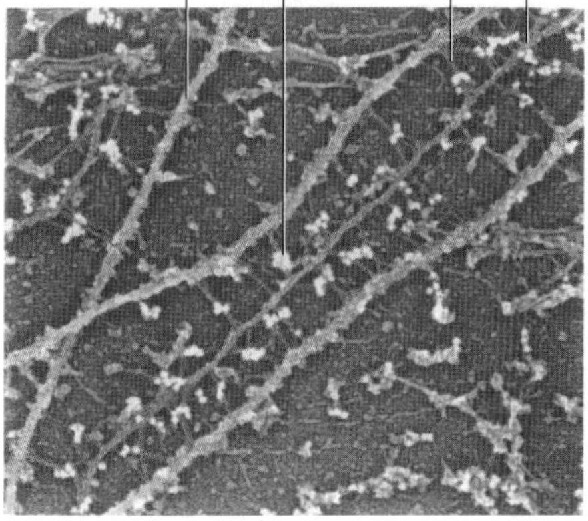

Figure 6-25 Connections between intermediate filaments and other components of the cytoskeleton. Intermediate filaments are linked to both microtubules and actin filaments by a protein called plectin. Plectin(red) links IFs(green) to MTs(orange) and to MFs (not shown). In this electron micrograph, the yellow dots are gold particles linked to antibodies that recognize plectin. Here, IFs server as strong but elastic connectors between the different cytoskeletal filaments (From: Becker WM. The World of the Cell. 2006)

6.3.1 Cytosolic intermediate filaments

IFs are often anchored to the plasma membrane at cell-cell junctions such as desmosomes where the external face of the membrane connects with that of another cell(Figure 6-26).

IFs are strong, solid, unbranched, ropelike fibers that provide strength and mechanically support the cells that are subject to physical stress such as neurons, muscle cells and the epithelial cells that line the body's cavities. IFs within the nucleus are called the nuclear lamina. They are a mesh of filaments that support the nuclear envelope in all eukaryotic cells. IFs were so named because in the smooth muscle cells where they were first discovered, they were found to have a diameter of about 8~12 nm, which is between that of the thin actin-containing microfilaments (about 7 nm) and thicker microtubule filaments (24 nm). Of the three types of cytoskeletal filaments, intermediate filaments are the strongest and the most stable. They are the only intact cytoskeleton filaments left when cells are treated with solutions containing non-ionic detergents and a high concentration of salts.

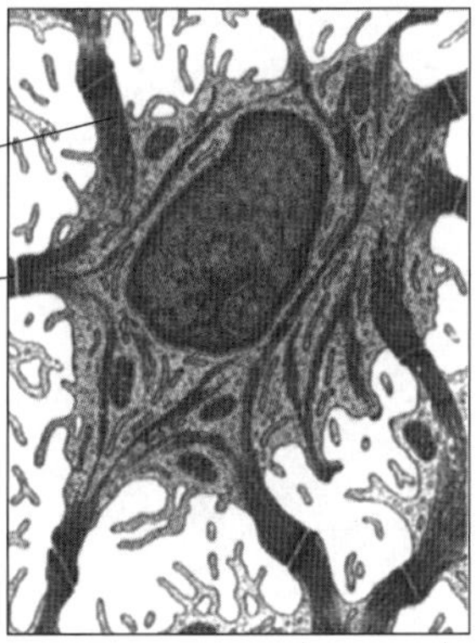

Figure 6-26 Intermediate filaments form a strong, durable network in the cytoplasm of the cell. Drawing from an electron micrograph of a section of epidermis showing the bundles of intermediate filaments that traverse the cytoplasm and that are inserted at desmosomes (From: Alberts B, et al. Essential Cell Biology (3rd edition). 2009)

Each subunit of intermediate filaments is an elongated fibrous polypeptide with an N-terminal globular head, a C-terminal globular tail and a center elongated rod domain. The rod domain has an extended α-helical region that enables pairs of intermediate filament proteins to form stable dimers by wrapping around each other in a coiled-coil configuration. Because the two polypeptides are aligned parallel to each other in the same direction, the dimer has polarity with one end defined by N-termini of the polypeptides and the other by C-termini. Two such dimmers bind together side-by-side in a staggered fashion with their N-and C-termini pointing in opposite (antiparallel) direction to form a tetramer via non-covalent bonding. Then tetramers bind to each other end-to-end and side-by-side also via non-cova-

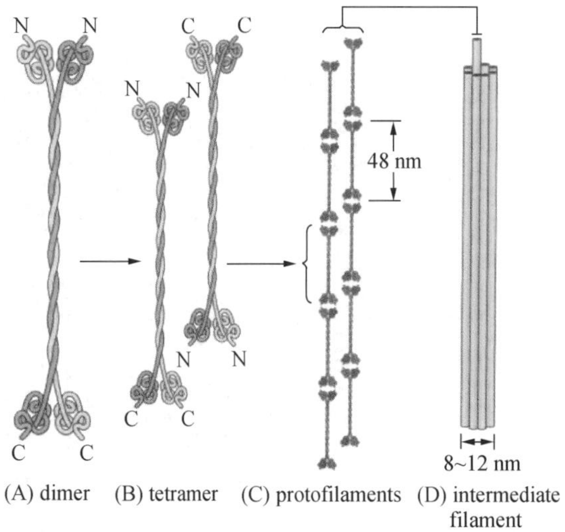

Figure 6-27 A model for intermediate filaments assembly in vitro. (A) The starting point for assembly is a pair to IFs polypeptides. The two polypeptides are identical for all IFs except kertatin filaments, which are obligate heterodimers with one each to the type I and type II polypeptides. The two polypeptides twist around each other to form a two-chain coiled coil, with their conserved center domain aligned in parallel. (B) Two dimmers align laterally to form a tetrame. (C) Tetramers aggtegate end-to-end into a protofilament. (D) The fully assembled intermediate filament is thought to be eight protofilaments thick at any point (From: Becker WM. The World to the Cell. 2006)

lent bonding to form the final ropelike intermediate filament. Because the tetremeric building blocks lack polarity, so do the intermediate filaments, which is another feature that distinguishes IFs from other cytoskeletal filaments (Figure 6-27).

The central rod domains of different IFs are all similar in size and amino acid sequence, therefore, different IFs are similar in diameter and internal structure. In contrast, the globular head and tail regions of different IFs vary greatly both in size and in amino acid sequence. The head and tail regions are exposed on the surface of the filament, allowing the filament to interact with other components of the cytoplasm. Unlike microtubules and microfilaments, IFs are a heterogeneous group of structures. The polypeptides subunits of IFs can be classified into six categories based on their tissue distribution, biochemical, genetic, and immunologic criteria (Table 6-2).

Table 6-2 Properties and distribution of the major mammalian intermediate filament proteins

IF protein	sequence type	average molecular mass ($\times 10^{-3}$)	estimated number of polypeptides	primary tissue distribution
keratin (acidic)	Ⅰ	40~64	>28	epithelia
keratin (basic)	Ⅱ	53~67	>28	epithelia
vimentin	Ⅲ	54	1	mesenchymal cells
desmin	Ⅲ	53	1	muscle
glial fibrillary acidic protein (GFAP)	Ⅲ	50	1	glial cells, astrocytes
peripherin	Ⅲ	57	1	peripheral neurons
neurofilament proteins				neurons of central and peripheral nerves
NF-L	Ⅳ	62	1	
NF-M	Ⅳ	102	1	
NF-H	Ⅳ	110	1	
Lamin proteins				all cell types (nuclear envelopes)
Lamin A	Ⅴ	70	1	
Lamin B	Ⅴ	67	1	
Lamin C	Ⅴ	60	1	
nestin	Ⅵ	240	1	heterogeneous

(From: Karp G, et al. Cell and Molecular Biology — concepts and experiments. 2005)

Because of the tissue specificity of intermediate filaments, animal cells from different tissues can be distinguished on the basis of the IF protein present, as determined by immunofluorescence microscopy. This intermediate filament typing serves as a diagnostic tool in medicine. IF typing is especially useful in the diagnosis of cancer, because tumor cells are known to retain the IF proteins characteristic of the tissue of origin, regardless of where the tumor occurs in the body.

6.3.2 Nuclear intermediate filaments

The nuclear envelope is supported by a meshwork of intermediate filaments. The inner nuclear membrane are organized as a two dimensional mesh called nuclear lamina which are composed of a class of intermediate filaments proteins called lamins (Figure 6-28).

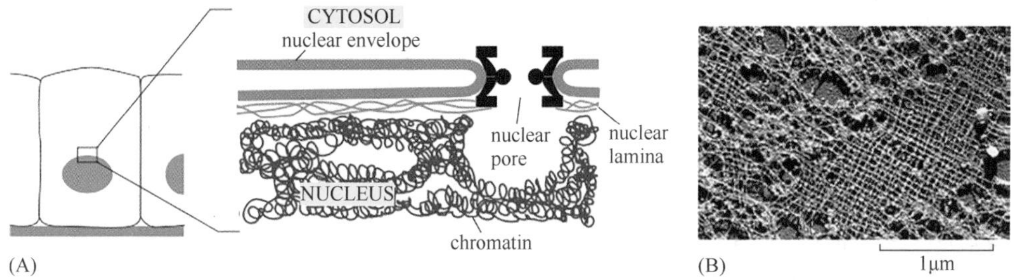

Figure 6-28 Intermediate filaments support and strengthen the nuclear envelope. (A) Schematic cross section through the nuclear envelope. The intermediate filaments of the nuclear lamina line the inner face of the nuclear envelope and are thought to provide attachment sites for the DNA-containing chromatin. (B) Electron micrograph of a portion of the nuclear lamina from a frog egg, or oocyte. The lamina is formed from a square lattice of intermediate filaments composed of lamins (Nuclear laminae from other cell types are not always as regularly organized as the one shown here) (From: Alberts B. et al. Essential Cell Biology (3rd edition). 2009)

Lamins are not as stable as cytoplasmic IFs. They disassemble and reassemble at each cell division when the nuclear envelope breaks down during mitosis and then reforms in each daughter cell. Disassembly and reassembly of the nuclear lamins are controlled by phosphorylation and dephosphorylation of the lamins by protein kinases. Phosphorylation of lamins causes conformational change in lamins that weakens the binding between IF tetramers and subsequently the filaments fall apart. On the other hand, dephosphorylation of lamins at the end of mitosis causes the lamins to reassemble (Figure 6-29).

Defects in a particular nuclear lamin are associated with certain types of progeria — rare disorders that cause affected individuals to appear to age prematurely. Children with progeria have wrinkled skin, lose their teeth and hair, and often develop severe cardiovascular disease by the time they reach their teens.

Summary

The cytoskeleton is an extensive system of fibers that are responsible for cell shape and cell motility. The cytoskeleton contains three principal types of filaments: microtubules, actin filaments and intermediate filaments. All three types of filaments undergo constant remodeling through the assembly and disassembly of their subunits. While tubulin and actin have been strongly conserved in eukaryotic evolution, the family of intermediate filaments is very diverse. There are a variety of tissue specific forms of intermediate filaments, including keratin, desmin and neurofilaments. In most animal cells, microtubules are nucleated at MTOC, whereas most actin filaments are nucleated near the plasma membrane. Through ATP hydrolysis, motor proteins convert chemical energy stored in ATP into movement along microtubules or actin filaments. They mediate the sliding of filaments relative to one

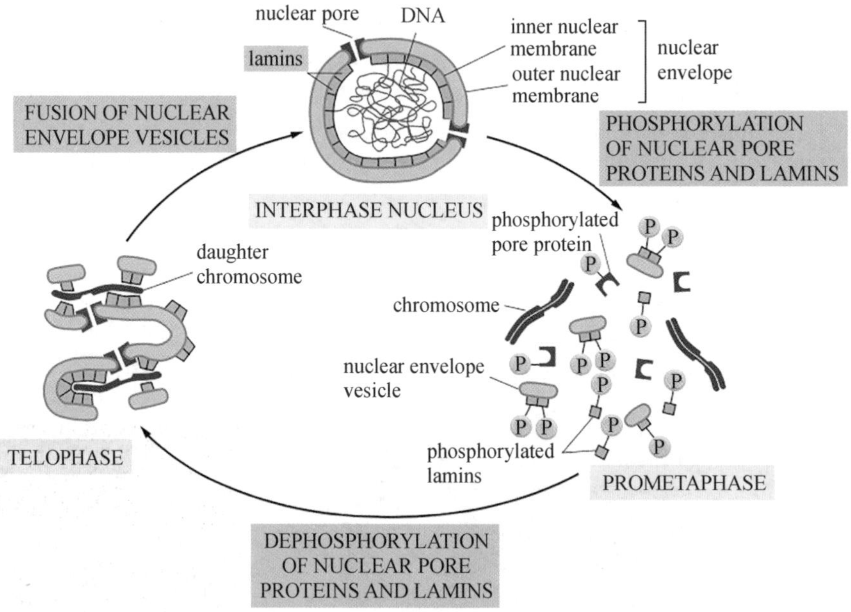

Figure 6-29 The nuclear envelope breaks down and reforms during mitosis. The phosphorylation of the lamins at prophase helps trigger the disassembly of the nuclear lamina, which in turn causes the nuclear envelope to break up into vesicles. Dephosphorylation of the lamins at telophase helps reverse the process (From: Alberts B, et al. Essential Cell Biology (3rd edition). 2009)

another and thus help transporting membrane-enclosed organelles along filament tracks. The motor proteins that move on actin filaments are members of the myosin superfamily. During muscle contraction, actin filaments slide against myosin filaments. The motor proteins that move on microtubules are members of either the kinesin family or the dynein family.

Questions

1. Define these key terms: actin filament, myosin, cell cortex, centriole, dynein, kinesin, intermediate filament, nuclear lamina, sarcomere, tubulin.
2. Why do eukaryotic cells, especially animal cells have such large and complex cytoskeletons?
3. There are no known motor proteins that move on intermediate filaments, why do you think that is the case?
4. Which of the following changes takes place when a skeletal muscle contracts:
   A. Z discs move farther apart
   B. Actin filaments contract
   C. Myosin filament contract
   D. Sarcomeres become shorter
5. Describe the functions of microtubule.
6. Describe the structure of the sarcomere of a skeletal muscle myofibril and the changes that occur during muscle contraction.
7. Why we say intermediate filament typing is a useful diagnostic tool?
8. Compare the structure of a fully assembled microtubule, actin filament, and intermediate filament.

（沈大棱，孙　媛修订）

Chapter 7

Cell communication and signaling

To carry out one or more specific biological functions, cells need to communicate and work with each other to coordinate their activities. **Cell communication** is also called **cell signaling**, which is one of the most basic characteristics of the cell. Cell communication (signaling) affects every aspect of cell structure and function. An understanding of how cells communicate with each other can tie together a variety of seemingly independent cellular processes.

7.1 Signaling components

Communication between cells is mediated mainly by extracellular signal molecules. The **signaling molecules** from physical and chemical changes of the environment, or produced by a **signaling cell** can be detected by a **target cell**, which in turn responds specifically to the signaling molecule.

Single and cells in multi-cellular organisms use hundreds of signaling molecules such as cytokines, growth factors, hormones, fatty acid derivatives, neurotransmitter, and even dissolved gases. The target cells responds by means of a **receptor**, which usually embedded in the plasma membrane of the target cell and specifically binds the signal molecule and then initiates a response through one or more intracellular signaling pathways mediated by a series of signaling proteins.

7.1.1 Cell communication styles

There are huge varieties of signaling molecules and receptors used in cell communication, however, relatively few basic styles of cell communication are existed. **Contact-dependent signaling** and non-contact dependent signaling are two main styles.

Contact-dependent signaling is the most intimate and short-range cell-cell communication. This happens between the cells that make direct contact through signaling molecules anchored in the plasma membrane of the signaling cell and bind to a receptor molecule embedded in the plasma membrane of the target cell. Lateral inhibition mediated by Notch and Delta during nerve cell production in Drosophila is a good example of contact-dependent communication (Figure 7-1A). The nervous system originates in the embryo from a sheet of epithelial cells. Isolated cells in this sheet begin to specialize as neurons, while their neighbors remain non-neuronal and maintain the epithelial structure of the sheet. The signals that control this process are transmitted via direct cell-cell contacts: each future neuron delivers an inhibitory signal molecule Delta and binds to North receptor proteins on the neighboring cells. The same mechanism, mediated by essentially the same molecules, controls the detailed pattern of differentiated cell types in various other tissues, in both vertebrates and invertebrates.

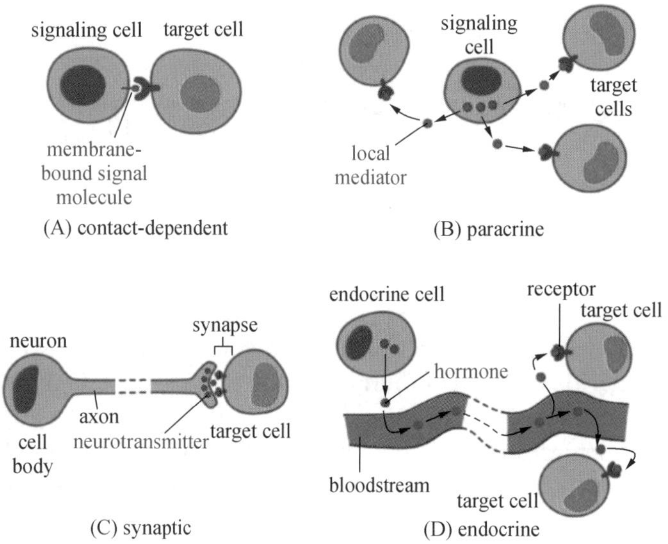

Figure 7-1 Four forms of intercellular signaling (From: Alberts B, et al. Molecular Biology of The Cell. 2008)

Non-contact mediated signaling is also referred as **chemical signaling**. Signaling cells secrete signal molecules into the extracellular fluid, and it is the most common style of communication. Depending on the distance over which the signal acts, chemical signaling can be classified into three sub-types.

(1) Paracrine signaling (Figure 7-1B) The signaling molecules stay in the neighborhood of the signaling cells and only affect nearby target cells. But cells may also produce signals that they themselves respond to, which is **autocrine signaling**. This is especially common in tumor cells in which over-produced growth factors stimulate over-growth of tumor cells themselves and adjacent non-tumor cells.

(2) Synaptic signaling It requires a special structure called the synapse between the cell originating and the cell receiving the signal to achieve rapid communication. Synaptic signaling only occurs between cells with the synapse; for example between a neuron and the muscle that is controlled by neural activity. For instance, neuronal signaling involves **neurotransmitter** release from nerve cells, then diffuse across the narrow (<100 nm) gap between the axon-terminal membrane and membrane of the target cell in less than 1 millisecond (Figure 7-1C).

(3) Endocrine signaling (Figure 7-1D) **Hormones** are used as signaling molecules and are secreted into the bloodstream (in an animal) or the sap (in a plant) from endocrine cells. This is the most "public" style of communication in which signals are broadcasted throughout the body, reaching distant target cells. For example, the endocrine gland of the pancreas secrets the hormone insulin into the bloodstream and regulates glucose uptake in cells all over the body.

In contrast with other modes of cell signaling, **gap junctions** generally allow communication to pass in both directions symmetrically. Gap junctions are channels that allow the diffusion of ions and small molecules (up to 1,000 Da) between closely apposed cells, and their typical effect is to homogenize conditions in the communicating cells, including the propagation of electrical signals, metabolic cooperation etc. Thus, gap junctions allow neighboring cells to share signaling information.

7.1.2 Types of receptors

Detection of signals arriving from outside of the cell is usually fulfilled by the presence of specific receptors. In most cases, the receptors are transmembrane proteins on the target cell surface. When these proteins bind an extracellular signal molecule (a ligand), they become activated and generate various intracellular signals that alter the behavior of the cell.

The number of different types of receptors is even greater than the number of signaling molecules because in many cases, there are multiple receptors for a single signaling molecule. Most cell-surface receptors belong to three large families: **ion-channel-coupled receptors**, **G-protein-coupled receptors**, or **enzyme-coupled receptors** (Figure 7-2).

(1) Ion-channel-coupled receptors Ion-channel-coupled receptors (also known as transmitter-gated ion channels) (Figure 7-2A) are linked to ion channels, and the conductance of the channels is modulated by the binding of signal molecules (agonists or antagonists). Ion-channel-coupled receptors can be found in the nervous system and other electrically excitable cells such as muscle, neuroendocrine cells etc., and often involved in the detection of neurotransmitter molecules. They function in the simplest and most direct way compared with how other cell-surface receptor types act. In most cases, **neurons** pass signals to other cells at **synapses** where a presynaptic neuron releases a chemical signal in the form of neurotransmitters. Neurotransmitters bind to ion-channel-coupled receptors on the postsynaptic target cells. These receptors alter their conformation to transiently open or close the efflux and influx of specific ions such as Na^{2+}, K^{+}, Ca^{2+}, or $Cl^{-}$and to allow them going across the cell membrane. The ion-flows are driven and directed by their **electrochemical gradient** (i. e. much higher concentration of ions on one side of the membrane than on the other side). Ions rushing into or out of the cell create a temporary change in the membrane electric potential due to the positive or negative nature of the ions. This conversion of chemical signals to electric signals happens within a milli-second or so. The sudden change in membrane potential can trigger or suppress a nerve impulse depending on what ion channels were involved. It may also affect the activity of other membrane protein such as voltage-gated ion channels, which in turn can trigger a nerve impulse, or stop an impulse from happening. Furthermore, this process can also change the activity of cytoplasmic proteins. For example, the opening of Ca^{2+} channels increases the intracellular concentration of Ca^{2+}, which in turn profoundly alter the activities of many proteins because Ca^{2+} is an intracellular messenger and has many substrates.

(2) G-protein-coupled receptors The majority of cell-surface receptors are G-protein-coupled receptors (GPCRs) (Figure 7-2B), with hundreds of members identified in different organisms ranging from yeast to flowering plants and mammals. GPCRs respond to a huge variety of extracellular signaling molecules including hormones, neurotransmitters, opium derivatives, chemoattractants, odorants and tastants, as well as photons. These signaling molecules (**ligands**) can be proteins, small peptides, or derivatives of amino acids or fatty acids; and for each of them there is a different receptor or set of receptors. As the name indicated, all GPCRs interact with G proteins. G proteins are a group of proteins that were discovered and characterized in the early 1970s and so named because they all bind guanine nucleotides, either GDP or GTP. All GPCRs have a similar structure: each is made of single polypeptide chain that passes through plasma membrane seven times. That is why GPCRs are also called seven-transmembrane (7TM) receptors. A **trimeric GTP-binding protein** (**G protein**) mediates the interaction between the activated receptor and the target protein.

(3) enzyme-coupled receptors Like G-protein-coupled receptors, enzyme-coupled

receptors(Figure 7-2C) are also transmembrane proteins that have a ligand-binding domain on the outer surface of the plasma membrane and a cytosolic domain. Unlike G-protein-coupled receptors, enzyme-coupled receptors usually have only one transmembrane segment, trespassing plasma membrane by a single α helix. Unlike G-protein-coupled receptors, the cytoplasmic domain of enzyme-coupled receptors acts as an enzyme or complex with another protein that acts as an enzyme.

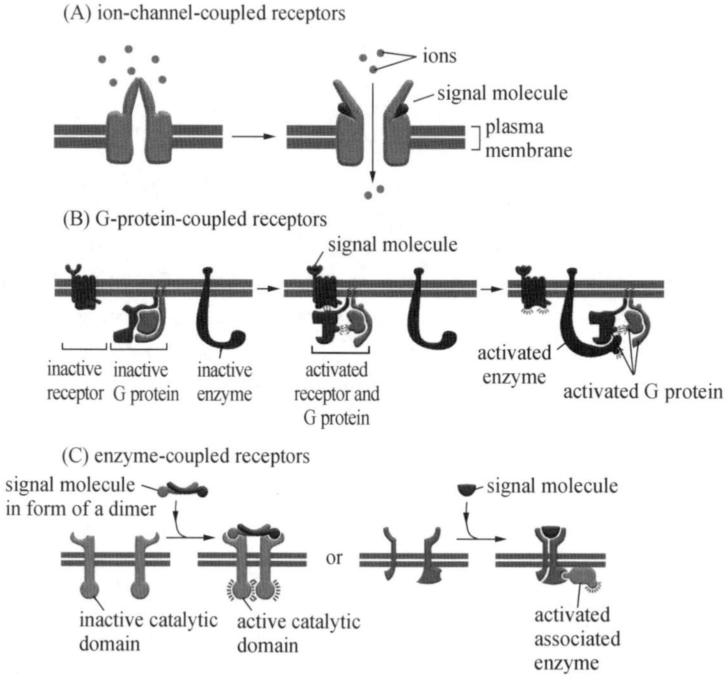

Figure 7-2 Cell-surface receptors fall into three basic classes. (A) An ion-channel-coupled receptor opens (or close, not shown) in response to binding of its signal molecule. (B) When a G-protein-coupled receptor binds its extracellular signal molecule, the signal is passed first to a GTP-binding protein (a G protein) that is associated with the receptor. The activated G protein then leaves the receptor and turns on a target enzyme (or ion channel, not shown) in the plasma membrane. (C) An enzyme-coupled receptor binds its extracellular signal molecule, switching on an enzyme activity at the other end of the receptor. Although many enzyme-coupled receptors have their own enzyme activity (*left*), others rely on associated enzymes (*right*) (From: Alberts B, et al., Molecular Biology of The Cell. 2008)

However, not all extracellular signaling molecules are detected on the surface of the cell by plasma membrane receptors. Many very important signals are released by cells but the receptors for their perception are inside the target cells, either in the cytoplasm or in the nucleus. The signaling molecule has to enter the cell to bind to them. Intracellular ligand binding for these extracellular signaling molecules means that they have to move through the plasma membrane, and commonly these signals are small and hydrop-hobic. Various small hydrophobic signal molecules diffuse directly across the plasma membrane of the target cells and bind to intracellular receptors that are gene regulatory proteins.

The signal molecules such as steroid hormones, thyroid hormones, retinoids, and vitamin D, they all act by a similar mechanism. When these signal molecules bind to their receptor proteins, they activate the receptors, which bind to DNA to regulate the transcription of specific genes. The receptors are **nuclear receptors**, that is, the ligand-activated gene regulatory proteins. Some nuclear receptor proteins may be activated by intracellular metabolites rather than by secreted signal molecules. Some receptor proteins are located in the cytosol and enter the nucleus only after ligand binding; others are bound to DNA in the nucleus even in the absence of the ligand.

7.1.3 The basic cell signaling process

An overview of the basic cell signaling process is described in figure 7-3. Cell signaling usually starts with extracellular messenger molecules (**first messenger**) produced by signaling cells. The first messenger reaches the target cells and binds to its receptor of the target cells. This binding causes a signal to be relayed across the membrane to the cytoplasmic domain of the receptor. Subsequently, most signals are transmitted to the cell interior using one of the two pathways. One way is through a **second messenger** (a. k. a. effector), which is a small substance that typically activates or inactivates specific proteins. The other way is through making the cytoplasmic domain a recruiting site for docking the intracellular signaling proteins. Either way, a protein positioned at the upstream of the pathway is activated, which in turn activates a series of distinct downstream proteins one by one, relaying the signal downstream.

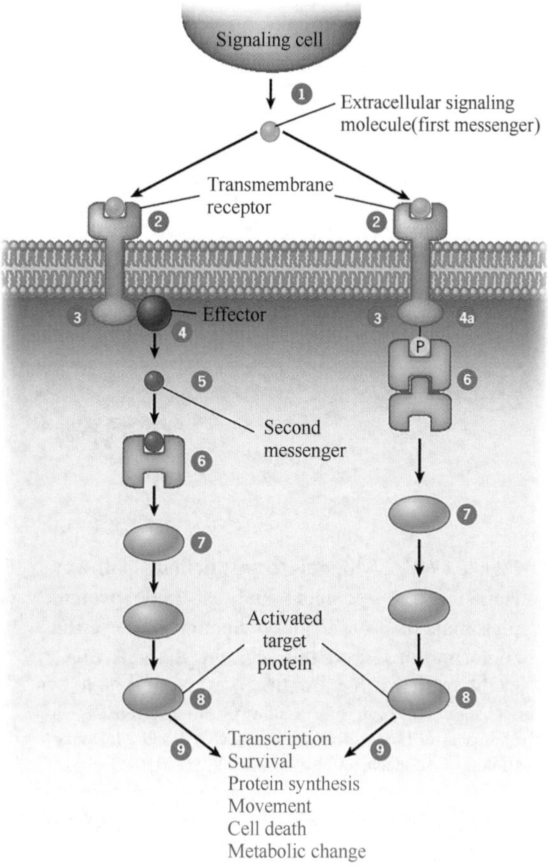

Figure 7-3 An overview of the signaling pathways by which extracellular messenger molecules can elicit intracellular responses. One signaling pathway is activated by a diffusible second messenger, and another pathway is activated by recruitment of proteins into the plasma membrane. Most signal transduction pathways refer to a combination of these two mechanisms. It should also be noted that signaling pathways are not typically linear tracks as depicted here, but are branched and interconnected to form a complex web (From: Karp G, et al. Cell and Molecular Biology — concepts and experiments. 2010)

The activation of proteins in the pathways usually involves alterations in signaling protein conformation (Figure7-4). This is usually done by protein kinases and protein phosphatases that add (**phosphorylate**) or remove (**dephosphorylate**) phosphate groups from target proteins respectively. Approximately one third of all proteins of a cell are thought to be subjected to phosphorylation. Phosphorylation can change protein activities in many ways. It can activate or inactivate an enzyme, increase or decrease protein-protein interaction, induce a protein to move from one subcellular compartment to another, or initiate protein degradation. Many signaling proteins controlled by phosphorylation are often organized into phosphorylation cascade, which is a sequence of events where one enzyme phosphorylates another, causing a chain reaction leading to the phosphorylation of different proteins.

The other class of proteins that function by gaining and losing phosphate groups consists of GTP-binding proteins. The protein can be further classified to G proteins and **monomeric GTPases** (also called **monomeric G proteins**). Both proteins are activated by GTP-binding and inactivated by GDP-binding. The figure 7-5 illustrates the regulation of monomeric GTPases.

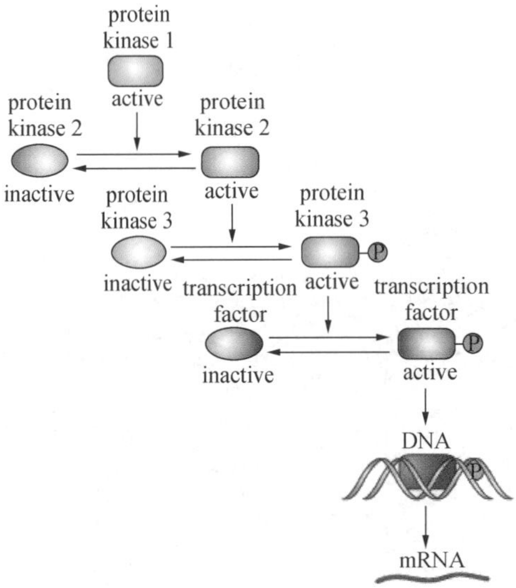

Figure 7-4 Signal transduction pathway consisting of protein kinases and protein phosphatases whose catalytic actions change the conformations, and thus change the activities, of the proteins they modify. Note that each of these activation steps in the pathway is reversed by a phosphatase(From: Karp G, et al. Cell and Molecular Biology — concepts and experiments. 2010)

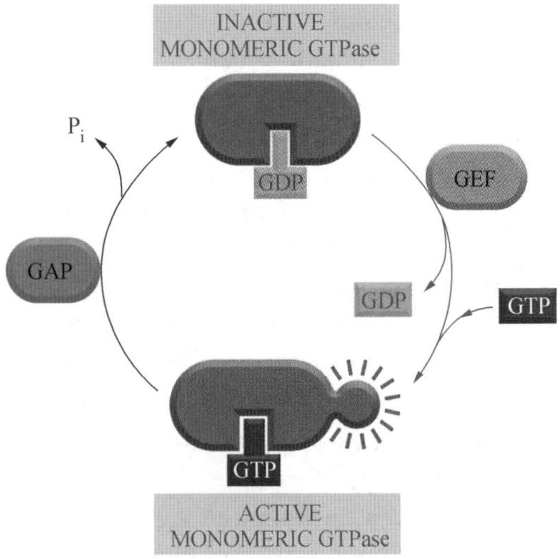

Figure 7-5 The regulation of a monomeric GTPases. GTPase-activating proteins (GAPs) inactivate the protein by stimulating it to hydrolyze its bound GTP to GDP. Guanine nucleotide exchange factors (GEFs) activate the inactive protein by stimulating it to release its GDP (From: Alberts B, et al. Molecular Biology of The Cell. 2008)

Ultimately, signals reach target proteins. Depending on the type of cells and messages relayed, these target proteins may regulate gene expression, alter the activity of metabolic enzymes, reconfigure the cytoskeleton, change ion permeability, activate DNA synthesis, or even cause cell death. Virtually all cell activity is regulated by signals originated from the cell surface. The whole process, where a signal carried by the extracellular messenger molecules is relayed into changes that occur inside a cell, is called **signal transduction.**

Finally, all signaling process must be turned off in order for cells to be responsive to other messages or next wave of the same message that they may receive. This can be accomplished in many ways. Some cells produce extracellular enzymes that destroy specific extracellular messenger/signaling molecules. Some cells internalize activated receptors along with ligands(signaling molecules). Once inside the cell, the receptors and the ligands are either both degraded, or the ligands are degraded and the receptors are returned to the cell surface. Both processes decrease cells' ability to respond to subsequent stimuli. In addition, cells have enzymes in the cytosol that de-activate signaling proteins in the pathway. Some cytosolic signaling proteins are degraded and new signaling proteins are synthesized to maintain signaling capabilities.

7.2 The role of intracellular receptor: signaling of nitric oxide

Nitric Oxide (NO), an inorganic gas, was found to be an intracellular messenger in 1986. How it works is best illustrated by dissecting the mechanism of acetylcholine-induced smooth muscle cell relaxation (Figure7-6). Endothelial cells are the flattened cells that line every blood vessel. Binding of acetylcholine to the endothelial cell membrane gives a rise in cytosolic Ca^{2+} concentration which in turn activates nitric oxide synthase (NOS), an enzyme that catalyzes the amino acid L-arginine into nitric oxide (NO). NO formed in the endothelial cell diffuses across the plasma membrane into the adjacent smooth muscle cells where it binds and activates **guanylyl cyclase**, which in turn synthesizes **cyclic GMP** (**cGMP**).

One of the main cellular receptors of NO is the enzyme guanylyl cyclase, the enzyme responsible for the production of cGMP. It is usually the soluble form of the cyclase that is the receptor of NO, but some data have been reported that suggest the membrane form may also be NO regulated to some extent. cGMP is an important intracellular second messenger, which resembles **cyclic AMP** (**cAMP**), a much more commonly used intracellular signaling molecule. cGMP leads to a decrease in cytosolic Ca^{2+} concentration through activation of cGMP-dependent kinase (PKG), causing the muscle cells to relax and blood vessels to dilate (Figure7-6).

The discovery of NO acting as an activator of guanylyl cyclase sheds light on the therapeutic effect of nitroglycerine, a commonly used drug in treatment of the pain of angina caused by an inadequate flow of blood to the heart muscle. In the body, nitroglycerine is metabolized into NO, which rapidly relaxes the smooth muscles lining the blood vessels of the heart, increasing blood flow into the heart. Many nerve cells also use NO to signal neighboring cells. For instance, during sexual arousal, nerve terminals in the penis release NO, which triggers the local blood-vessel dilation and penis erection.

NO acts only locally because NO is converted to nitrates and nitrites within seconds by reaction with oxygen and water outside of the cells, thus terminating the NO signaling. NO signaling is a great example of how an extracellular signal can directly activate an enzyme and alter a cell within a few seconds or minutes.

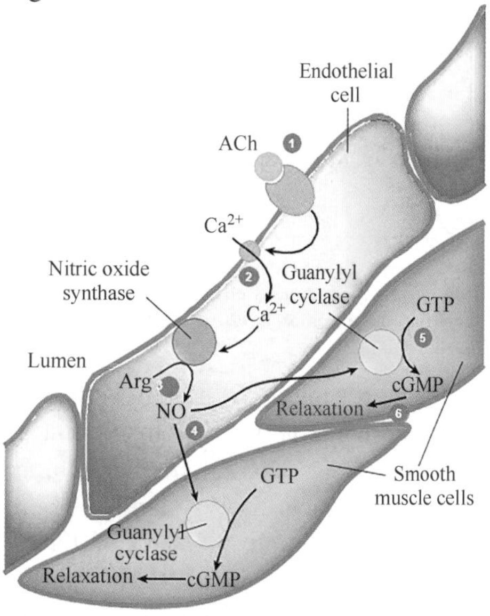

Figure 7-6 A signal transduction pathway that operates by means of NO and cyclic GMP that leads to the dilation of blood vessels (From: Karp G, et al. Cell and Molecular Biology — concepts and experiments. 2010)

Recent research shows that NO can also function without involving production of cGMP. For example, in a post-translational modification process called S-nitrosulation, NO is added to the SH group of certain cysteine residues in a number of proteins causing the alteration of protein properties and activity. Hemoglobin, ras, ryanodine channels and caspases are some good examples of these proteins.

7.3 Signaling through G-protein-coupled cell-surface receptors

The typical topology of GPCRs includes three loops of amino-terminus presented outside of the cell and forming the ligand-binding site; the seven α-helices that traverse the membrane are connected by loops of varying length; another three loops of carboxyl-terminus on the cytoplasmic side of the cell provide the binding sites for intracellular signaling molecules. G proteins bind to the third intracellular loop (Figure 7-7). Arrestins, an inhibitor of G protein actions, compete the same binding site of GPCR with G proteins. More and more proteins acting as molecular scaffolds that link receptors to various intracellular signaling proteins and effectors are found to bind to the carboxyl-termini of GPCRs.

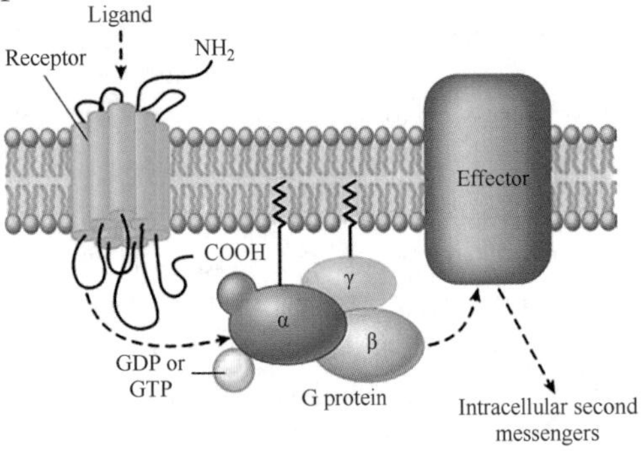

Figure 7-7 The membrane-bound machinery for transducing signals by means of a seven transmembrane receptor and a heterotrimeric G protein. When bound to their ligand, GPCRs interact with a trimeric G protein, which activates an effector, such as adenylyl cyclase. Note that the α and γ subunits of the G protein are linked to the membrane by lipid groups that are embedded in the lipid bilayer. (From: Karp G, et al. Cell and Molecular Biology — concepts and experiments. 2010)

Binding of a ligand to the extracellular ligand-binding site of a GPCR causes a conformational change in the extracellular ligand-binding site. This conformational change is transferred across the plasma membrane and changes the conformation of cytoplasmic loops of GPCR, which in turn increase the affinity of GPCR to a cytoplasmic G protein. Consequently, the G protein binds to the activated receptor forming a receptor-G protein complex. The G protein thus becomes active.

There are different kinds of G proteins; each is specific for a particular set of receptors and a particular set of downstream target proteins. Most of G proteins are called **heterotrimeric G proteins** to distinguish them from small, **monomeric G proteins**, such as Ras, which are discussed later.

Heterotrimeric G proteins share similar structure. The physiologic processes mediated by heterotrimeric G proteins include glycogen breakdown, behavioral sensitization and learning, visual excitation, chemotaxis, etc. They all have three protein subunits — α, β, γ, two of which are covalently tethered to the plasma membrane short lipid tails. In idle state, the α subunit of G protein has GDP bound to it. The interaction between a G protein with its receptor activates G protein by inducing a conformation change in the α subunit of the G protein, causing it to lose some of its affinity for GDP, which releases GDP in exchange for the binding of a molecule of GTP. The activated G protein breaks into two free activated subunits: the GTP-bound α subunit and the activated βγ complex unit. Both of these two subunits can bind and interact directly with effector proteins located in the plasma membrane, which in turn may relay signals to other destinations. As long as these effector proteins have a α or a βγ subunit bound to them, they can amplify the relayed signals.

To regain sensitivity to future stimuli, the receptor, the G protein, and the effector protein must all return to their inactive state. This is a two-step process called desensitization. The first step is the phosphorylation of the cytoplasmic domain of the activated GPCR by a **G protein-coupled receptor kinase** (**GRK**). GRKs are a small family of serine-threonine protein kinases. Most of them are located next to the cytoplasmic surface of the plasma membrane. Some of them that located in the cytoplasm are recruited to the membrane upon activation of certain G proteins. The conformation changes that occur for GPCRs to activate G proteins also make G proteins good substrates for GRKs. Therefore, following ligand binding, GPCRs become sensitive to phosphorylation by GRKs. The second step involves protein arrestins that have the same binding sites on GPCRs with G proteins. As a result, **arrestins** binding to the phosphorylated GPCRs prevent the further activation of G proteins. When desensitized, the cell stops responding to the stimulus even though the stimulus is still acting on the outer surface of the cell. Desensitization is an important mechanism that allows a cell to respond to changes in its environment rather than continue to be in a state of activation endlessly in the presence of an unchanging environment. The importance of desensitization can be illustrated in a disease called retinitis pigmentosa. People who suffer from this disease have mutations that interfere with phosphorylation of rhodopsin by a GRK, leading to the death of the **photoreceptor** cells in the retina causing retinal death and subsequent blindness.

Arrestins can also bind to **clathrin** molecules that are located in **clathrin-coated pits**. The bound arrestins and clathrin promotes a process called receptor internalization where phosphorylated GPCRs are taken into the cell by endocytosis. Under certain circumstances, endocytosed receptors are dephosphorylated and returned to the plasma membrane, while other times receptors are degraded in the lysosomes. When the receptors are returned, the cells remain sensitive to ligands; when receptors are degraded, the cells lose, at least temporarily, sensitivity to the ligands. Interestingly, both desensitization and receptor internalization appears to be initiated by a large variety of extracellular signals activating GRKs.

Termination of signaling via activated α subunit of the G protein goes through another type of process, which depends on the behavior of the α subunit. Following its interaction with an effector, the α subunit goes through an intrinsic GTP-hydrolyzing (GTPase) process, which hydrolyzes its bound GTP back to GDP and Pi. GTP hydrolysis is accelerated by regulators of G protein signaling (RGSs). GTP hydrolysis induces a conformational change that causes α subunit to dissociate from the effector protein and reunite with the βγ complex unit, which thus inactivates G protein. The reconstituted G protein is now ready to be reactivated by another activated receptor, thus the system is returned to a resting state.

The target proteins for activated G protein subunits are either ion channels or membrane-bound enzymes. The most popular G-protein-coupled receptor-initiated enzyme-mediated signaling pathways are **the cyclic AMP pathway** and **the inositol phospholipid pathway.**

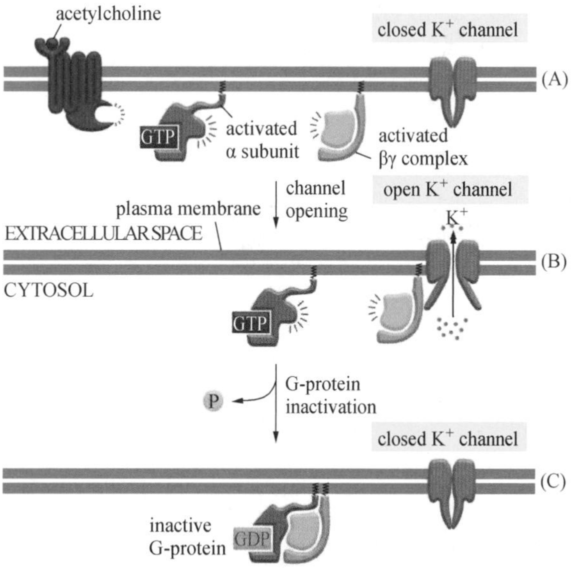

Figure 7-8 G-proteins-couple receptor activation to the opening of K^+ channels in the plasma membrane of heart muslce cells. (A) Binding the neurotransmitter acetylcholine to its G-protein-coupled receptor results in the dissociation of the G protein into an activated βγ complex and an activated α subunit. (B) The activated βγ complex binds to and opens a K^+ channel. (C) Inactivation of the α subunit by hydrolysis of bound GTP causes it to reassociate with the βγ complex to form an inactive G protein, allowing the K^+ channel to close(From: Alberts B, et al. Essential Cell Biology. 2004)

7.3.1 G-protein regulation of ion channels

G protein-gated ion channels are associated with a specific type of G protein-coupled receptor.

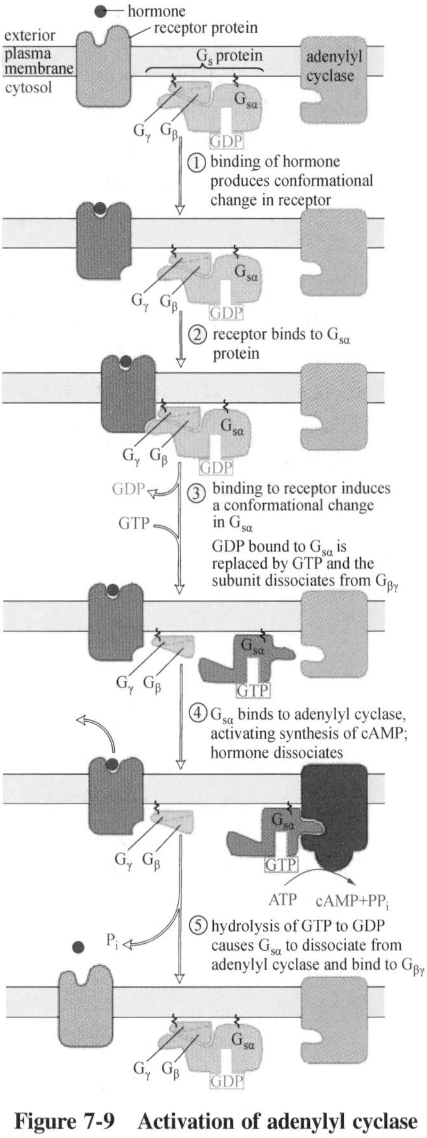

Figure 7-9 Activation of adenylyl cyclase following binding of an appropriate hormone (e.g., epinephrine, glucagon) to a Gs protein-coupled receptor. Following ligand binding to the receptor, the G_s protein relays the hormone signal to the effector protein adenylyl cyclase. G_s cycles between an inactive form with bound GDP and an active form with bound GTP. Dissociation of the active form yields the $G_{s\alpha}$, GTP complex, which directly activates adenylyl cyclase. Activation is short-lived because GTP is rapidly hydrolyzed. This terminates the hormone signal and leads to reassembly the inactive Gβγ. GDP form, returning the system to the resting state. Binding the activated receptor to $G_{s\alpha}$ promotes dissociation of GDP and its replacement with GTP(From: Lodish H, et al. Molecular Cell Biology. 2000)

These ion channels are transmembrane ion channels with selectivity filters and a G protein binding site. G-proteins directly activate these ion channels using effector proteins or the G protein subunits themselves. Unlike most effectors, not all G protein-gated ion channels have their activity mediated by G_α of their corresponding G proteins. For instance, the opening of inwardly rectifying K^+ channels is mediated by the binding of $G_{\beta\gamma}$. Two sets of nerve fibers control the heartbeat in animals: one speeds up the heart rate, the other slows it down. The nerves that slow the heartbeat do so by releasing neurotransmitter acetylcholine (Ach), which binds to a G-protein-coupled receptor (muscarinic receptor) on the surface of heart muscle cells. When Ach binds to this receptor, a G protein (Gi, "i" stands for inhibitory) is activated by dissociating into a α subunit and a βγ complex. βγ complex in turn binds to the cytoplasmic side of a type of K^+ channels in the heart muscle cell membrane and opens the K^+ channels. K^+ ions flow out of the cell following K^+ electrochemical gradient causing a hyperpolarization of the heart muscle cell, which slows heartbeat down. When the α subunit inactivates itself by hydrolyzing its bound GTP and reassociates with the βγ complex, G protein is no longer active. Consequently, the K^+ channels reclose, and the signal shuts down (Figure. 7-8).

7.3.2 GPCR-initiated cyclic AMP pathway

Many extracellular signals acting via GPCRs go through the cyclic AMP pathway. The activated G protein α subunit ($G_{s\alpha}$) usually switches on adenylyl cyclase, causing a sudden increase in the concentration of cyclic AMP (cAMP) synthesized from ATP in the cell. Because this G protein stimulates **adenylyl cyclase**, it is also called G_s. A second enzyme that is continuously active inside the cell called cyclic AMP **phosphodiesterase** can rapidly convert cAMP back to AMP, thus shutting down the signaling process (Figure 7-9). Because cAMP can be broken down very quickly, the concentration of this second messenger can rise or fall ten-fold within seconds, thus responding to extracellular signals rapidly.

cAMP is water-soluble, it can carry its signal throughout the cell, traveling from cell membrane (its synthetic site) to target proteins located in the cytosol, the nucleus, or other organelles. Cyclic AMP exerts its effect mainly through activating an enzyme **cyclic-AMP-dependent protein kinase (PKA)**. PKA is a normally inactive, as it is a heterotetramer made of two regulatory (R) and two catalytic (C) subunits and the regulatory subunits normally inhibit the catalytic activity of the enzyme. When binding to cAMP, PKA goes through a conformational change and releases its active catalytic subunits. The activated PKA then moves into the nucleus and phosphorylates specific gene regulatory proteins. Once phosphorylated, these proteins stimulate the transcription of a whole set of target genes. This type of signaling pathway controls many processes in cells, ranging from hormone synthesis in endocrine cells to the production of proteins involved in long-term memory in the brain. Figure 7-10 shows the process of such a pathway. Activated PKA can also catalyze the phosphorylation of particular serines or threonines on certain intracellular proteins in the cytosol, and changing their activity.

Some cAMP responses are very fast; such as the ones that happen in skeletal muscles, which only takes a few seconds to complete and does not depend on changes in gene transcription. Some cAMP do depend on changes in the transcription of specific genes and take minutes or even hours to develop.

Many cells signal through cAMP. However, different target cells respond very differently to the same extracellular signals that change intracellular cAMP concentrations. Although activation of PKA in a liver cell in response to epinephrine leads to the breakdown of glycogen, activation of the same enzyme in a kidney tubule cell in response to vasopressin causes an increase in the permeability of the membrane to water, and the secretion of thyroid hormone. The response of a given cell to cAMP is typically determined by the specific proteins phosphorylated by PKA. Different cell types have different sets of target proteins available to be phosphorylated; that is why the effects of cAMP vary with different target cells. Some of the known PKA target proteins are phosphorylase kinase α, protein phosphatase-1, pyruvate kinase, CREB, protein phosphatase inhibitor-1, etc.

7.3.3 GPCR-initiated inositol phospholipid pathway

Some extracellular signals exert their effects via activating a type of G protein that activates membrane-bound enzyme phospholipase C instead of adenylyl cyclase. The activated phospholipace C cleaves a lipid molecule that is a component of the cell membrane called **inositol phospholipid**. Inositol phospholipid is a phospholipid that has the sugar inositol attached to its head and is present in small quantities in the inner half of the plasma membrane lipid bilayer. Because inositol phospholipid is involved, this signaling pathway through the activation of phospholipase C is called the inositol phospholipid pathway (Figure 7-11). This pathway occurs in almost all eukaryotic cells and affects many different types of target proteins in those cells. Some examples of cell responses mediated by phospholipase C activation are glycogen breakdown, secretion of amylase, smooth muscle contraction and blood platelets aggregation, etc.

The activated phospholipase C chops the sugar-phosphate head off the inositol phospholipids, producing two small second messenger molecules — inositol 1, 4, 5-triphosphate (IP_3) and diacylglycerol (DAG) (Figure 7-11). Both molecules play a crucial part in signaling inside the cell. Therefore, phospholipase C activates IP_3 and DAG, two

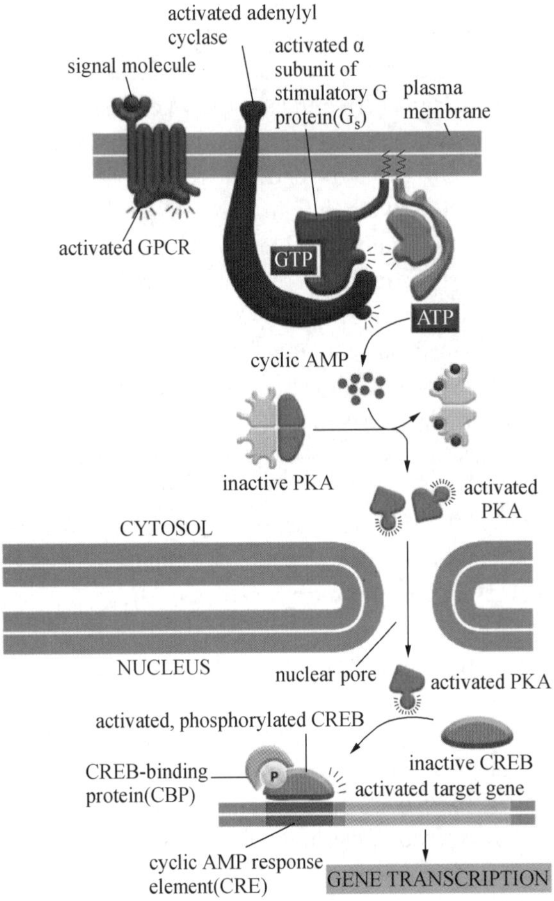

Figure 7-10 A rise in intracellular cyclic AMP can activate gene transcription. Binding a signal molecule to its G-protein-coupled receptor can lead to the activation of adenylyl cyclase and a rise in the concentration of intracellular cyclic AMP. In the cytosol, cyclic AMP activates PKA, which then moves into the nucleus, phosphorylates the gene regulatory protein CREB, and then stimulates a whole set of target gene transcription (From: Alberts B, et al. Molecular Biology of the Cell. 2008)

signaling cascades.

IP_3 is a small, hydrophilic sugar phosphate. It easily diffuses into the cytosol, where it binds to IP_3 receptor embedded in the endoplasmic reticulum membrane. This IP_3 receptor also functions as a tetrameric Ca^{2+} channel. Binding of IP_3 opens the Ca^{2+} channels so that Ca^{2+} stored inside the endoplasmic reticulum rushes out to the cytosol and triggers a sharp rise in the cytosolic concentration of free Ca^{2+} (Figure 7-11). Calcium ion is an intracellular secondary messenger that binds to various target proteins, triggering different responses. Some of IP_3 mediated cellular responses are muscle contraction, cyclic GMP formation, actin polymerization, blood platelets shape change and aggregation, calcium mobilization, membrane depolarization, etc.

DAG is a lipid that remains embedded in the plasma membrane. With Ca^{2+}, DAG helps to recruit and activate an enzyme called protein kinase C (PKC), translocating PKC from the cytosol to the plasma membrane. Ca^{2+} is needed to bind to PKC for PKC to be activated (Figure 7-11). Once activated, PKC phosphorylates different sets of intracellular target proteins in different types of cells. PKC works in a similar way to PKA, although they have different target proteins. Some PKC-activated responses are serotonin release, histamine release, secretion of epinephrine, secretion of insulin, secretion of calcitonin, dopamine release, glycogen hydrolysis, etc.

7.4 Signaling through enzyme-coupled cell-surface receptors

Enzyme-coupled receptors usually have only one transmembrane segment, and the cytoplasmic domain of enzyme-coupled receptors acts as an enzyme or complex with another protein that acts as an enzyme. Some of these protein kinases and protein phosphatases are hydrophilic, while others are integral membrane proteins. Some of these enzymes have multiple proteins as substrates; others have only a single protein substrate. Many of the substrates of these enzymes are enzymes themselves, most often they are other protein

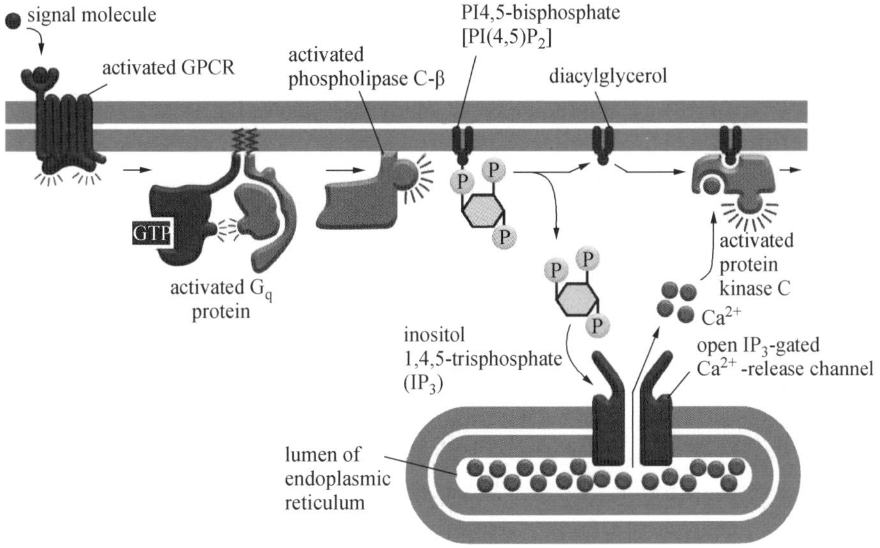

Figure 7-11 Phospholipase C activates two signaling pathways (From: Alberts B, et al. Molecular Biology of the Cell. 2008)

kinases and phosphatases, or ion channels, transcription factors and various types of regulatory proteins.

7.4.1 Receptor tyrosine kinases

The largest class of enzyme-coupled receptors is made up of those with a cytoplasmic domain that acts as a tyrosine protein kinase, phosphorylating tyrosine side chains on selected proteins. These receptors are called **receptor tyrosine kinases (RTKs)**. We will use receptor tyrosine kinase as an example to illustrate how external signals work through enzyme-coupled receptors(Figure 7-12).

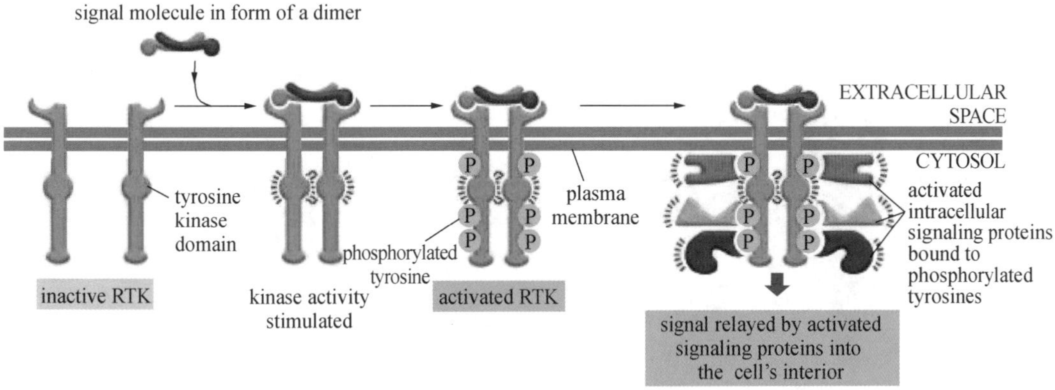

Figure 7-12 Activation of a receptor tyrosine kinase stimulates the assembly of an intracellular signaling complex (From: Alberts B, et al. Essential Cell Biology. 2004)

In many cases, an external signal first causes two enzyme-coupled receptors to come together in the membrane and to form a dimer. In figure 7-12 however, the signaling molecule itself is a dimer and thus can physically cross-link two receptor molecules. Within the dimer, contact with the two adjacent intracellular receptor tails activates their kinase function, causing each receptor to phosphorylate the other. In the case of receptor tyrosine kinase, the phosphorylations occur on specific tyrosine located on the cytosolic tail of the receptors. The phosphorylation then triggers the assembly of an elaborate intracellular signaling complex on the receptor tails. The newly phosphorylated tyrosines serve as binding sites for many kinds of intracellular signaling proteins, which become active upon binding. As long as the receptor is activated, the intracellular signaling complex transmits signals along several routes simultaneously to many destinations inside the cell, thus activating and coordinating the numerous biochemical changes that are required to trigger a complex response such as cell proliferation. The activated receptors are usually turned off by protein tyrosine phosphatases in the cell. These phosphatases remove the phosphates that were added to receptors previously in response to extracellular signal. In some cases, the activated receptors are endocytosed into the interior of the cell and then destroyed by digestion in lysosomes.

Different receptor tyrosine kinases recruit different sets of intracellular signaling molecules, producing different effects. The more widely used ones include a phospholipase that functions the same way as phospholipase C to activate the inositol phospholipid signaling pathway. They also include an enzyme called phosphatidyl-inositol 3-kinase (PI3-kinase), which phosphorylates inositol phospholipids in the plasma membrane making them docking sites for other intracellular signaling proteins. One of these is **protein kinase B (PKB)**, which phosphorylates target proteins on their serines and threonines and is essential in cell survival and growth. Some of these intracellular signaling molecules function solely as physical adaptors; they help to build a larger signaling complex by coupling receptors to other proteins, which in turn may bind to and activate yet other proteins to pass the signals along. One of the key players in these adaptor-assembled signaling complexes is **Ras**, a small protein that has a lipid tail bound to the cytoplasmic side of the plasma membrane.

Virtually all receptor tyrosine kinases activate Ras. The Ras protein is a member of a large family of small, single-subunit GTP-binding proteins called the **monomeric GTP-**

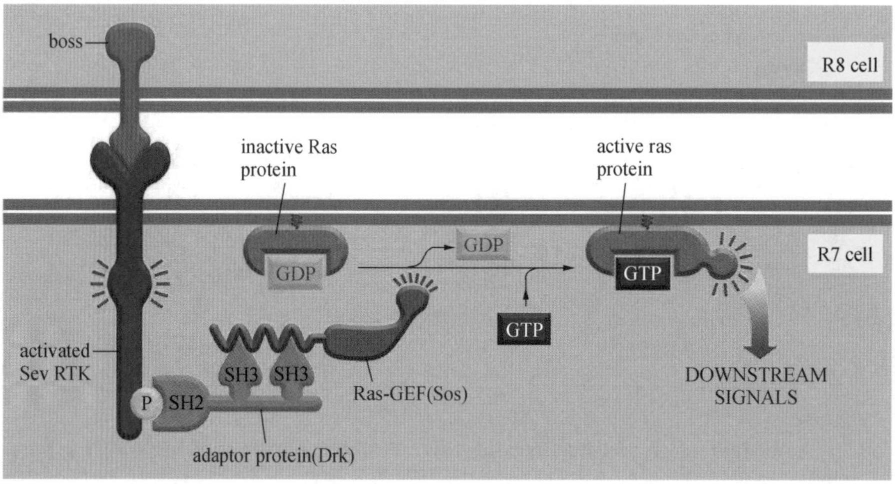

Figure 7-13 Sev RTK activates Ras in the fly eye. Sev (on the surface of the R7 precursor cell) is activated by the Boss protein (on the surface of R8). The adaptor protein Drk docks on a particular phosphotyrosine on the activated receptor. Drk recruits and stimulates a protein that functions as a Ras-activating protein Sos. Sos stimulates the inactive Ras protein to replace its bound GDP by GTP, which activates Ras to relay the signal downstream, inducing the R7 precursor cell to differentiate into a UV-sensing photoreceptor cell. Note that the Ras protein contains a covalently attached lipid group (*red*) that helps anchor the protein to the plasma membrane(From: Alberts B, et al. Molecular Biology of the Cell. 2008)

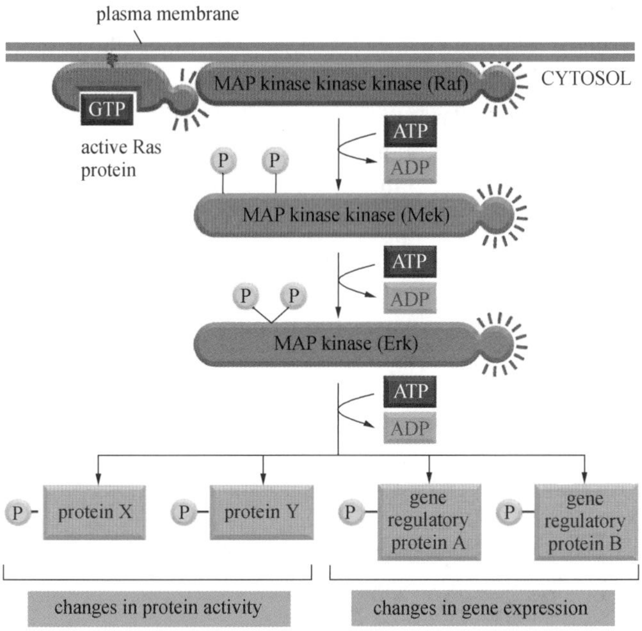

Figure 7-14 The steps of a generalized MAP kinase cascade. Activated Ras recruits the protein Raf to the membrane, where it is phosphorylated and thus activated. In the pathway depicted here, Raf (MAPKKK) phosphorylates and activates another kinase named MEK (MAPKK), which in turn phosphorylates and activates still another kinase termed ERK (MAPK). Erk in turn phosphorylates a variety of downstream proteins. The resulting changes in gene expression and protein activity cause complex changes in cell behaviour (From: Alberts B, et al. Molecular Biology of the Cell. 2008)

binding proteins. Ras is similar to subunit of a G protein in both its structure and function: it has GTP-bound active state and GDP-bound inactive state; it is activated when Ras interacts with an activating protein and exchanges its GDP for GTP; Ras switches itself off by hydrolyzing its GTP to GDP. Figure 7-13 shows how the Sev RTK activates Ras in the fly eye. The assembly of photoreceptor cells in a developing *Drosophila* ommatidium begins with R8 and ending with R7, which is the last photoreceptor cell to develop.

When activated, Ras triggers a phosphorylation domino cascade in which a series of protein kinases phosphorylate and activate one another in turn, relaying messages along the way from the plasma membrane all the way to the nucleus. This relay system is called a **MAP-kinase cascade** (Figure 7-14). It takes its name from the final kinase in the cascade MAP-kinase (**mitogen-activated protein kinase**, **MAPK**). In this cascade, MAP-kinase is phosphorylated and activated by an enzyme called **MAP-kinase-kinase** (**MAPKK**), which itself is activated by an enzyme called **MAP-kinase-kinase-kinase** (**MAPKKK**), which is activated by Ras. MAPKKs are dual-specificity kinases, a term denoting that they can phosphorylate tyrosine as well as serine and threonine residues. All MAPKs have a tripeptide near their catalytic site with the sequence Thr-X-Tyr. MAPKK phosphorylates MAPK on both the threonine and tyrosine residue of this sequence, thereby activating the enzyme.

Once activated, the MAP-kinase relays the signal downstream by phosphorylating various proteins in the cell, including gene regulatory proteins and other protein kinases. The activated MAP-kinase **phosphorylates** certain **gene regulatory proteins** on serines and threonines, altering their ability to control gene transcription and thereby to change the pattern of gene expression. The final outcome is determined by what signals the cell receives and which genes are active in the cell.

7.4.2 Tyrosine-kinase-associated receptor

Not all enzyme-coupled receptors work through activating a sequence of protein kinases to pass messages to the nucleus. Some use a more direct way to control gene expression. **Cytokines** can bind to receptors that activate the latent gene regulatory proteins at the plasma membrane. Once activated, these regulatory proteins head straight for the nucleus where they stimulate the transcription of certain genes (Figure 7-15). For example, interferon, the cytokine that instructs cells to produce proteins that will make them more resistant to viral infection, uses such a pathway. Unlike the receptor tyrosine kinases, cytokine receptors have no intrinsic enzyme activity but are associated with **cytoplasmic tyrosine kinases** called JAKs (Janus kinases). Cytokine binding to **cytokine receptor** causes receptor dimerization and brings two JAKs into close proximity so that they transphosphorylate each other, thereby increasing the activity of their tyrosine kinase domain. JAKs then phosphorylate and activate **cytoplasmic gene regulatory proteins** called STATs (signal transducers and activators of transcription). STATs migrate to the nucleus where they stimulate transcription of specific **target genes**. Different cytokine receptors activate different STATs, thus have different effects. As any pathway that is turned on by phosphorylation, the cytokine signal is also shut off by removal of the phosphate groups from the activated signaling proteins with phosphatases.

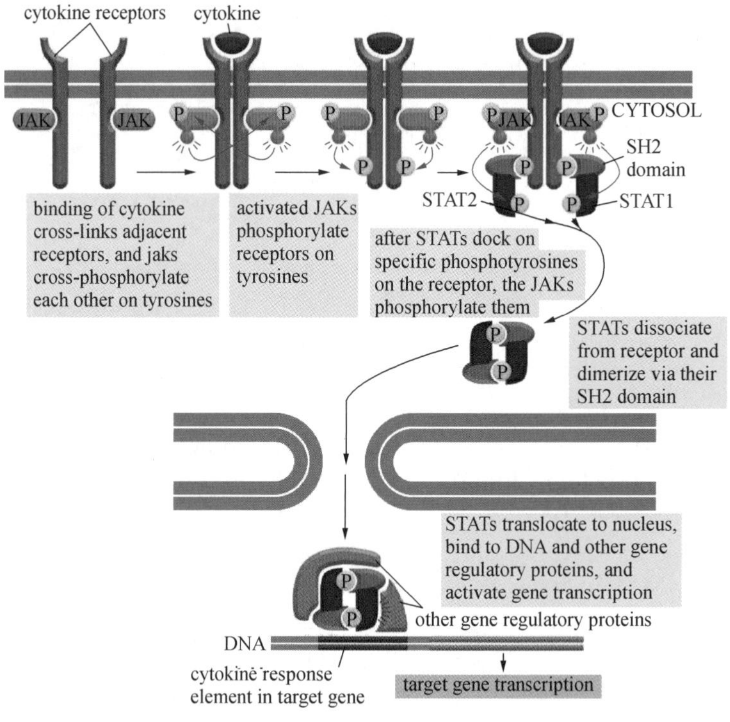

Figure 7-15 Cytokine receptors are associated with cytoplasmic tyrosine kinases. Binding a cytokine to its receptor causes JAKs to cross-phosphorylate and activate one another. The activated kinases then phosphorylate the receptor proteins on tyrosines. STATs present in the cytosol then attach to the phosphotyrosines on the receptor, and the JAKs phosphorylate and activate these proteins as well. The STATs then dissociate from the receptor proteins, dimerize, migrate to the nucleus, and activate the transcription of specific target genes(From: Alberts B, et al. Molecular Biology of the Cell. 2008)

7.4.3 Receptor serine/threonine kinases

There are other enzyme-coupled receptors — receptor serine/threonine kinases — use an even more direct signaling pathway. When activated, they directly phosphorylate and activate cytoplasmic gene regulatory proteins called SMADs. The hormones and local mediators that activate these receptors belong to the TGF-β (transforming growth factor-β) superfamily of extracellular proteins, which are essential in animal development, tissue repair and in immune regulation, as well as in many other processes.

Members of the TGF-β superfamily act through receptor serine/threonine kinases that are single-pass transmembrane proteins with a serine/threonine kinase domain on the cytosolic side of the plasma membrane. There are two classes of receptor serine/threonine kinases, type I and type Ⅱ. Typically, the ligand first binds to and activates a type-Ⅱ receptor, which recruits, phosphorylates, and activates a type-Ⅰ receptor, forming an active tetrameric receptor complex.

Once activated, the receptor complex uses a strategy for rapidly relaying the signal to the nucleus. The type-I receptor directly binds and phosphorylates a latent gene regulatory protein of the Smad family. Once one of these Smads has been phosphorylated, it dissociates from the receptor and binds to Smad 4, which can form a complex with the phosphorylated receptor-activated Smads. The Smad complex then moves into the nucleus, where it associates with other gene regulatory proteins, binds to specific sites in DNA, and activates a particular set of target genes (Figure 7-16).

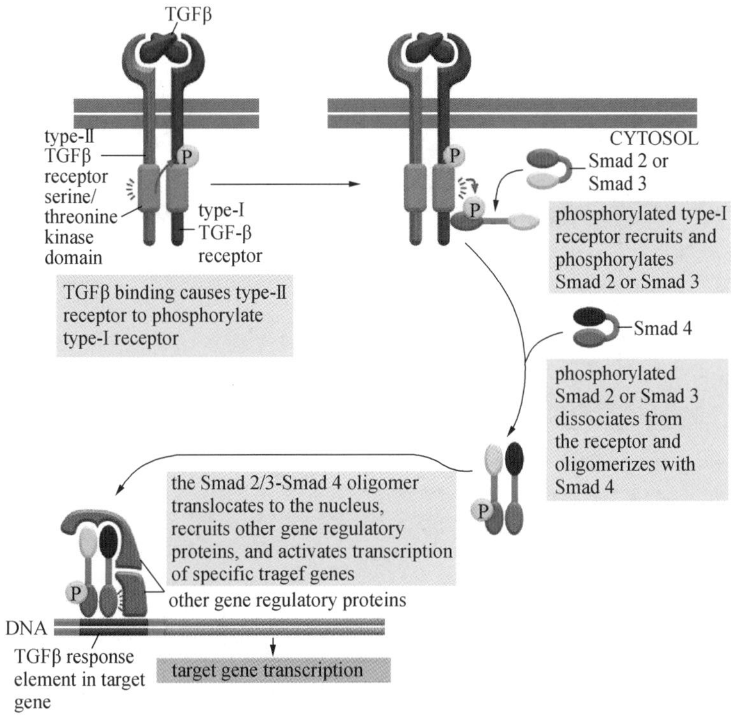

Figure 7-16 TGF-β receptors activate gene regulatory proteins directly at the plasma membrane. These receptor serine/threonine kinases phosphorylate themselves and then recruit and activate cytoplasmic SMADs. The SMADs then dissociate from the receptors and bind to other SMADs, and the complexes then migrate to the nucleus, where they stimulate transcription of specific target genes(From: Alberts B, et al. Molecular Biology of the Cell. 2008)

7.5 Signal network system

Cells communicate with each other via signaling pathways. A cell receives information from its environment through the activation of various receptors. These receptors can bind only to specific ligands, thus neglecting a large variety of unrelated signaling molecules in the environment. A single cell may have many different receptors receiving and passing through many different signals into the cell simultaneously. Once transmitted into the cell, these signals are routed through selected pathways reaching their final target proteins to cause different effects in a cell such as causing the cell to divide, differentiate, change shape, relocate, turn on a particular metabolic pathway, send out a signal of its own, or even commit suicide. In previous sections, we reviewed some signaling pathways as if they acted linearly and independently. In real life, however, signaling pathways in the cell form a complex signal network system, where players in one pathway can participate in events of other pathways. Signals from different ligands binding to different unrelated receptors can converge to activate a common effector, such as Ras, leading to the same cellular responses (Figure 7-17).

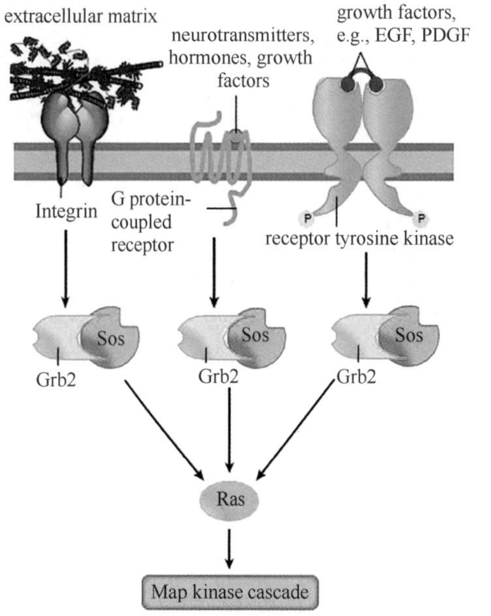

Figure 7-17 Signals transmitted from a G protein-coupled receptor, an integrin, and a receptor tyrosine kinase all converge on Ras and are then transmitted along the MAP kinase cascade (From: Karp G, et al. Cell and Molecular Biology — concepts and experiments. 2010)

Signals from the same ligand can diverge to activate different effectors, leading to diverse cellular responses. For example, insulin is a hormone that induces both immediate and long-term cellular responses. Within minutes, 10^{-9} to 10^{-10} M insulin may cause an increase in the rate of glucose uptake from the blood or fat cells into muscle cells, as well as modulate the activity of various enzymes of glucose metabolism. The long-term insulin effects include increasing expression of liver enzymes that synthesize **glycogen** and of adipocyte enzymes that synthesize **triacylglycerols**. Higher concentration of insulin (10^{-8}M) can also function as a growth factor for many cells, the effects of which are manifested in hours and require continuous exposure. Signals can be passed back and forth among different pathways; this is called **cross-talk** within the **signal network system**, and this cross-talk occurs in virtually all of the control systems of the cell. Figure 7-18 shows how five pathways that we discussed previously are interconnected with each other. It is through the complex signal network system that a cell integrates and responds to information from different sources accordingly.

Most cross-talks are mediated by the protein kinases that are present in each of the pathways. Protein kinases often phosphorylate -and hence regulate -components in other pathways. Hundreds of distinct types of protein kinases are thought to be present in a single mammalian cell. One example of this type of crosstalk involves cAMP. cAMP initiates the cascade leading to glucose mobilization. However, cAMP can also inhibit the growth of a variety of cells including fibroblasts and fat cells. This is achieved by cAMP activating PKA, the **cAMP-dependent kinase**, which can phosphorylate and inhibit Raf (a player in MAP kinase cascade), thus blocking signals transmitted through the MAP kinase cascade. These two pathways also intersect at another important signaling effector, the transcription factor CREB, which can be phosphorylated by PKA of cAMP-mediated

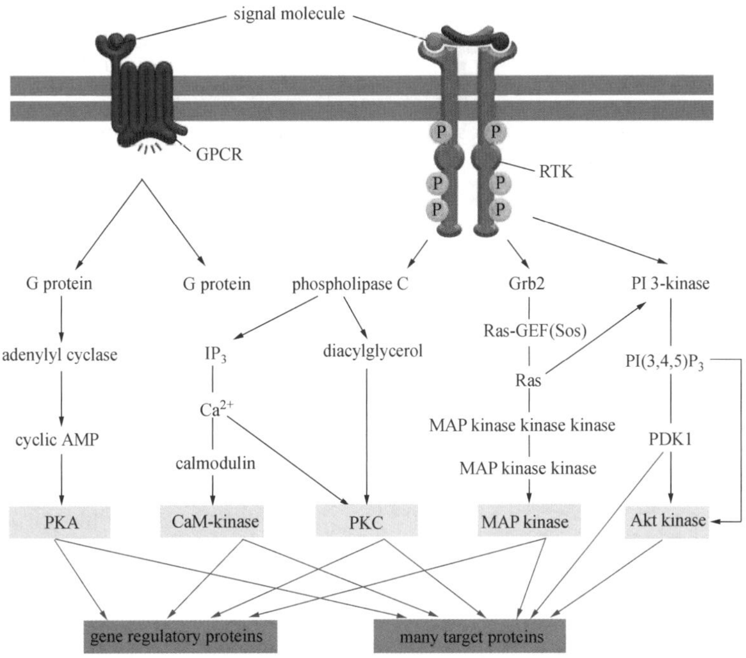

Figure 7-18 Five parallel intracellular signaling pathways activated by G-protein-coupled receptors, receptor tyrosine kinases, or both. In this schematic example, the five kinases (shaded yellow) at the end of each pathway phosphorylate target proteins (shaded red), some of which are phosphorylated by more than one of the kinases. The specific phospholipase C activated by the two types of receptors is different: G-protein-coupled receptors activate PLC-β, whereas receptor tyrosine kinases activate PLC-γ (not shown) (From: Alberts B, et al. Molecular Biology of the Cell. 2008)

pathways as well as by Rsk-2, a substrate of MAPK of the MAP kinase pathway. Both PKA and RSK-2 phosphorylate CREB on the same amino acid residue, Ser133, which endow the transcription factor with the same potential in both pathways (Figure 7-19).

In reality, the complexity of cell signaling is much greater than we are able to describe. We are still in the continuous quest for new links in the chains, new signaling molecules, new connections and even new pathways.

7.6 Cell signaling and the cytoskeleton

In most cells, cell adhesion and the organization of cytoskeleton directly affect cell functions. The receptors responsible for cell adhesion and cytoskeleton organization can also initiate intracellular signaling pathways that regulate other cell behavior such as gene expression. On the other hand, signaling molecules such as growth factors can induce cytoskeletal changes resulting in cell shape modification or cell movement. Thus components of cytoskeleton can act as both receptors and targets in cell signaling pathways.

7.6.1 Integrin signal transduction through FAK

As briefly mentioned in chapter 6, the integrins are major adhesion proteins (receptors) that are responsible for the attachment of cells to the extracellular matrix. Integrins does so by

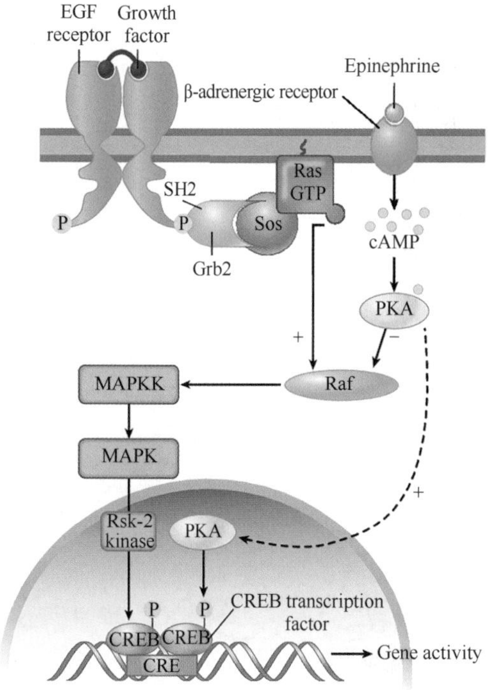

Figure 7-19 An example of crosstalk between two major signaling pathways. Cyclic AMP acts in some cells, by means of the cAMP-dependent kinase PKA, to block the transmission of signals from Ras to Raf, which inhibits the activation of the MAP kinase cascade. In addition, both PKA and the kinases of the MAP kinase cascade phosphorylate the transcription factor CREB on the same serine residue, activating the transcription factor and allowing it to bind to specific sites on the DNA (From: Karp G, et al. Cell and Molecular Biology — concepts and experiments. 2010)

interacting with components of the cytoskeleton to build a linkage between the extracellular matrix and adherent cells at the two types of cell-matrix junctions (focal adhesions and hemidesmosomes). Meanwhile, the integrins are also receptors that receive stimuli from extracellular matrix, activate intracellular signaling pathway, and subsequently control gene expression and other cell behavior.

Nonreceptor protein-tyrosin kinases, particularly **FAK** (**focal adhesion kinase**), play an important role in **integrin signal transduction**. Upon binding of integrin with extracellular matrix, FAK gets activated by autophosphorylating itself. In addition, members of the Src family of nonreceptor protein-tyrosin kinases bind to the autophosphorylation site of FAK and further phosphorylate some additional sites on FAK. Tyrosine-phosphorylated FAK provides binding sites for many downstream signaling molecules such as Grb2-Sos complex, which in turn activates Ras and MAP kinase cascades (Figure 7-20).

7.6.2 Rho family regulation of the actin cytoskeleton

Cell movement and changes in cell shape are two common cell responses to its extracellular signals. As discussed in Chapter 6, these cell

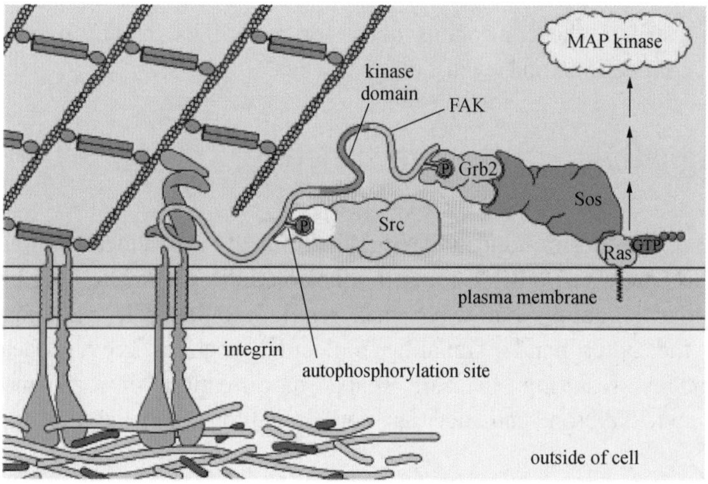

Figure 7-20 Model for signaling from the FAK protein-tyrosine kinase. Binding integrins to the extracellular matrix stimulates FAK activity, leading to its autophosphorylation. Src then binds to the FAK autophosphorylation site and phosphorylates FAK on additional tyrosine residues. These phosphotyrosines serve as binding sites for the Grb2-Sos complex, leading to activation of Ras and the MAP kinase cascade, as well as for additional downstream signaling molecules, including pI3-kinase (From: Geoffrey MC. The Cell — A Molecular Approach. 2000)

behaviors are governed by the actin cytoskeleton modeling and re-modeling underlying the plasma membrane. Therefore, assembly and disassembly of the actin filaments are crucial parts of cell response to extracellular signaling molecules.

Members of the **Rho family** of small GTP-binding proteins (including Rho, Rac and Cdc42) serve as universal regulators of the dynamic assembly and disassembly of the actin filaments, thus connecting extracellular signals to a variety of cell processes. Early studies of fibroblasts responding to growth factor stimulation first reveal the role of Rho family in different aspects of actin remodeling: Different members of the Rho family regulate the organization of actin to form filopodia (Cdc42), lamellipodia (Rac) and focal adhesion and stress fibers (Rho), respectively. Further studies show that the Rho family plays similar roles in regulating the actin cytoskeleton in all types of eukaryotic cells. Moreover, Rho family members are found to activate MAP kinase signaling pathways. So these small GTP-binding proteins regulate not only the cytoskeletal remodeling but also gene expression.

Several experiments have elucidated some of the pathways by which the Rho family members regulate cytoskeletal changes (Figure 7-21). A key target of Rho is a protein-serine/threonine kinase called Rho kinase. When Rho kinase is activated by Rho, it phosphorylates the light chain of myosin II as well as inhibits myosin light chain phosphatase. The resulting increase in myosin II light chain phosphorylation leads to the assembly of actin-myosin filaments, thus the formation of stress fibers and focal adhesions. On the contrary, Rac and Cdc42 stimulate the PAK (another protein-serine/threonine kinase), which in turn phosphorylates and inhibits myosin light chain kinase, causing a decrease in myosin II phosphorylation. Inactivated myosin II decreases its ability to interact with actin. As a result, cell surface protrusions occur. Both Rho and Rac also stimulate LIM-Kinase which regulates actin remodeling by phosphorylating cofilin, a key regulator of actin disassembly.

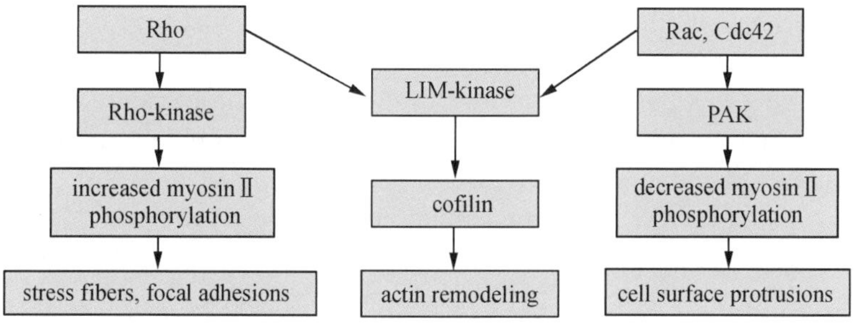

Figure 7-21 Targets of Rho family members (From: Geoffrey MC. The Cell — A Molecular Approach. 2000)

7.7 Cell communication in Plants

Summary

Cell communication is a universal and fundamental biological process that mediates interactions between a cell and its surroundings. Cell signaling requires not only extracellular signal molecules, but also a complementary set of receptor proteins in each cell that enable it to bind and respond to the signal molecules in a specific way. Some small hydrophilic signal molecules, including steroid or thyroid hormones and the dissolvable gas nitric oxide, signal by diffusing across the plasma membrane of target cell and activating intracellular receptor proteins or enzymes. However, most extracellular signal molecules are hydrophobic and can activate receptor proteins only on the surface of the target cells. These receptors act as signal transducers converting the extracellular signal to intracellular proteins (messengers) or other downstream target(s), which relay the signal by changing their level or activity, and ultimately effecting specific cell function(s). There are three main families of cell-surface receptors. Ion-channel-coupled receptors are transmitter-gated ion channels that open or close briefly in response to the binding of a signal molecule. G-protein-coupled receptors indirectly activate or inactivate plasma-membrane-bound enzymes via trimeric G proteins. Enzyme-coupled receptors either act directly as enzymes or act indirectly via their association with other enzymes. These enzymes are usually protein kinases that phosphorylate specific proteins in the target cell. Signaling pathways are highly interconnected into a complicated communication network system in which signals can diverge or converge, and they can go back and forth between two or more different signaling pathways.

In addition to the structural role, the integrins serve as receptors that activate intracellular signaling pathways, a nonreceptor protein-tyrosine kinase called FAK plays a key role in integrin signaling.

Cell signaling is not only important for the understanding of the functioning of a normal cell, but also is of vital importance to understand the growth and activity of an aberrant cell, or that of a cell that is combating adverse conditions.

Questions

1. Please define the following terms: signal molecule, cell signaling, signal transduction, A-kinase (PKA), C-kinase (PKC), G-protein, nitric oxide, phospholipase C, Ras, receptor tyrosine kinase, signaling cascade.
2. Describe the basic types of signal molecules and second messengers.
3. Describe the role that the inositol-lipid signaling pathway plays in the activation of protein kinase C.
4. What are the similarities and differences between the reactions that lead to the activation of G proteins and the reactions that lead to the activation of Ras?
5. ①Compare and contrast signaling by neurons which secrete neurotransmitters at synapses to signaling carried out by endocrine cells, which secrete hormones into the blood. ②Discuss the relative advantages of the two mechanisms.
6. Compare animal cells and plant cells intracellular signaling mechanisms.

(沈大棱,黄　燕修改)

Chapter 8

Nucleus and chromosomes

The presence of membrane-enclosed organelles such as the nucleus is a hallmark feature of eukaryotic cells. The nucleus encloses genomic DNA within a double lipid bilayer termed the nuclear envelope, which separates genomic DNA from the rest of the cell. This architecture serves not only to isolate the genome from sources of damage, but also to provide opportunities for gene regulation. Although some DNA is present within the mitochondria of animals, plants, and fungi and within the chloroplasts of plants, nearly all the DNA in a eukaryotic cell is sequestered in a nucleus, which in many cells occupies about 10% of the total cell volume. It is usually the most prominent organelle in most plant and animal cells. In this chapter, first we begin by describing the structure and properties of the eukaryotic cell nucleus. We then consider how eukaryotic cell fold long DNA molecules into compact chromosomes. Finally, we discuss how about nucleolus and ribosome biogenesis.

8.1 The nucleus of a eukaryotic cell

Considering its importance in the storage and utilization of genetic information, the nucleus of a eukaryotic cell has a rather undistinguished morphology (Figure 8-1A). The nucleus is surrounded by two concentric lipid bilayer membranes. These membranes are connected at each nuclear pore, through which molecules move between the nucleus and the cytosol. The nuclear envelope is directly connected to the extensive system of intracellular membranes called the endoplasmic reticulum, which extend out from it into the cytoplasm. And it is mechanically supported by a network of intermediate filaments called the nuclear lamina — a thin feltlike mesh just beneath an inner nuclear membrane (Figure 8-1B). The space between the inner and outer nuclear membranes is continuous with the lumen of the rough endoplasmic reticulum.

8.1.1 The nucleus as an organized organelle

The nucleus is composed of several substructures and highly dynamic intranuclear regions. At its periphery, the nucleus possesses a double membrane, i. e., the nuclear envelope, which serves to separate the nuclear contents from the cytoplasm. Deeper inside the nucleus resides the DNA, which usually exists in the form of interphase chromatin, as highly extended nucleoprotein fibers. Another prominent structure found in the nucleus is the nucleolus. This is often seen as a distinctly dense body and is sometimes referred to as a suborganelle, although it is not bound by membrane. It functions in the synthesis of rRNA

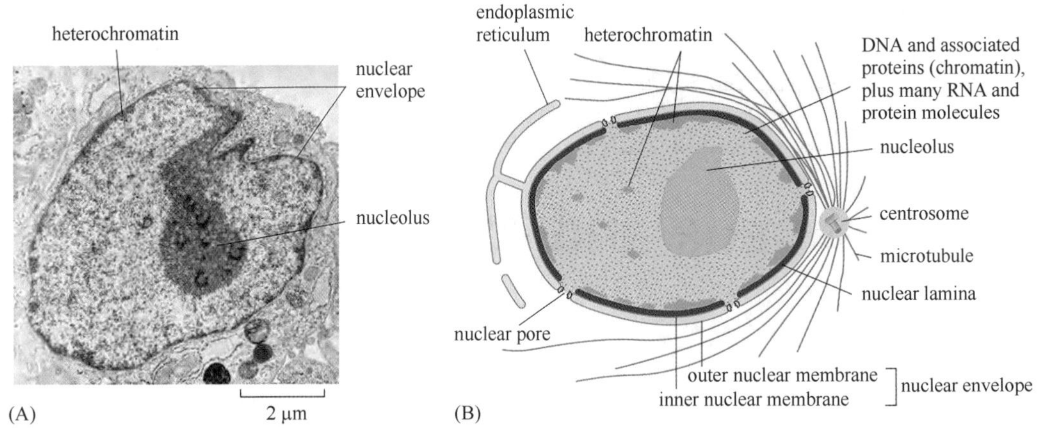

Figure 8-1 A cross-sectional view of a typical cell nucleus. (A) Electron micrograph of a thin section through the nucleus of a human fibroblast. (B) Schematic drawing, showing that the nuclear envelope consists of two membranes, the outer one being continuous with the endoplasmic reticulum membrane. The lumen inside the endoplasmic reticulum is colored yellow. A sheetlike network of intermediate filaments (brown) inside the nucleus forms the nuclear lamina. The dark-staining heterochromatin contains specially condensed regions of DNA (From: Alberts B, et al. Molecular Biology of the Cell. 2015)

and the assembly of ribosomes.

Examination of the nucleus under the electron microscope typically reveals little more than scattered clumps of chromatin and one or more irregular nucleoli. As a result, researchers were left with the impression that the nucleus is largely a "sack" of randomly positioned components. With the development of new, powerful strategies to probe for the spatial organization of nuclear activities in both fixed and live cells, it became possible to localize specific gene loci within the interphase nucleus. It became evident from these studies that the nucleus maintains considerable order. For example, the chromatin fibers of a given interphase chromosome are not strewn through the nucleus like a bowl of spaghetti, but are concentrated into a distinct territory that does not overlap extensively with the territories of other chromosomes.

Along with these major structural and functional components of the nucleus, several other types of smaller assemblies (e. g., cajal bodies, PML bodies, and nuclear speckles) are often seen under the microscope. Each of these contains large numbers of proteins that move in and out of the structure in a dynamic manner. Because none of these are enclosed by a membrane, no special transport mechanisms are required for these large-scale movements. While the functions of these structures remain unclear, their presence clearly indicates a high level of functional compartmentalization and organization within the nucleus.

Overall, the nucleus is a rich environment with a rich inner life. Its constituents are numerous and diverse, ranging from polymers to colloids, from small molecules to macromolecules, giving rise to a highly heterogeneous system. Strikingly, the nucleus lacks any internal boundaries, yet its content is evidently functionally organized. This reflects a changing view of the nucleus as an organelle; not a passive repository for genetic information, but an active, dynamic superstructure whose processes dictate how that information is organized, accessed and used. As a result, three major principles have emerged: first, the nucleus is not just a bag filled with nucleic acids and proteins. Rather, many distinct functional domains, including the chromatin fibers, resides within the confines of the nuclear envelope. Second, all these nuclear domains are highly dynamic, with components exchanging rapidly between the nuclear compartments and the surrounding nucleoplasm. Finally, the motion of molecules within the nucleoplasm appears to be mostly driven by random diffusion.

8.1.2 Structure of the nuclear envelope

The formation of the NE and the subsequent compartmentalization of the cell nucleus is a defining feature of eukaryotes. The NE consists of several distinct components. The core of the NE consists of two cellular membranes arranged parallel to one another and separated by 10 to 50 nm. The membrane of the NE, acting as a barrier that keeps ions, solutes, and macromolecules from passing between the nucleus and cytoplasm, maintain the nucleus as a distinct biochemical compartment. Ruptures of the NE and exposure of chromatin threaten cell viability and cause genome instability. Despite its essential boundary function, the NE undergoes remarkable morphological changes, most noticeable during mitosis.

The nuclear envelope is the defining structure of the eukaryotic cell and includes three interconnected domains with morphological differences: the inner nuclear membrane (INM) and outer nuclear membrane (ONM) and the pore membranes. These two individual lipid bilayers are separated by a luminal space of 30 ~ 50 nm in human cells, named the lumen or perinuclear space (Figure 8-2). For the transport of macromolecules in and out of the nucleus, the two membranes are fused at sites forming circular pores, i. e., the **nuclear pore complexes** (NPCs) that contain complex assemblies of proteins. Indeed, INM and ONM form discrete domains of a single membrane system separated by the NPCs. The ONM is contiguous with the rough endoplasmic reticulum (ER) and contains ribosomes on its outer surface. Furthermore, the ONM is connected to the cytoskeleton through its integral proteins, and in turn, these proteins connect to the luminal parts of INM proteins at the luminal space, all in all connecting the cytoskeleton to the nucleoskeleton and

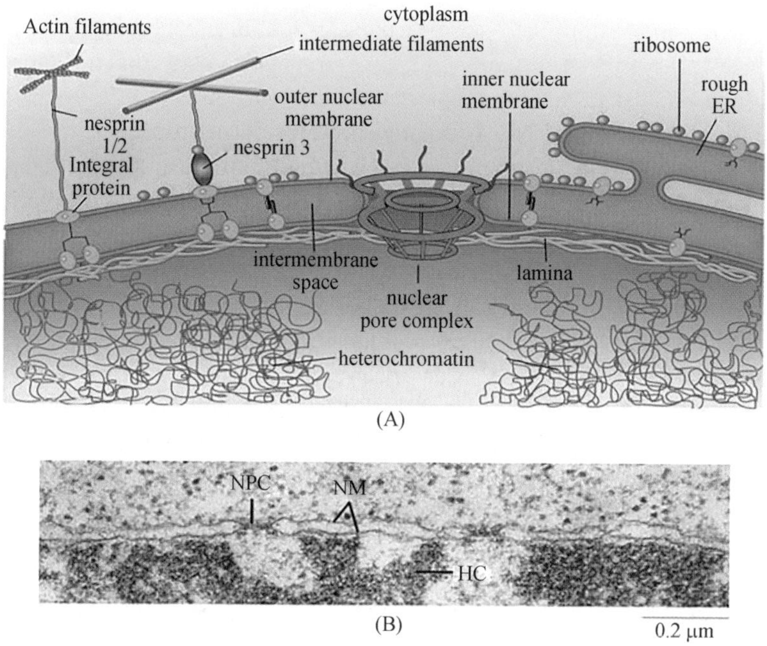

Figure 8-2 The nuclear envelope. (A) Schematic diagram showing the double membrane (NM), NPC, nuclear lamina, and the continuity of the outer membrane with the rough endoplasmic reticulum. (B) Electron micrograph of a section through a portion of the NE of an onion root tip cell. Note the NM with intervening space, NPC, and associated heterochromatin (HC) that does not extend into the region of the nuclear pores (From: Karp G. 2016)

chromatin. Although INM, ONM, pore membranes, and ER originate from a continuous structure, they maintain their identities to a large extent through unique profiles of integral-membrane and other associated proteins, together with specific compositions of lipids and cholesterol. Moreover, the protein composition of the NE is variable across different mammalian tissues.

In metazoan organisms, the primary scaffold of NE is provided by a fibrous layer called the nuclear lamina. This thick protein layer that underlies the INM has been described in studies with isolated NEs, since nuclear lamina is resistant to most chemical extractions used in biology. The most abundant proteins of the nuclear lamina are, by far, three polypeptides of around 65~70 kDa that correspond to type V intermediate filament proteins and are named lamins. It is estimated that there are roughly 3,000,000 copies of lamins in a typical mammalian nucleus. The specific lamin nucleoskeleton is distinct from the nuclear matrix that supports chromatin inside the nucleus. Four lamin proteins are expressed in mammalian somatic cells. A-type lamins (LamA/C) are produced by alternative mRNA splicing of *LMNA* gene, whereas the most abundant B-type lamins (LamB1/B2) are encoded, respectively, by *LMNB*1 and *LMNB*2 genes. Lamins C2 and B3 are germ-cell-specific isoforms produced by alternative splicing of *LMNA* and *LMNB*1, respectively. B-type lamin expression occurs early in embryonic development and persists ubiquitously through adult life. In contrast, the A-type lamins are expressed in an asynchronous and developmentally regulated manner and are only detected after tissue differentiation. Indeed, before day 10 of murine embryonic development, no A-type lamins are detected in the embryo proper. The last constituent of the nuclear lamina is a collection of integral and associated proteins of the INM.

8.1.3 Functions of the nuclear envelope

As a compartment border, the NE encircles the chromatin and physically separates it from the cytoplasm, while still facilitating directed transport between the nucleus and the cytoplasm through nuclear pores that connect the inner and outer membranes. In the nuclear interior, an intricate network of NE-associated proteins forms the nuclear lamina in animal cells. Besides its pivotal role in separating nuclear and cytoplasmic functions, the NE and its associated lamina function in key biological processes, including chromatin tethering to the nuclear periphery, monitoring DNA repair progression, regulating gene expression and transmembrane signaling.

Despite its important structural role in sheltering the genome, the NE is a dynamic and highly adaptable boundary that changes composition during differentiation, deforms in response to mechanical challenges, can be repaired upon rupture and even rapidly disassembles and reforms during open mitosis. NE remodeling is fundamentally involved in cell growth, division and differentiation, and if perturbed can lead to devastating diseases such as muscular dystrophies or premature ageing. Therefore, the NE is also a key cellular hub that plays a dynamic role in the control of cell cycle regulation, apoptosis, ageing, nuclear architecture, chromatin organization, and cell migration.

Unlike the partitioning of the plasma membrane during cell division, the NE is distinct in that it is formed and disassembled during the course of the cell cycle, and is likely to involve similar mechanisms during mitosis and meiosis. Especially during meiosis, the NE takes on multiple essential functions beyond separating the nucleoplasm from the cytoplasm. These include associations with meiotic chromosomes to mediate pairing, being a sensor for recombination progression, and supportive of enormous nuclear growth during oocyte formation.

8.2 The nuclear pore complex

Numerous nuclear pores perforate the NE in all eukaryotic cells. Each nuclear pore is formed from an elaborate structure termed the **nuclear pore complex** (NPC), which is one of the largest protein assemblages in the cell. The total mass of the pore structure is 60~80 mDa in vertebrates, which is about 16 times larger than a ribosome. It has a diameter of 100~150 nm and a depth of 50~70 nm, depending on the organism. Despite its enormous dimensions, the NPC is built from a surprisingly small number of proteins called **nucleoporins** (Nups) which are largely conserved between yeast and vertebrates, are found in these structures (see below for details).

The gatekeeper function necessary for the translocation of proteins in and out of nucleus is controlled principally by the NPC. By controlling the traffic of molecules between the nucleus and cytoplasm, the NPC plays a fundamental role in the physiology of all eukaryotic cells. NPCs are the only channels through which small polar molecules, ions, and macromolecules (proteins and RNAs) are able to travel between the nucleus and the cytoplasm. For example, small metabolites, and globular proteins up to about 40 kDa can diffuse passively through the central aqueous region of the NPC. However, large proteins and ribonucleoprotein complexes cannot diffuse in and out of the nucleus. Rather, these macromolecules are actively transported through the NPC with the assistance of soluble transport proteins that bind macromolecules and also interact with Nups. The capacity and efficiency of the NPC for such active transport is remarkable. In one minute, each NPC is estimated to import 60,000 protein molecules into the nucleus, while exporting 50~250 mRNA molecules, 10~20 ribosomal subunits, and 1000 tRNAs out of the nucleus.

8.2.1 The structure of the nuclear pore complex

(1) The overall architecture of the human NPC Much of our understanding of the overall shape and architecture of the NPC comes from studies performed with electron microscopy (EM). Early studies revealed that the NPC approximates a hollow cylindrical shape with eight-fold rotational symmetry along the axis perpendicular to the NE. More recently, cryo-electron tomography (cryo-ET) and subtomogram averaging have been utilized to generate 3D reconstructions of intact NPCs. The overall architecture emerging from current cryo-ET studies is one in which the most central NPC proteins typically form a symmetric core, which possesses not only eight-fold rotational symmetry but also two-fold rotational symmetry along an axis parallel to the NE. This symmetric core surrounds the central transport channel and serves as the scaffold onto which asymmetric Nups attach on the cytoplasmic and nuclear faces to form the cytoplasmic filaments and nuclear basket, respectively (Figure 8-3). Morphologically, the nuclear basket is a flexible fish-trap-like structure that has been shown to establish a heterochromatin exclusion zone beneath the nuclear face of the NPC. The nuclear basket has important functions in the disassembly of import complexes, interactions with chromatin, and NPC assembly. It is also a very important docking site for various molecular factors that function in nuclear proximity to the NPC. Although the symmetric core accounts for a major portion of the NPC, decorations specific to the cytoplasmic or nuclear side formed by the asymmetric cytoplasmic filament and nuclear basket Nups provide much of the functionality associated with the NPC. In some cases, these Nups are much more flexible, and parts or all of the proteins may not be visible in EM

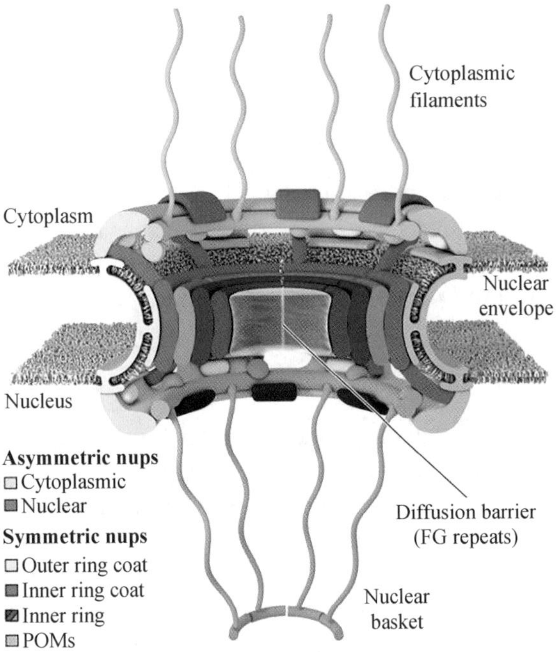

Figure 8-3 Schematic representation of the NPC architecture. A cutaway view depicting half of an NPC is shown (From: Annu. Rev. Biochem. 2019)

reconstructions owing to the averaging necessary to acquire high-resolution reconstructions. On both the cytoplasmic and nuclear sides, these Nups contain FG-repeats (Phe and Gly residues) and bind sites for transport factor-cargo complexes as well as other cellular machinery. In addition to their structural flexibility, these asymmetric Nups also display substantial divergence in composition and domain organization between humans and *S. cerevisiae*.

(2) Nucleoporins and their properties

Despite their large size, a NPC is made up of multiple copies of only ~34 unique proteins, i. e., Nups, with the exact number depending on the species. The majority of Nups are conserved throughout eukaryotes, with recent surveys of eukaryotic genomes consistent with ancestral NPCs being present in the last eukaryotic common ancestor. However, Nups generally display poor sequence conservation despite strong structural conservation, which prevents robust identification of Nups in distantly related genomes. FG-repeat regions are intrinsically disordered, typically span hundreds of residues, and are often enriched in polar amino acids and depleted of charged amino acids. The exact composition and specific sequence motifs of FG repeats vary between Nups and species.

Nups range in size from a few hundred to several thousand residues and possess molecular masses of 30 ~ 358 kDa. A unifying Nup nomenclature does not exist, and proteins are most often named with Nup followed by a number that refers to their molecular mass. In general terms, the Nups are of three types: architectural Nups, transmembrane Nups, and FG-Nups. Each category has unique structural features that are essential to execute specific functions. The architectural Nups form the stable scaffold of the nuclear pore, which is a ring of eightfold rotational symmetry that traverses both membranes of the NE, creating an annulus. Of all Nups, the transmembrane Nups, also known as the **pore membrane proteins** (POMs), are the least conserved and structurally characterized. The roles of the various POMs have been difficult to decipher in part owing to their apparent functional redundancy. Most POMs possess large regions that either face the NE lumen or project toward the pore side, the latter of which could facilitate interactions with the soluble Nups. The FG-Nups, which line the channel of the NPC and are also found associated with the nuclear basket and the cytoplasmic filaments, contain multiple repeats of short hydrophobic sequences that are rich in FG-repeats. The FG-Nups are essential for the function of the NPC and in particular, play the most relevant role in defining the NPC diffusion limit and the list of nuclear transport receptors that might be shuttled through the NPC; however, the NPC remains functional even if up to half of the FG-repeats have been deleted. The FG-Nups are thought to form a flexible gel-like matrix with bulk properties that allow the diffusion of small molecules while excluding unchaperoned hydrophilic proteins larger than 40 kDa. Almost half of the Nups in any given species possess FG repeats, including Nups in the symmetric core, the cytoplasmic filaments, and the nuclear basket. These unique and critical sequences not only form the NPC's diffusion barrier but also serve as specific binding sites for nuclear transport factors, which facilitates rapid, selective transport through the diffusion barrier (Figure 8-4B).

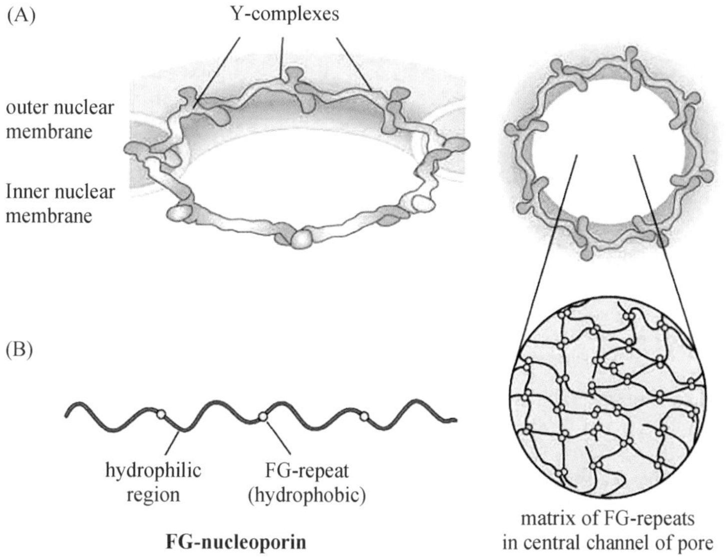

Figure 8-4 Y-complexes and FG-Nups. (A) Sixteen copies of the Y-complex form a major part of the structural scaffold of the NPC. The three-dimensional structure of the Y-complex is modeled into the pore structure. Note the twofold symmetry across the double membrane of the nucleus (left) and the eightfold rotational symmetry around the axis of the pore (right). (B) The FG-Nups have extended disordered structures that are composed of repeats of the sequence FG interspersed with hydrophilic regions (left). The FG-Nups are most abundant in the central part of the pore, and the FG-repeat sequences are thought to fill the central channel with a gel-like matrix (right) (From: Lodish H, et al. 2016)

(3) Composite structure of the human NPC symmetric core The human NPC symmetric core is formed by three sandwiched rings: an inner ring that resides at the midplane of the NE and lines the central transport channel and two outer rings, each of which resides above the NE on either the nuclear or cytoplasmic side. The inner ring and outer rings are sparsely connected. The symmetric core comprises eight distinct spokes, which are the units related by the NPC's eight-fold rotational symmetry. Each spoke also possesses two-fold rotational symmetry relating its cytoplasmic and nuclear halves (Figure 8-5C).

One of the main architectural elements of the NPC is the Y-shaped complex (also referred to as coat nucleoporin complex), which is the largest and best-characterized NPC subcomplex (Figure 8-5A). It contains six universally conserved Nups, but can have up to ten members depending on the species. The Y-complex is an essential scaffolding unit within the NPC, depletion of which abolishes NPC formation. Cryo-ET analysis indicates that the Y-complex comprises the majority of the nuclear and cytoplasmic ring structures, respectively. The single Y-complex has two arms: long and short, leading to a stem region (Figure 8-5A). The human Y-complexes form a basic element dimer of two shifted vertices (cytoplasmic side: red/blue dimer and yellow/green dimer) (Figure 8-5B). The Y-complexes, arranged in a head-to-tail conformation (Figure 8-4A), assemble into two imbricated rings of the NPC, one at the cytoplasmic side and the other at the nuclear side of the

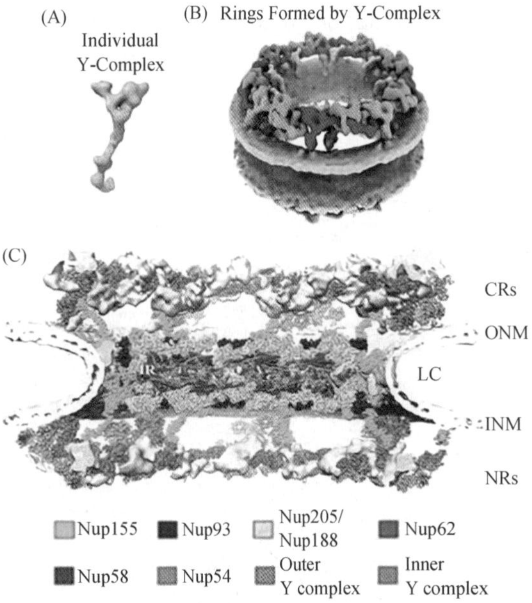

Figure 8-5 Human double ring structure and cross-section of the NPC. (A) Structural model of individual Y-complex (yellow). (B) Human double ring structure formed by collections Y-complex dimers (red/blue and yellow/green). (C) NPC viewed from the cytosol side (top) to the nuclear side (bottom), structure is based on cryo-ET (23Å). Abbreviations: luminal curve: LC; inner ring: IR; cytoplasmic ring: CR; nuclear ring: NR (From: Glavy J. Protein J. 2019)

NPC (Figure 8-5C).

Indeed, NPCs are not a mere gateway to the nucleus. First, the size and maturation of nuclear pores is a crucial event to the nuclear import and nuclear growth and size at the end of mitosis, and during interphase. Second, NPCs might alter nuclear morphology through the physical link with lamins. Third, Nups are a model of long-term protein endurance: NPCs are sustained over the lifetime of a cell by means of a slow but finite interchange of its steadily more stable subcomplexes. In tumorigenesis, the incidental finding of Nup88 as a biomarker of cancer opened the door to detection of high levels of Nup88 in several types of tumors. Beyond this protein, only a small number of Nups have been associated with tumorigenesis.

8.2.2 Selective transport of proteins to and from the nucleus

The basis for selective traffic across the NE is best understood for proteins that are imported from the cytoplasm to the nucleus. Such proteins are responsible for all aspects of genome structure and function; they include histones, DNA polymerases, RNA polymerases, transcription factors, splicing factors, and many others. These proteins are targeted to the nucleus by specific amino acid sequences, called **nuclear localization signals** (NLS) that direct their transport through the NPC (Figure 8-6).

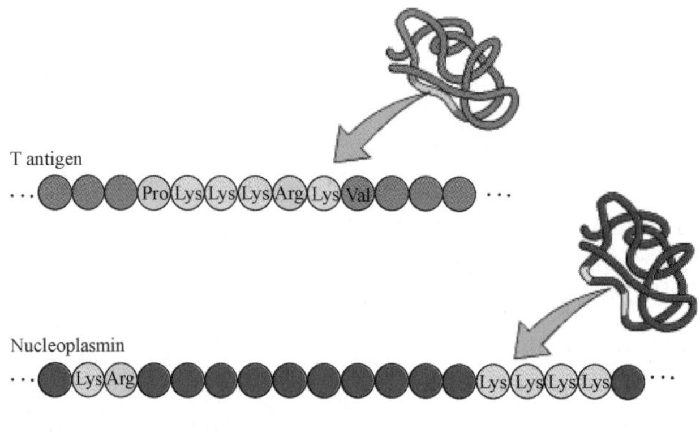

Figure 8-6 Classical nuclear localization signal (cNLS). The T antigen NLS is a single stretch of amino acids. In contrast, the NLS of *Xenopus* nucleoplasmin is bipartite, consisting of a Lys-Arg sequence, followed by a Lys-Lys-Lys-Lys sequence located ten amino acids farther downstream.

(1) Composition of the nuclear localization signals The first **NLS** to be mapped in detail was characterized by Robert Laskey and his coworkers in 1982. They found that nucleoplasmin (a protein involved in chromatin assembly), one of the more abundant nuclear proteins of amphibian oocytes, contains a stretch of amino acids near its C-terminus that functions as an NLS. There is considerable diversity in the type and number of NLS. The best-studied, or classical NLSs (cNLS) were described in the SV40 large antigen (^{126}PKKKRKV^{132}) and *Xenopus* nucleoplasmin (^{155}KRPAATKKAGQAKKKK^{170}). These motifs contain either one or two short stretches of basic amino acids (Figure 8-6), and so they are called monopartite or bipartite cNLS, respectively. The two clusters of basic amino acids within the bipartite cNLS type are usually separated by a linker element of 10~12 amino acids, but its length has also been found to functionally tolerate additional residues. NLSs that do not conform to these structures are globally termed non-classical NLS.

(2) Nuclear transport receptors escort proteins containing NLSs into the nucleus

Assisted transport of macromolecules across NPC is an energy-dependent process that requires nuclear transport factors (NTFs). These factors mainly include transport receptors, adaptor molecules and constituents of the RanGTPase system. The mechanism for the import of cytoplasmic cargo proteins mediated by a soluble nuclear transport receptor (NTR) known as importin is shown in Figure 8-7. A protein bearing a cNLS (cargo protein) is recognized in the cytoplasm by importin α (Imp α). A trimeric complex is formed with the inclusion of importin β (Imp β), which is the actual translocator. Imp α and β interact with each other via a protein domain present near the N-terminus of Imp α, which is the so-called Imp-β-binding domain (IBB). This import trimeric complex docks the cytoplasmic fibrils of the NPCs. Imp β directs the transit of the complex through the NPC channel. This passage is carried out by weak interactions with the innermost layer of FG-Nups. Once in the nucleus, the imported complex attaches to the nuclear basket Nups, such as Nup50. The complex is dissociated by RanGTP, which is found in high concentrations near the nuclear basket structure.

A key player in the translocation process is a protein, **Ran**, which locates predominantly in cell nuclei, a small GTP-binding protein that exists in two conformations, one when complexed with GTP and an alternative one when complexed with GDP. Therefore, the RanGTP-Imp β complex can be translocated back to the cytoplasm. Nup50 connects to Imp α, and the cargo is released. Imp α cytoplasmic re-export requires the participation of CAS (the cellular apoptosis susceptibility gene product), which is a shuttling exportin. CAS participates as a complex with RanGTP. RanGTP-responsive transport accounts for a majority of the nucleocytoplasmic exchange of macromolecules; however, Ran-independent transport receptors also exist (Figure 8-9).

Once in the cytoplasm, the exported complexes of Imp α and β are released due to RanGTP hydrolysis. Free CAS can return to the nucleus by itself. Notably, cytoplasmic and nuclear differential binding or dissociation of cargos and transporters ultimately depends on the differential RanGTP concentrations between these two cellular compartments. The conformation and activity of Ran is regulated by GTP binding and hydrolysis, like other GTP-binding proteins. Enzymes that stimulate GTP binding to Ran are localized to the nuclear side of the NE whereas enzymes that stimulate GTP hydrolysis are localized to the cytoplasmic side. Consequently, there is a gradient of RanGTP across the NE, with a high concentration of RanGTP in the nucleus and a high concentration of RanGDP in the cytoplasm. This gradient of RanGTP is thought to determine the directionality of nuclear transport, and GTP hydrolysis by Ran appears to account for most (if not all) of the energy required for nuclear import. How is this RanGTP gradient preserved? GTP hydrolysis by Ran is a very slow reaction, which explains the high concentration of GTP in the

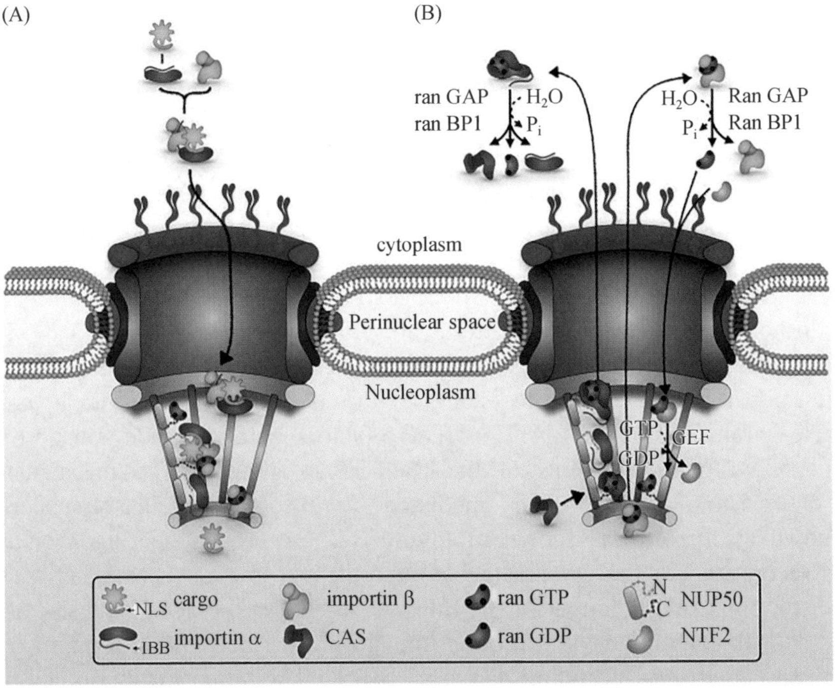

Figure 8-7 The major events in the classical nuclear transport pathway. (A) Main molecular players for cNLS-based import of proteins in eukaryotic cells. A protein bearing a functional cNLS is identified in the cytoplasm by the mediator protein Imp α, and then a ternary complex is formed with the subsequent binding of Imp β. This ternary complex is transported through the NPC channel to the nucleus, where RanGTP binds to the translocated complex and causes dissociation of the cargo and Imp α. Imp β remains bound to RanGTP. (B) Recycling of factors. Nuclear Imp β-RanGTP can be transported back through the NPC channel. Imp α returns to the cytoplasm with the aid of the protein CAS and newly added RanGTP as a tripartite complex. Both Imp α and β are liberated in the cytoplasm as distinct complex after RanGTP hydrolysis. See the main text for details (From: Molecular & Biochemical Parasitology. 2019)

nucleus. The intrinsic GTPase activity of Ran is stimulated by a Ran-specific GTPase-activating protein (RanGAP1). Ran-binding protein 1 (RanBP1) or RanBP2 further stimulate RanGAP1-mediated GTPase activation. Consistent with a functional role in the cytoplasmic disassembly of transport-factor-RanGTP complexes, RanBP1 and RanGAP1 are excluded from the nucleus. The consequence of the location of these proteins is that RanGTP depletion occurs in the cytoplasm.

Altogether, cycles of nuclear export involving complexes containing RanGTP along with the import of complexes that do not include Ran would deplete Ran from the nucleus. Although Ran is small enough to diffuse through the NPC, efficient nuclear import of RanGDP is aided by nuclear transport factor 2 (NTF2). During its transport through the NPC channel, NTF2 interacts with FG repeats. Nuclear release of NTF2 and reconversion of RanGTP from RanGDP occurs with participation of the chromatin-bound guanine nucleotide exchange factor (GEF), which is identified as the product of gene *RCC*1 (regulator of chromosome condensation). GEF is a solely nuclear protein, which indicates RanGTP formation is a nuclear process. Especially, the cytoplasmic hydrolysis of RanGTP is aided by the cytoplasmic proteins RanGAP and RanBP1, and the nuclear conversion of RanGDP to RanGTP is stimulated by the nuclear protein GEF. As a consequence of the asymmetric distribution of these three proteins (RanGAP1, RanBP1, and GEF), there is a cellular gradient of RanGTP with a higher RanGTP concentration in the nucleus (Figure 8-7).

Hundreds of individual nuclear proteins or proteins that shuttle between the nucleus and the cytoplasm are thought to use this classic Imp α/β nuclear import pathway. Over the last couple of years, however, a number of proteins have been described that do not depend on the adapter protein Imp α. Some of these bind directly to the classic import receptor, Imp β, without requiring any adapter protein.

(3) A second type of NTR escorts proteins containing NESs out of the nucleus The main mechanism by which proteins exit the nucleus differs mechanistically from nuclear protein import. This mechanism was initially elucidated by studies of certain ribonuclear protein complexes that shuttle between the nucleus and the cytoplasm. Such shuttling proteins contain a Leu-rich **nuclear export signal** (NES) that stimulates their export from the nucleus to the cytoplasm through nuclear pores, in addition to an NLS that results in their uptake into the nucleus.

The mechanism whereby shuttling proteins are exported from the nucleus is best understood for those containing NES. According to the current model, shown in Figure 8-8, a specific NTR, called exportin 1, first forms a complex with RanGTP in the nucleus and then binds the NES in a cargo protein. Binding of exportin 1 to RanGTP causes a conformational change in exportin 1 that increases its affinity for the NES, so that a trimolecular cargo complex is formed. Like other NTRs, exportin 1 interacts transiently with FG-repeats in FG-Nups and diffuses through the NPC. The cargo complex dissociates when it encounters the RanGAP associated with the NPC cytoplasmic filaments, which stimulates Ran to hydrolyze the bound GTP, shifting it into a conformation that has low

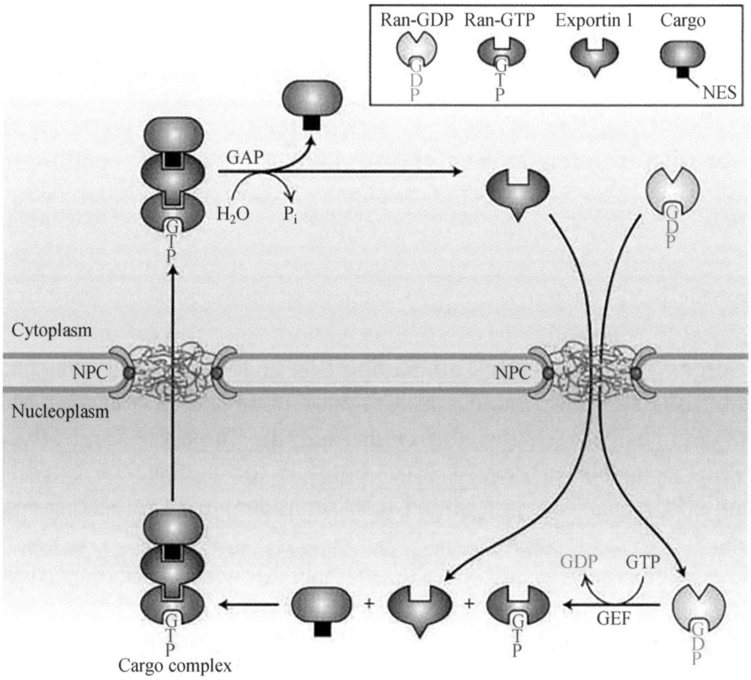

Figure 8-8 Ran-dependent nuclear export. In the nucleoplasm, exportin 1 binds cooperatively to the NES of a cargo protein and to RanGTP. After the resulting cargo complex diffuses through an NPC via transient interactions with FG-repeats, the GAP associated with the NPC cytoplasmic filaments stimulates GTP hydrolysis, converting RanGTP to RanGDP. The accompanying conformational change in Ran leads to dissociation of the complex. The cargo protein is released into the cytosol, whereas exportin 1 and RanGDP are transported back into the nucleus. RanGEF in the nucleus then stimulates conversion of RanGDP to RanGTP (From: Lodish H et al. 2016)

affinity for exportin 1. After RanGDP dissociates from the trimolecular cargo complex, exportin 1 changes its conformation to one that has low affinity for the NES, releasing the cargo into the cytosol. The direction of the export process is driven by this dissociation of the cargo from exportin 1 in the cytoplasm, which causes a concentration gradient of the cargo complex across the NPC that is high in the nucleoplasm and low in the cytoplasm. Exportin 1 and RanGDP are then transported back into the nucleus through the NPC (Figure 8-8).

By comparing this model for nuclear export with that in Figure 8-7 for nuclear import, we can see one obvious difference: RanGTP is part of the cargo complex during export, but not during import. Apart from this difference, the two transport processes are remarkably similar. In both processes, association of an NTR with RanGTP in the nucleus causes a conformational change that affects its affinity for the transport signal. During import, the interaction causes release of the cargo, whereas during export, the interaction promotes association with the cargo. In both export and import, stimulation of RanGTP hydrolysis in the cytoplasm by RanGAP produces a conformational change in Ran that releases the NTR. During nuclear export, the cargo is also released. Localization of RanGAP and RanGEF to the cytoplasm and nucleus, respectively, is the basis for the unidirectional import and export of cargo proteins across the NPC.

In keeping with their similarity in function, the two types of NTRs — importins and exportins are highly homologous in sequence and structure. The family of NTRs has 14 members in yeast and more than 20 in mammalian cells. The NESs or NLSs to which they bind have been determined for only a fraction of them. Some individual NTRs function in both import and export.

8.2.3 Transport of RNA between nucleus and cytoplasm

Like protein import, the export of an RNA species through the NPC is an active, energy-dependent process that requires the RanGTPase protein. A similar shuttling mechanism has been shown to export cargos from the nucleus. Exportins are also dedicated to RNA transit. For example, exportin-t functions to export fully mature tRNA with the 5′- and 3′-ends correctly processed. Exportin-t binds the processed tRNAs in a complex with RanGTP that diffuses through NPCs and dissociates when it interacts with RanGAP in the NPC cytoplasmic filaments, releasing the tRNA into the cytosol. The exportin is recycled back to the nucleus. Cargos containing NLS bind importins at low RanGTP concentrations in the cytoplasm and translocate to the nucleus as a dimeric complex. Association of RanGTP with the importin in the nucleus releases the cargo and the importin-RanGTP complex is re-exported. A Ran-dependent process is also required for the nuclear export of ribosomal subunits through NPCs once the protein and RNA components have been properly assembled in the nucleolus.

Likewise, certain specific mRNAs that associate with particular ribonu-cleoprotein (RNP) proteins can be exported by a Ran-dependent mechanism, but the majority of mRNAs are exported from the nucleus by a Ran-independent mechanism. Once the processing of an mRNA is completed in the nucleus, it remains associated with specific proteins in a messenger ribonuclear protein complex (mRNP). The principal transporter of mRNPs out of the nucleus is the mRNP exporter, a heterodimeric protein composed of a large subunit called nuclear export factor 1 (NXF1) and a small subunit called nuclear export transporter 1 (NXT1). NXF1 is the major driver of interaction between the export mRNP and the NPC. Multiple NXF1/NXT1 dimers bind to nuclear mRNPs through cooperative interactions with the RNA and other mRNP adapter proteins that associate with nascent pre-mRNAs during transcriptional elongation and pre-mRNA processing. In many respects, the subunits of NXF1/NXT1 act like an NTR that binds to an NLS or NES, since both subunits interact with the FG-repeats of FG-Nups, and this interaction allows them to diffuse through the central channel of the NPC. At the cytoplasmic face, the cargo mRNPs are released and the export factors are recycled (Figure 8-9).

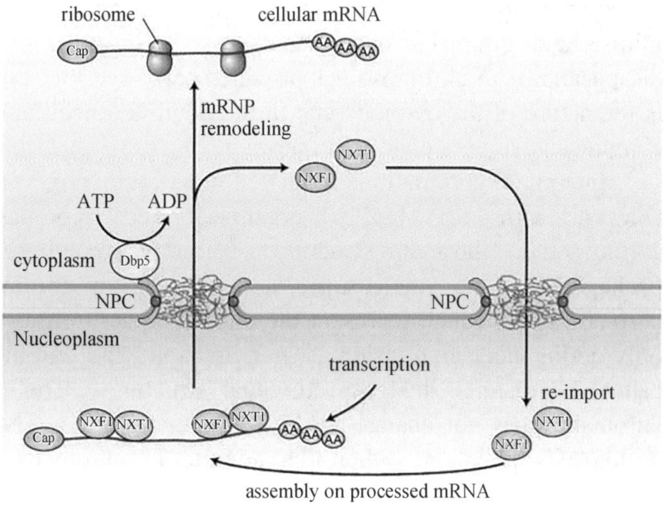

Figure 8-9 Ran-independent nuclear export of mRNAs. The heterodimeric NXF1/NXT1 complex binds to mRNPs in the nucleus. NXF1/NXT1 acts as an NTR and directs the associated mRNP to the central channel of the NPC by transiently interacting with FG-Nups. The Dbp5 located on the cytoplasmic side of the NPC removes NXF1 and NXT1 from the mRNA in a reaction that is powered by ATP hydrolysis (From: Lodish H, et al. 2016)

The process of mRNP export does not require Ran, and thus the unidirectional transport of mRNA out of the nucleus requires a source of energy other than GTP hydrolysis by Ran. Once the mRNP-NXF1/NXT1 complex reaches the cytoplasmic side of the NPC, NXF1 and NXT1 dissociate from the mRNP with the help of the RNA helicase Dbp5, which is associated with cytoplasmic NPC filaments. The RNA helicases use the energy derived from hydrolysis of ATP to move along RNA molecules, separating dsRNA chains and dissociating RNA-protein complexes. This leads to the simple idea that Dpb5, which is associated with the cytoplasmic side of the NPC, acts as an ATP-driven motor to remove NXF1/NXT1 from the mRNP complexes as they emerge on the cytoplasmic side of the NPC. The assembly of NXF1/NXT1 onto mRNPs on the nucleoplasmic side of the NPC and the subsequent ATP-dependent removal of NXF1/NXT1 from mRNPs on the cytoplasmic side of the NPC creates a concentration gradient of mRNP-NXF1/NXT1, which drives unidirectional export. After being removed from the mRNP, the free NXF1 and NXT1 subunits that have been stripped from the mRNA by Dbp5 helicase are imported back into the nucleus by a process that depends on Ran and an NTR (Figure 8-9).

In Ran-dependent nuclear export, hydrolysis of GTP by Ran on the cytoplasmic side of the NPC causes dissociation of the NTR from its cargo. In basic outline, the Ran-independent nuclear export discussed here operates by a similar mechanism, except that Dbp5 on the cytosolic side of the NPC uses hydrolysis of ATP to dissociate the mRNP exporter from mRNA.

8.3 Chromatin and chromosomes

Eukaryotic DNA is complexed with an array of acidic and basic histone proteins into thin fibers. During nondivisional phases of the cell cycle, these fibers are uncoiled and dispersed into **chromatin**. During cell division, chromatin fibers coil and condense into structures called **chromosomes**. Chromosomes seem to appear out of nowhere at the beginning of mitosis and disappear once again when cell division has ended. The appearance and disappearance of chromosomes provided early cytologists with a challenging question: What is the nature of the chromosome in the nonmitotic cell? We are now able to provide a fairly comprehensive answer to this question.

Interphase chromatin is a tangled mass occupying a large part of the nuclear volume, in contrast with the highly organized and reproducible ultrastructure of mitotic chromosome. Chromatin structure is hierarchic, ranging from the two lowest levels of DNA packaging — the **nucleosome** and the 30-nm chromatin fiber — to the metaphase chromosomes, which represent the most compact form of chromatin in eukaryotes and occur only during nuclear division. After division, the chromosomes become less compact and cannot be distinguished as individual structures. The global structure of the interphase chromatin does not change visibly between divisions. No disruption is evident during the period of replication, when the amount of chromatin doubles. Chromatin is fibrillar, although the overall configuration of the fiber in space is hard to discern in detail. The fiber itself, however, is similar or identical to that of the mitotic chromosomes.

8.3.1 Nucleosomes: the lowest level of chromosome organization

In eukaryotic chromatin, a substantial amount of protein is associated with the acidic sugar-phosphate backbone of DNA in all phases of the cell cycle. The associated proteins are divided into basic, positively charged histones and less positively charged nonhistones. The major proteins of chromatin are the histones — small proteins containing a high proportion of basic amino acids (Arg and Lys) that facilitate binding to the negatively charged DNA molecule. There are five major types of histones — called H1, H2A, H2B, H3, and H4 — which are very similar among different species of eukaryotes. In addition, chromatin contains an approximately equal mass of a wide variety of nonhistone chromosomal proteins.

In the early 1970s, it was found that when chromatin was treated with nonspecific nucleases, most of the DNA was converted to fragments of approximately 200 bp in length. In contrast, a similar treatment of naked DNA produced a randomly sized population of fragments. This finding suggested that chromosomal DNA was protected from enzymatic attack, except at certain periodic sites along its length. It was presumed that the proteins associated with the DNA were providing the protection. In 1974, using the data from nuclease digestion and other types of information, Roger Kornberg proposed an entirely new structure for chromatin. He proposed that DNA and histones are organized into repeating subunits, called **nucleosome**, the central unit of chromatin. When interphase nuclei are broken open very gently with nuclease and their contents examined under electron microscope, most of the chromatin is in the form of a fiber with a diameter of about 30 nm (Figure 8-10A). If this chromatin is subjected to treatments that cause it to unfold partially, it can then be seen under the electron microscope as a series of "beads on a string". The string is connecting (linker) DNA, and each bead is a nucleosome core particle (10 nm in diameter) that consists of DNA wound around a core of proteins formed from histones. Each nucleosome particle is connected by ~60 bp of linker DNA and forms a fiber with structurally repetitive motifs of ~200 bp that is referred to as the 10-nm fiber or the beads on a string (Figure 8-10B). The complex of the 10-nm fiber associated with various nonhistone proteins is together known as chromatin.

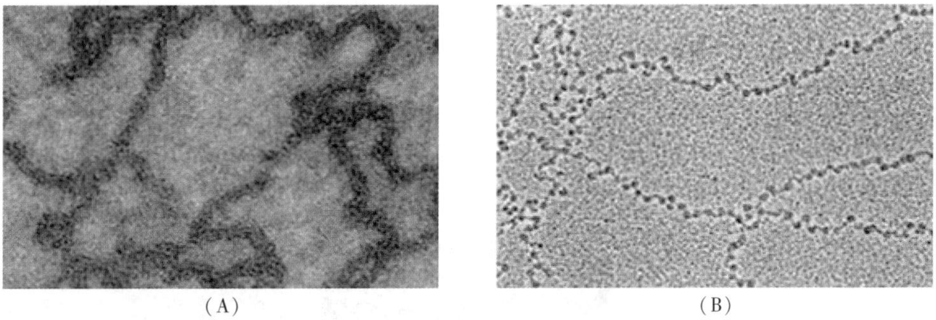

Figure 8-10 The condensed and extended forms of extracted chromatin have very different appearances in electron micrographs. (A) Chromatin isolated in buffer with a physiological ionic strength (0.15 M KCl) appears as a condensed fiber 30 nm in diameter. (B) Chromatin isolated in low-ionic-strength buffer has an extended beads-on-a-string appearance (From: Lodish H. 2016)

We now know that each nucleosome consists of a protein core with DNA wound around its surface like thread around a spool. The core is an octamer containing two copies each of histones H2A, H2B, H3, and H4. Nucleosomes from all eukaryotes contain about 147 bp of DNA wrapped one and two-thirds turns in a left-handed coil around the globular histone core (Figure 8-11). The length of the linker DNA is more variable among species, and even between different cells of the same organism, ranging from about 10 to 90 bp. During cell replication, DNA is assembled into nucleosomes shortly after the replication fork passes. This process depends on specific chaperone molecules that bind to histones and assemble them, together with newly replicated DNA, into nucleo-somes.

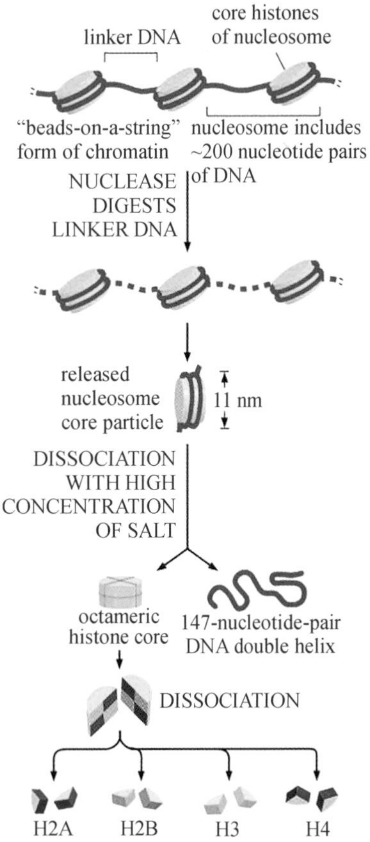

Figure 8-11 Structural organization of the nucleosome. In biochemical experiments, the nucleosome core particle can be released from isolated chromatin by digestion of the linker DNA with a nuclease. After dissociation of the isolated nucleosome into its protein core and DNA, the length of the DNA that was wound around the core can be determined (From: Alberts B, et al. Molecular Biology of the Cell. 2015)

Our understanding of DNA packaging has been greatly advanced by portraits of the nucleosome core particle obtained by X-ray crystallography. The eight histone molecules that comprise a nucleosome core particle are organized into four heterodimers: two H2A-H2B dimers and two H3-H4 dimers (Figure 8-12A, B). Dimerization of histone molecules is mediated by their C-terminal domains, which consist largely of α helices (represented by the cylinders in Figure 8-12C) folded into a compact mass in the core of the nucleo-some. In contrast, the N-terminal segments from each core histone (and also the C-terminal segment of H2A) take the form of a long, flexible tail (represented by the dashed lines of Figure 8-12C) that extends out through the DNA helix that is wrapped around the core particle. For many years, histones were thought of as inert, structural molecules, but the extending N-terminal segments are targets of key enzymes that play a role in making the chromatin accessible to proteins. In this way, chromatin is a dynamic cellular component in which histones, regulatory proteins, and a variety of enzymes move in and out of the nucleoprotein complex to facilitate the complex tasks of DNA transcription, compaction, replication, recombination, and repair.

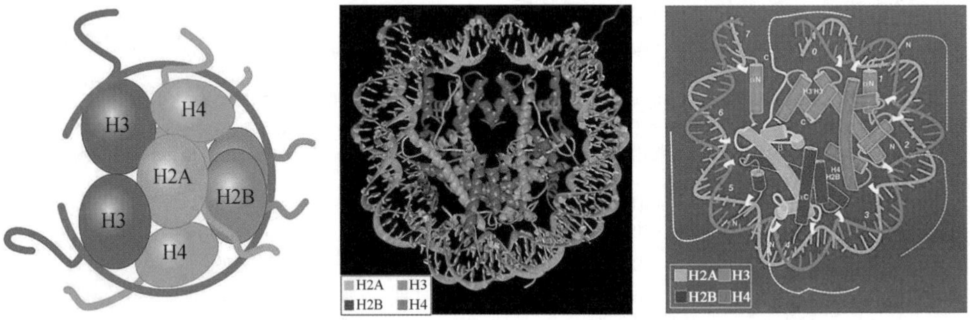

(A) (B) (C)

Figure 8-12 The structure of a nucleosome. (A) Schematic representation of a nucleosome core particle with its histone octamer composed of four histone heterodimers. (B) X-ray crystallographic structure of a nucleosome core particle viewed down the central axis of the DNA superhelix, showing the position of each of the eight histone molecules of the core octamer. The histones are organized into four dimeric complexes. Each histone dimer binds 27 to 28 bp, with contacts occurring where the minor groove of the DNA faces the histone core. (C) A simplified, schematic model of half of a nucleosome core particle, showing one turn (73 bp) of the DNA superhelix and four core histone molecules. The four different histones are shown in separate colors, as indicated by the key. Each core histone is seen to consist of (1) a globular region, called the histone fold, consisting of three α helices (represented by the cylinders) and (2) a flexible, extended N-terminal tail (indicated by the letter N) that projects out of the histone disk and out past the DNA double helix. The intermittent points of interaction between the histone molecules and the DNA are indicated by white hooks. The dashed lines indicate the outermost portion of the histone tails, which are sites of modification (From: Karp G, et al. 2016)

As a reflection of their fundamental role in DNA function through controlling chromatin structure, the histones are among the most highly conserved eukaryotic proteins. For example, the amino acid sequence of histone H4 from a pea differs from that of a cow at only 2 of the 102 positions. This strong evolutionary conservation suggests that the functions of histones involve nearly all of their amino acids, so that a change in any position is deleterious to the cell. But in addition to this remarkable conservation, eukaryotic organisms also produce smaller amounts of specialized variant core histones that differ in amino acid sequence from the main ones. These variants, combined with the surprisingly large number of covalent modifications that can be added to the histones in nucleosomes, give rise to a variety of chromatin structures in cells.

8.3.2 Chromatin remodeling complexes alter nucleosome structure

Chromatin remodeling complexes include a subunit that hydrolyzes ATP. This subunit binds both to the protein core of the nucleosome and to the dsDNA that winds around it. By using the energy of ATP hydrolysis to move this DNA relative to the core, the protein complex changes the structure of a nucleosome temporarily, making the DNA less tightly bound to the histone core. Through repeated cycles of ATP hydrolysis that pull the nucleosome core along the DNA double helix, the remodeling complexes can catalyze nucleosome sliding. In this way, they can reposition nucleosomes to expose specific regions of DNA, thereby making them available to other proteins in the cell. In the example depicted in Figure 8-13, path 2, the DNA has formed a transient loop or bulge on the surface of the histone octamer, making that site more accessible for interaction with DNA-binding regulatory proteins. In addition, by cooperating with a variety of other proteins that bind to histones and serve as histone chaperones, some remodeling complexes are able to remove either all or part of the nucleosome core from a nucleosome — catalyzing either an exchange of its H2A-H2B histones, or the complete removal of the octameric core from the DNA. As a result of such processes, measurements reveal that a typical nucleosome is replaced on the DNA every one or two hours inside the cell (Figure 8-13).

Eukaryotic cells contain dozens of different chromatin remodeling complexes that are specialized for different roles. Most are large protein complexes that can contain 10 or more subunits, some of which bind to specific modifications on histones. The activity of these complexes is carefully controlled by the cell. As genes are turned on and off, remodeling complexes are brought to specific regions of DNA where they act locally to influence chromatin structure.

Although some DNA sequences bind more tightly than others to the nucleosome core, the most important influence on nucleosome positioning appears to be the presence of other tightly bound proteins on the DNA. Some bound proteins favor the formation of a nucleosome adjacent to them. Others create obstacles that force the nucleosomes to move elsewhere. The exact positions of nucleosomes along a stretch of DNA therefore depend mainly on the presence and nature of other proteins bound to the DNA. And due to the presence of ATP-dependent chromatin remodeling complexes, the arrangement of nucleosomes on DNA can be highly dynamic, changing rapidly according to the needs of the cell.

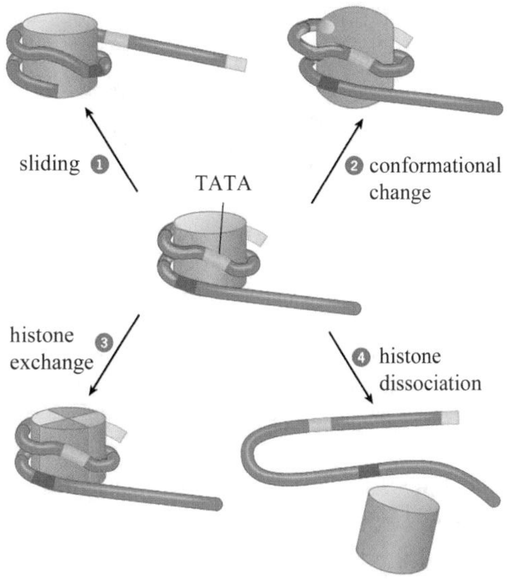

Figure 8-13 Chromatin remodeling. In pathway 1, a key nucleosome slides along the DNA, thereby exposing the TATA binding site and allowing the preinitiation complex to assemble. In pathway 2, the histone octamer of a nucleosome has been reorganized. Although the TATA box is not completely free of histone association, it is now able to bind the proteins of the preinitiation complex. In pathway 3, the standard H2A/H2B dimers of a nucleosome have been exchanged with histone variants (e. g., H2A. Z/H2B dimers) that associate with active chromatin. In pathway 4, the histone octamer has been disassembled and is lost from the DNA entirely (From: Karp G, et al. 2016)

8.3.3 Flexibility of nucleosome and chromatin structures

The high-resolution structure of a nucleosome core particle, solved in 1997, revealed that the histone core is a roughly disk-shaped structure made of interlocking histone subunits. Nucleosomes are connected by linker DNA to form nucleosomal arrays like a "beads-on-a-string." The arrayed nucleosomes are further condensed by the linker histone into the 30-nm chromatin fiber, which is typically regarded as the secondary structure of chromatin. The dynamic change and regulation of chromatin fiber, e. g., the transitions between the accessible form of nucleosomal arrays and the highly compacted form of 30-nm chromatin fiber, play a critical role in gene regulation.

Despite intense efforts during the past three decades, the existence of the 30-nm chromatin fiber *in vivo* remains controversial. Recently, Chinese scholars developed the chromatin *in vitro* reconstitution and structural analysis system with EM/cryo-EM techniques, and obtained cryo-EM structures of 30-nm chromatin fibers reconstituted in the presence of linker histone H1 and with different nucleosome repeat lengths (Figure 8-14). The asymmetric binding and the location of histone H1 in chromatin play a role in the formation of the 30 nm fiber. These results provide mechanistic insights into how nucleosomes compact into higher-order chromatin fibers.

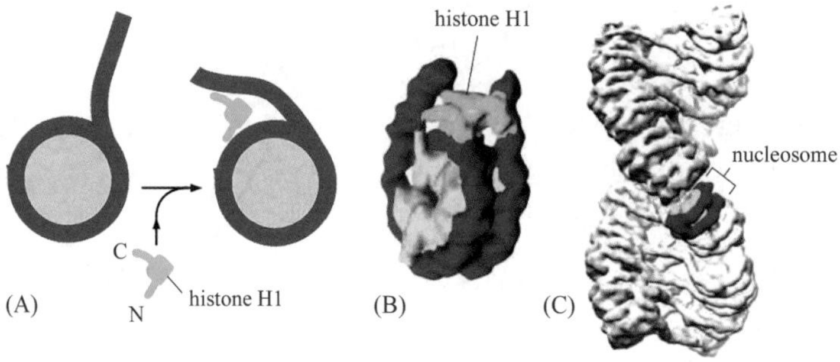

Figure 8-14 How the linker histone binds to the nucleosome. The position and structure of histone H1 is shown. The H1 core region constrains an additional 20 nucleotide pairs of DNA where it exits from the nucleosome core and is important for compacting chromatin. (A) Schematic, and (B) structure inferred for a single nucleosome from a structure determined by high-resolution electron microscopy of a reconstituted chromatin fiber (C) (From: Song F. et al. Science, 2014)

One of the powerful techniques in cryo-EM is to computationally separate heterogeneous or flexible regions of proteins. Using this strategy, a mono-nucleosome was reconstituted and reconstructed into several structurally district dynamic states (Figure 8-15A). These structures show the rearrangements of histones during the unwrapping of the nucleosomal DNA. For example, the αN helix of Histone H3 moves toward unwrapped DNA while still being bound to the DNA (Figure 8-15B). This arrangement is expected to stabilize nucleosomal DNA during unwrapping to prevent further detachment and DNA opening. Although DNA unwrapping occurs near the αN helix of Histone H3, the helices of H2A-H2B are also rearranged to stabilize the altered nucleosome structure, probably to prevent further DNA unwrapping. In accordance with the findings from this study, the nucleosome is not a solid and static structure, but flexible enough to accommodate some degree of alteration in the structure, such as DNA unwrapping.

In addition to the mono-nucleosome structure, the structures of higher order nucleosomes, such as di-nucleosome and nucleosome fibers, were also investigated using cryo-EM (Figure 8-15C). In the cryo-EM structure of HP1 dimer, two HP1 proteins bridge two adjacent nucleosomes with major contacts with histone H3. Unexpectedly, a linker DNA between di-nucleosome in the HP1 di-nucleosome complex does not seem to interact with HP1α or histones and remains exposed to the solvent (crosslinking). In the HP1α-dinucleosome complex with a 15-bp linker DNA, the MNase susceptibility of the linker DNA was maintained in the presence of HP1α. In contrast, the linker histone H1 completely protected the linker DNA from the MNase attack in the dinucleosome and in the HP1α-dinucleosome complex. These results indicate that the HP1α binding did not conceal the linker DNA in the HP1α-dinucleosome complex, in contrast to the histone H1 binding. The linker DNA exposure detected in the present structure is consistent with the

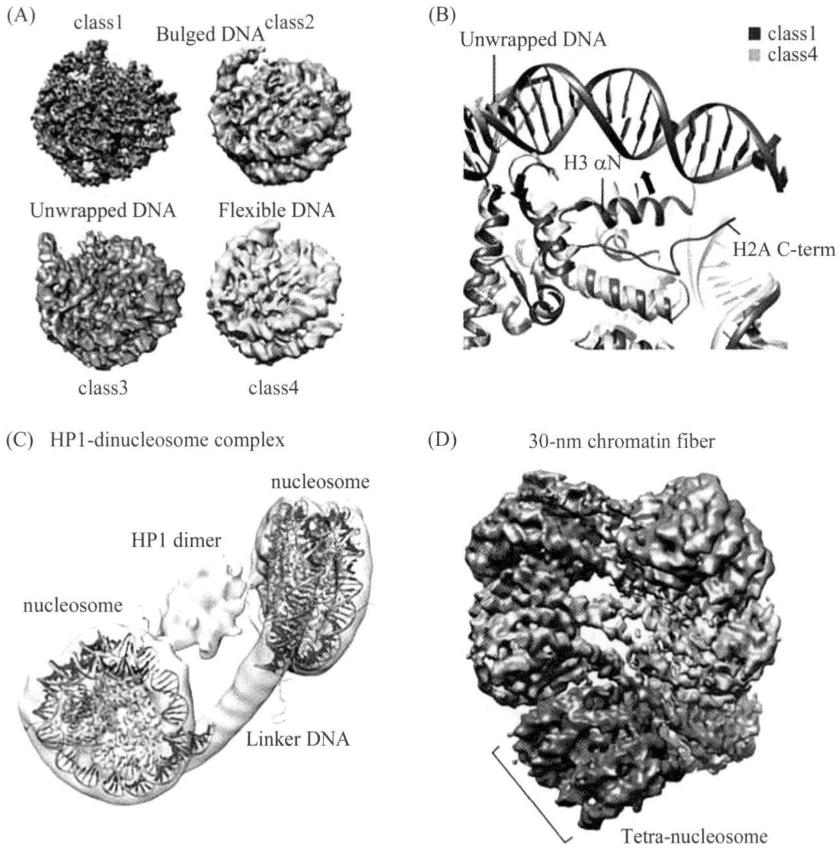

Figure 8-15 Dynamics of nucleosome and its higher order structures. (A) Four cryo-EM maps of nucleosome core particle showing dynamic nature of nucleosome. Class 1 presents canonical structure of nucleosome core particle. Classes 2, 3, and 4 show the destabilization of nucleosome core particles. (B) A comparison of nucleosome structures of the class 1 and the class 4 in (A). The C-terminal region of H2A is delocalized in the class 4. The αN helix of histone H3 moved toward the detached DNA. (C) Cryo-EM map of HP1 dimer bound to di-nucleosome. (D) Cryo-EM map of 30-nm chromatin fiber including 12 nucleosomes. Each tetranucleosomal unit is presented in green, orange, and pink (From: Current Opinion in Structural Biology. 2019)

finding that the linker histone H1 efficiently binds to nucleosomes bound to HP1. The configuration of the HP1-dinucleosome structure may change when the linker histone H1 binds to the complex. Such a structural change by histone H1 may allow the HP1 dimer to bind the linker DNA in the HP1-dinucleosome complex.

The structure obtained by X-ray crystallography and high-resolution EM of reconstituted chromatin shows a histone H1-dependent left-handed twist of the repeating tetranucleosomal structural units, within which the four nucleosomes zigzag back and forth with a straight linker DNA. Four nucleosomes move in a zigzag manner to make two stacks of two nucleosomes, consistent with the two-start model based on the crystal structure of the tetranucleosome (Figure 8-15D). Each tetranucleosomal unit is stacked on top of another through an internucleosomal electrostatic interaction, where the positively charged N-terminus of histone H4 from one tetranucleosome contacts the H2A-H2B acidic patch from the other. This electrostatic interaction suggests that the structure of the chromatin fiber might be altered in a manner depending on local ionic strength *in vivo*.

What causes nucleosomes to stack so tightly on each other? Nucleosome-to-nucleosome linkages that involve histone tails, most notably the H4 tail, constitute one important factor. Another important factor is an additional histone that is often present in a 1-to-1 ratio with nucleosome cores, i. e., histone H1. This so-called linker histone is larger than the individual core histones and it has been considerably less well conserved during evolution. A single histone H1 molecule binds to each nucleosome, contacting both DNA and protein, and changing the path of the DNA as it exits from the nucleosome. This change in the exit path of DNA is thought to help compact nucleosomal DNA (Figure 8-16).

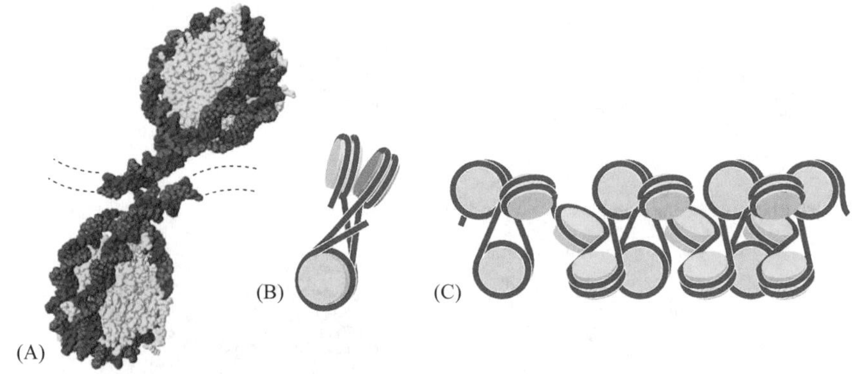

Figure 8-16 A zigzag model for the 30-nm chromatin fiber. (A) The conformation of two of the four nucleosomes in a tetranucleosome. (B) Schematic of the entire tetranucleosome; the fourth nucleosome is not visible, being stacked on the bottom nucleosome and behind it in this diagram. (C) Diagrammatic illustration of a possible zigzag structure that could account for the 30-nm chromatin fiber (From: Alberts B, et al. 2015)

Despite the many structural studies of chromatin, we are far from understanding the whole picture of the organization of higher order DNA structures and many outstanding questions remain, such as how the chromatin structure affects gene expression, how the chromatin structure itself is regulated and how DNA forms a higher-order structure without any tangling problems. Integrated structural approaches combined with cell biology and genomic analysis are needed to address these questions.

8.3.4 The structure of a mitotic chromosome

Having discussed the dynamic structure of interphase chromosomes, we now turn to mitotic chromosomes. The chromosomes from nearly all eukaryotic cells become readily visible by light microscopy during mitosis, when they coil up to form highly condensed structures. This condensation reduces the length of a typical interphase chromosome only about tenfold, but it produces a dramatic change in chromosome appearance. Mitotic chromosome condensation can thus be thought of as the final level in the hierarchy of chromatin packaging; 1 μm of mitotic chromosome length typically contains approximately 1 cm of DNA, which represents a packing ratio of 10,000 : 1. This compaction occurs by a poorly understood process. An overview of the various levels of chromatin organization, from the nucleosomal filament to a mitotic chromosome, is depicted in Figure 8-17. The chromatin in chromosomal regions that are not being transcribed or replicated exists predominantly in the condensed, 30-nm fiber form and in higher-order folded structures whose detailed conformation is not currently understood. The regions of chromatin actively being transcribed and replicated are thought to assume the extended beads-on-a-string form.

As the result of compaction, the chromosomes of a mitotic cell appear as distinct, rod-like structures (Figure 8-18A). Close examination of mitotic chromosomes reveals each of them to be composed of two mirror-image, sister chromatids, which are a result of replication in the previous interphase. Prior to replication, the DNA of each interphase

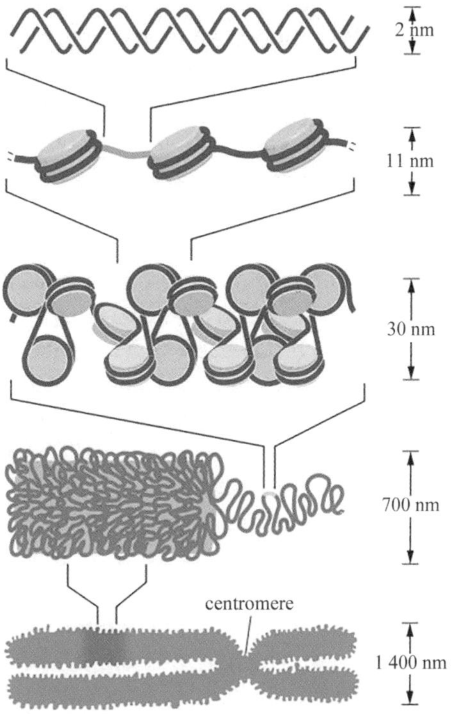

Figure 8-17 Chromatin packing. This model shows some of the many levels of chromatin packing postulated to give rise to the highly condensed mitotic chromosome (From: Alberts B, et al. 2015)

chromosome becomes associated at sites along its length with a multiprotein complex called cohesin (Figure 8-18B). Just after replication, the DNA helices of a pair of sister chromatids would be held in association by cohesion molecules that encircled the sister DNA helices. Cohesin molecules would continue to hold the DNA of sister chromatids together when they are ultimately separated.

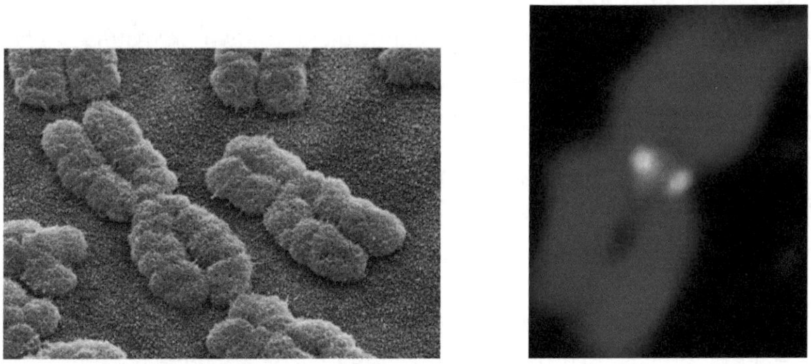

(A) (B)

Figure 8-18 Each mitotic chromosome is comprised of a pair of sister chromatids connected to one another by the protein complex cohesin. (A) Colorized scanning electron micrograph of several metaphase chromosomes showing the paired identical chromatids associated loosely along their length and joined tightly at the centromere. The chromatids are not split apart from one another until anaphase. (B) Fluorescence micrograph of a metaphase chromosome in a cultured human cell. The DNA is stained blue, the kinetochores are green, and cohesin is red. At this stage of mitosis, cohesin has been lost from the arms of the sister chromatids but remains concentrated at the centromeres where the two sisters are tightly joined (From: Karp G, et al. 2016)

The compaction of chromosomes during mitosis is a highly organized and dynamic process that serves at least two important purposes. First, when condensation is complete (in metaphase), sister chromatids have been disentangled from each other and lie side by side. Thus, the sister chromatids can easily separate when the mitotic apparatus begins pulling them apart. Second, the compaction of chromosomes protects the relatively fragile DNA molecules from being broken as they are pulled to separate daughter cells.

Although eukaryotic chromosomes differ in length and number among species, cytogenetic studies have shown that they all behave similarly at the time of cell division. Moreover, any eukaryotic chromosome must contain three functional elements in order to replicate and segregate correctly: **Centromere**, the constricted region required for proper segregation of daughter chromosomes; **Telomeres**, special telomeric sequences at the ends function in preventing chromosome shortening; and **Replication origins** at which DNA polymerases and other proteins initiate synthesis of DNA.

(1) Centromere The centromere is a specialized region of the chromosome that plays a critical role in ensuring the correct distribution of duplicated chromosomes to daughter cells during cell division. Chromosome fragments that lack a centromere (acentric fragment) do not become attached to the spindle, and so fail to be included in the nuclei of either of the daughter cells. The centromeres thus serve both as the sites of association of sister chromatids and as the attachment sites for microtubules of the mitotic spindle. They consist of specific DNA sequences to which a number of centromere-associated proteins bind, forming a specialized structure called **kinetochore.** Kinetochores mediate attachment between chromosomes and microtubules, assemble on each sister chromatid at their centromere. Proteins associated with the kinetochore then act as "molecular motors" that drive the movement of chromosomes along the spindle fibers, segregating the chromosomes to daughter nuclei (see Chapter 9).

In eukaryotic cells, the centromere is comprised of a complex of specific DNA sequences. These sequences can vary enormously in size; in budding yeast it is about 125 bp, whereas in plants and animals, centromeres are megabases in length and are composed of multiple repeats of simple-sequence DNA. In humans, centromeres contain 2 ~ 4 Mb arrays of a 171 bp simple-sequence DNA called α-satellite DNA, which appears to play a role in centromere structure and function.

S. cerevisiae has by far the simplest centromere known in nature. The centromere sequences of the well-studied yeast consist of three regions (Ⅰ, Ⅱ, and Ⅲ), which are conserved among the centromeres on different yeast chromosomes. Short, fairly well-conserved nucleotide sequences are present in regions Ⅰ and Ⅲ. Region Ⅱ does not have a specific sequence, but is AT-rich with a fairly constant length, probably so that regions Ⅰ and Ⅲ will lie on the same side of a specialized centromere-associated histone octamer. This specialized centromere-associated histone octamer contains the usual histones H2A, H2B, and H4, but a variant form of histone H3. Centromeres from all eukaryotes similarly contain nucleosomes with a specialized, centromere-specific form of histone H3, called CENP-A in mammals. Chromosomes lacking CENP-A fail to assemble a kinetochore and are lost during cell division.

(2) Telomere Another essential feature in all chromosomes is telomeres, which are the physical ends of eukaryotic linear chromosomes. Telomeres form special structures that cap chromosome ends to prevent degradation by nucleolytic attack and to distinguish chromosome termini from DNA double-strand breaks. With few exceptions, telomeres are composed primarily of repetitive DNA associated with proteins that interact specifically with double- or single-stranded telomeric DNA or with each other, forming highly ordered and dynamic complexes involved in telomere maintenance and length regulation. Linear chromosomes present two major biological hurdles at chromosome ends: the end replication and end protection problems. The end-replication problem exists, because DNA polymerases are unable to fully replicate chromosome ends. This results in the loss of DNA at the very ends of chromosomes during each replication cycle. The specialized ribonucleoprotein enzyme telomerase counteracts this DNA attrition by synthesizing new telomeric repeats at chromosome ends. The protein subunit of telomerase, called telomerase reverse transcriptase (TERT), contains a catalytic reverse transcriptase domain for DNA synthesis. The RNA subunit of telomerase called telomerase RNA component (TR) contains, amongst other elements, the template for telomeric repeat addition.

The composition of shelterin complex. Telomerase is expressed in both germ line and somatic stem cells to facilitate their continued cell division. It is, therefore, not surprising that mutations in telomere- and telomerase-associated genes result in inherited stem cell-dysfunction diseases collectively known as "telomeropathies". Normally, nondividing cells lacking telomerase enter a non-proliferative state called senescence once their telomere length falls below a certain threshold. This is an anti-tumorigenic mechanism to prevent uncontrolled cell division. However, if a rare cell escapes senescence and aberrantly resumes telomerase expression to re-establish telomere length maintenance, it could attain "replicative immortality", a hallmark of cancer. In fact, an overwhelming majority of cancers (~80%) show telomerase expression, and thus, this enzyme is a promising target for anti-cancer drug discovery.

While telomerase solves the end-replication problem, linear chromosomes must also solve the end protection problem. This problem occurs when the natural ends of linear chromosomes are misrecognized by the DNA damage response and repair machinery, as double strand breaks requiring repair. The six-protein complex **shelterin**, consisting of proteins POT1, TPP1, TIN2, TRF1, TRF2, and Rap1, solves the end protection problem by coating telomeric DNA (Figure 8-19). While TRF1 and TRF2 recognize ds telomeric DNA, POT1 binds the ss overhang with high specificity and affinity. By coating telomeric DNA, the shelterin complex sequesters it away from the ATM and ATR-mediated DNA damage response pathways. Although continuously dividing cells must protect chromosome ends from the DNA damage response machinery, they must also allow recruitment of telomerase to those ends. This is the essence of the telomerase recruitment problem, which is primarily solved by the telomere protein TPP1. Finally, telomeres also perform an essential function in meiosis. During prophase I in meiosis, homologous chromosomes pair together and undergo DNA recombination, a process central to generating genetic diversity. Telomeres play an essential role in this process by attaching chromosomes to the INM, and facilitating homolog pairing and genetic crossover. Telomere-INM engagement is achieved by interaction of the shelterin component TRF1 with the meiosis-specific complex that localizes at the INM.

Protection of ssDNA at chromosome ends. The very end of any eukaryotic chromosome is single-stranded and thus a potential substrate for illicit homology-driven recombination or repair. Thus, a major challenge in end protection involves preventing this DNA from participating in such processes. The G-rich 3′ ss overhang of telomeric DNA is also involved in end replication, as it provides the site for telomerase to bind and extend chromosome ends. Both of these functions are either directly dictated by (in case of end protection) or facilitated by (in case of end replication) proteins that bind the ss overhang

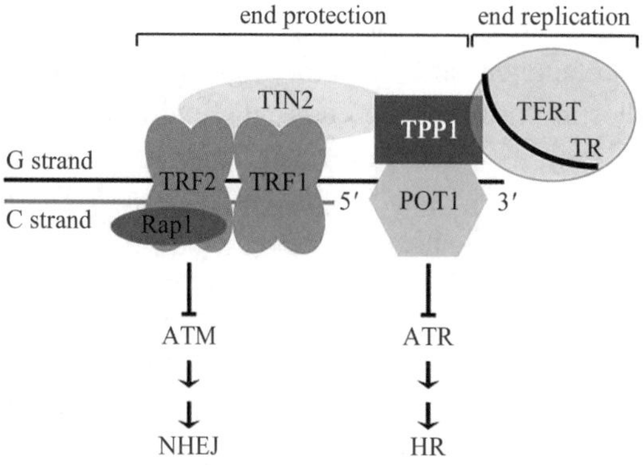

Figure 8-19 Schematic showing the composition of complexes involved in chromosome end protection and end replication. The shelterin complex protects chromosome ends from ATM kinase and ATR kinase-mediated DNA damage response pathways. The telomerase RNP adds telomeric DNA *de novo* at the 3′ ends of chromosomes (From: Cellular and Molecular Life Sciences. 2020)

(Figure 8-19).

The high affinity and specificity of POT1 for ss telomeric DNA provide an elegant mechanism for protection against an ATR kinase-mediated response and subsequent homologous recombination (HR) at chromosome ends. An essential initial step for HR is binding of the participating ssDNA with replication protein A (RPA), a heterotrimer composed largely of oligosaccharide/oligonucleotide binding (OB) domains. By binding ss telomeric DNA with high affinity and sequence specificity, POT1 blocks access of RPA to prevent ATR activation at chromosome ends.

Faithful protection of ss chromosome ends not only mandates POT1 to outcompete RPA, but also requires POT1 to avoid binding to TERRA, a non-coding ss G-rich RNA that is transcribed from and localized to the telomeres. TERRA contains many potential POT1 binding sites and is more abundant than the ss overhang. Yet, POT1 is able to avoid this RNA decoy and selectively bind to telomeric DNA. The preference for DNA over RNA of POT1 is further increased by binding of TPP1, although the structural basis for this remains unknown.

Double-stranded chromosome end protection. While sequestration of the ss G-rich overhang is an essential part of end protection, telomeric DNA is mostly double-stranded in nature. Therefore, a major task of shelterin is to coat and protect ds telomeric DNA. TRF1 and TRF2 constitute the dsDNA-binding proteins of shelterin. Both proteins have a similar domain layout, with each having a dimerization domain and a C-terminal DNA-binding myb domain. At the N-terminus, TRF1 contains a region rich in acidic residues, while TRF2 contains a highly basic stretch of amino acids known as the basic domain or GAR (Gly/Arg-rich) region. TRF1 and TRF2 assemble the remaining shelterin proteins (TIN2, TPP1, POT1, and Rap1) qualifying these as critical determinants of chromosome end protection. Given their importance in maintaining genome stability, it is not surprising that knock out of either TRF1 or TRF2 results in mouse embryonic lethality. TRF1 aids DNA replication through the G-rich repetitive telomeric DNA sequences. As a result, TRF1 deficiency leads to the characteristic "fragile telomere" phenotype. TRF2 plays a pivotal role in end protection by preventing activation of the ATM kinase DNA damage response (Figure 8-19). TRF2 also facilitates formation of telomeric structures known as t-loops, which bury the ss telomeric DNA end presumably to help prevent its recognition by the DNA damage response machinery.

Chromosome end replication by telomerase. Although telomerase does not play a direct role in end protection, it does prevent the severe erosion of telomeres (in continuously dividing cells) that could ultimately result in genome instability. Telomerase is a processive enzyme, adding multiple telomeric repeats per replication cycle, which is a property unique to telomerase and absent in any other known DNA/RNA polymerase and its activity is tightly controlled at multiple levels, from the expression of telomerase components to the appropriate functioning of assembly factors required to recruit and activate telomerase at the telomeres. Defects in any of these process lead to loss of telomeric repeats or the maintenance of short telomcres.

Telomerase is recruited to the telomeres via the shelterin or other telomere located proteins (TPP1-TIN2 in mammals). The recruited telomerase relocates to the end of a single-stranded G-overhang by hybridizing with its RNA template, allowing the reverse transcriptase component to add new repeats. The interaction between telomerase and the shelterin and/or the formation of the telomeric DNA/telomerase RNA hybrid retain telomerase at the telomere, permitting the processive extension of telomeres. Telomerase preferentially targets the shortest telomeres for extension, which ensures that an average telomere length is maintained across all chromosomes. Telomere elongation is observed only during S phase. At the end of S phase, telomerase access is blocked by the DNA-replication complex CST (CTC1, STN1 and TEN1 in mammals). In humans, the CST complex is able to act as terminator of telomerase activity and is responsible for recruiting polymerase-α-primase complex for fill-in synthesis of the C-strand after telomerase extension of the G-strand. Outside S phase, telomeres form a closed structure, blocking telomerase access known as a non-extendible state. This is possibly via the accumulation of suppressive proteins that bind the ds telomeric DNA (the classical counting mechanism), the folding of the 3′ G-overhang by the shelterin bridge (between ss and ds telomeres), or via a t-loop formation, in which 3′ G-overhang end displaces and anneals to its ds telomere (Figure 8-20). All these closed formation models involve the shelterin proteins.

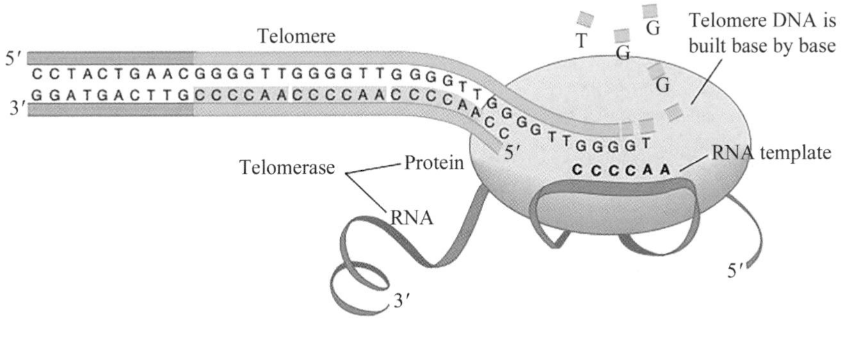

Figure 8-20 Telomerase operates at the end of chromosome. It is a specialized ribonucleoprotein enzyme consisting of a protein subunit and an RNA subunit. The RNA serves as a template for synthesizing telomere DNA.

In short, telomeres are considered so important that three scientists were awarded a Nobel Prize in Physiology or Medicine for the structure and function of telomeres in 2009.

(3) Replication origins DNA synthesis of daughter strands starts at discrete sites, termed replication origins, and proceeds in a bidirectional manner until all genomic DNA is replicated. From the single sequence-specific DNA replication origin of the small *E. coli* genome, up to thousands of origins that are necessary to replicate the large human genome, strict sequence specificity has been lost. Nevertheless, genome-wide analyses performed in recent years, using different mapping methods, demonstrated that there are precise locations along the metazoan genome from which replication initiates. These sites that are scattered throughout eukaryotic chromosomes, contain relaxed sequence consensus and epigenetic features. Early experimental findings suggested that between 30,000 and 100,000 potential DNA replication origins are activated in a mammalian cell cycle, and recent methods that allow characterizing origins at the genome-wide level gave similar numbers. Two important features were also confirmed. First, from yeast to human cells, the number of potential origins is large, but only 20~30% of them are activated each cell cycle. Moreover, there is a large flexibility in the choice of the origins to be activated in each cell of a given population, and this choice is apparently stochastic. Second, selected origins are not activated all at the same time, but according to a highly regulated timing of individual activations during the entire S phase. The actual selection of sites for initiation of replication is thought to be governed by local epigenetic factors, such as the positions of nucleosomes, the types of histone modifications, the state of DNA methylation, the degree of supercoiling, and the level of transcription.

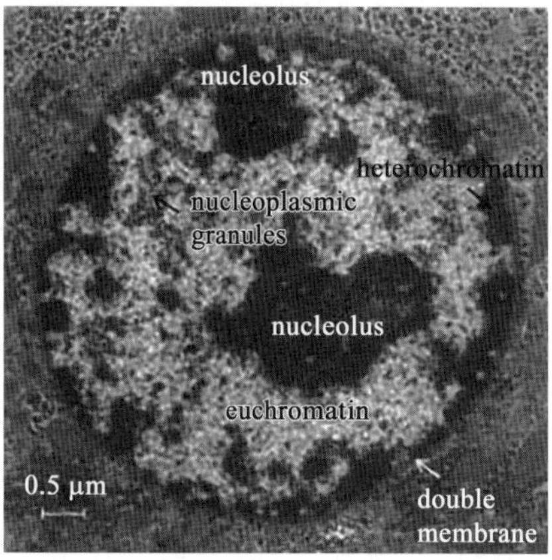

Figure 8-21 Transmission electron micrograph of a thin section of a rat liver cell nucleus. Heterochromatin is evident around the entire inner surface of the nuclear envelope. Two prominent nucleoli are visible, and clumps of chromatin can be seen scattered throughout the nucleoplasm.

Early studies in *S. cerevisiae* indicated that replication origins in eukaryotes might be recognized in a DNA-sequence-specific manner analogously to those in prokaryotes. In budding yeast, the search for genetic replicators lead to the identification of autonomously replicating sequences (ARS) that support efficient DNA replication initiation of bacterial DNA molecules. The ARS is necessary, but not sufficient for origin specification, because only about 3% of ARS sequences in the genome are used as origins.

8.3.5 Heterochromatin

After mitosis has been completed, most of the chromatin in highly compacted mitotic chromosomes returns to its diffuse interphase condition. Approximately 10 percent of the chromatin, however, generally remains in a condensed, compacted form throughout interphase. This compacted, densely stained chromatin is typically concentrated near the periphery of the nucleus, often in proximity with the nuclear lamina, as indicated in Figure 8-21. Chromatin that remains compacted during interphase is called **heterochromatin** (HC) to distinguish it from **euchromatin** (EC), which returns to a dispersed, active state. When a radioactively labeled RNA precursor such as[^{3}H] uridine is given to cells that are subsequently fixed, sectioned, and autoradiographed, the clumps of HC remain largely unlabeled, indicating that they have relatively little transcriptional activity. The peripheral regions of the nucleus are thought to contain factors that promote transcriptional repression, making it a favorable environment for the localization of HC.

HC is further classified into constitutive and facultative types, depending on whether the chromatin is permanently or transiently compacted. Constitutive HC (cHC) is a permanent feature of all cells and remains condensed and mostly transcriptionally silent throughout development and cell divisions. This fraction is found at subtelomeric regions and at pericentromeres, which surround repetitive centromeric DNA. In contrast, facultative HC (fHC) is not a permanent feature but is seen in some cells some of the time. fHC is thought to contain genes that are inactive in some cells or at some periods of the cell cycle. When these genes are inactive, their DNA regions are compacted into HC. A typical example of fHC is the inactive X chromosome in female mammals, but it also includes genomic regions that interact with specific nuclear structures, such as the lamina-associated domains located at the nuclear periphery and nucleolus-associated domains. Each of these HC domains is defined epigenetically by specific histone post-translational modifications, histone variants, and associated proteins, in addition to DNA methylation, which contributes to transcriptional silencing.

A crucial but poorly understood epigenetic component of eukaryotic genomes is cHC. Although highly enriched for repeated DNA sequences and containing few protein-coding genes, the cHC domain plays critical roles in safeguarding the genome, including chromosome segregation, telomere protection, suppression of transposon activity, and DNA repair.

8.4 Nucleolus and ribosome biogenesis

The **nucleolus** is the largest compartment of the interphase nucleus of a eukaryotic cell when viewed in light microscope. It is a membraneless substructure in the nucleus, where ribosome biogenesis takes place, a process that is responsible for the assembly of the translational machinery, the ribosome, and is tightly regulated according to cell state. Ribosome biogenesis is initiated by the transcription of rRNA genes, the genetic component of the nucleolus. In eukaryotic genomes, rRNA genes are generally present in high copy numbers and arranged in arrays of tandem repeats among different chromosomes at regions called **nucleolar organizer regions** (NORs). The number of NORs present in the genome varies between species, and the rDNA content of NORs can vary between individuals of the same species and even between the cells of an individual. Cells require large numbers of ribosomes to meet their needs for protein synthesis. Actively growing mammalian cells, for example, contain 5 million to 10 million ribosomes that must be synthesized each time the cell divides. The nucleolus is a ribosome production factory, designed to fulfill the need for large-scale production of rRNA and assembly of the ribosomal subunits.

8.4.1 Structural organization of the nucleolus

Morphologically, nucleoli consist of three distinguishable regions (Figure 8-22): a fibrillar center (FC), one or a few dense fibrillar components (DFCs), and a large granular component (GC). In addition, a layer of heterochromatin surrounds the nucleolus, forming the perinucleolar chromatin. These different regions are thought to represent the sites of progressive stages of rRNA transcription, processing, and ribosome assembly. The rRNA genes are located in the FCs, with transcription occurring primarily at the boundary of the FCs and DFC. Processing of the pre-rRNA is initiated in the DFC and continues in the GC, where the rRNA is assembled with ribosomal proteins to form nearly completed preribosomal subunits, ready for export to the cytoplasm.

Following each cell division, nucleoli form around the chromosomal regions that contain the 5.8S, 18S, and 28S rRNA genes. In humans, NORs are present in the short arm of five acrocentric chromosomes. The formation of nucleoli requires the transcription of 45S pre-rRNA, which appears to lead to the fusion of small prenucleolar bodies that contain processing factors and other components of the nucleolus. In most cells, the initially separate nucleoli then fuse to form a single nucleolus. The size of the nucleolus depends on the metabolic activity of the cell, with large nucleoli found in cells that are actively engaged in protein synthesis. This variation is due primarily to differences in the size of the GC, reflecting the levels of ribosome synthesis.

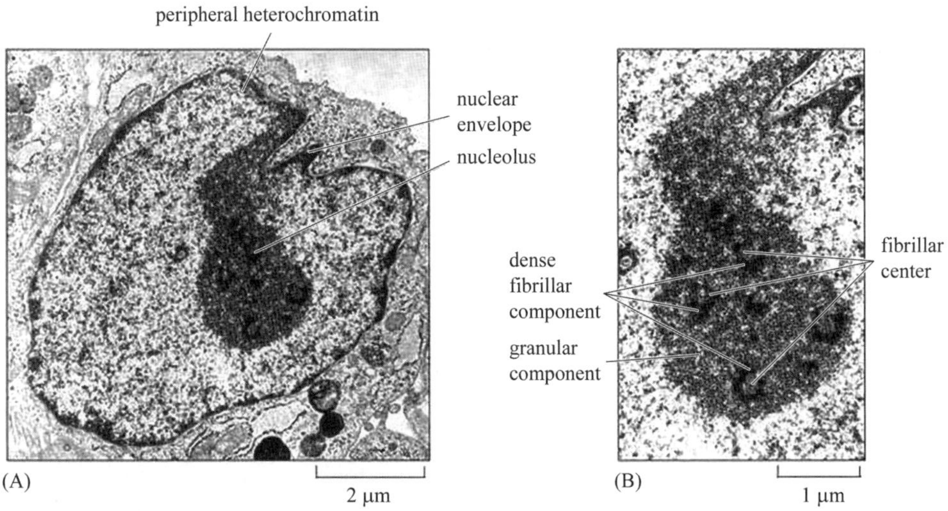

Figure 8-22 Electron micrograph of a thin section of a nucleolus in a human fibroblast, showing its three distinct zones. (A) View of entire nucleus. (B) High-power view of the nucleolus. It is believed that transcription of the rRNA genes takes place between FC and DFCs and that processing of the rRNAs and their assembly into ribosomes proceeds outward from the DFC to the surrounding GC (From: Alberts B, et al. 2015)

8.4.2 Features of the ribosomal DNA genes

In all organisms, ribosomal genes are present in multiple repeat units to satisfy the high demand for ribosomes. The human genome, for example, contains about 200 copies of the gene that encodes the 5.8S, 18S, and 28S rRNA and approximately 2000 copies of the gene that encodes 5S rRNA. Clustered arrays of human rDNA repeats are located at NORs on the short arm of five acrocentric chromosome pairs. The 5S rRNA genes are present in a single tandem array on the distal end of chromosome 1. The biological reason why 5S genes are localized in a separate genomic compartment and are transcribed by a different polymerase remains unknown (Figure 8-23).

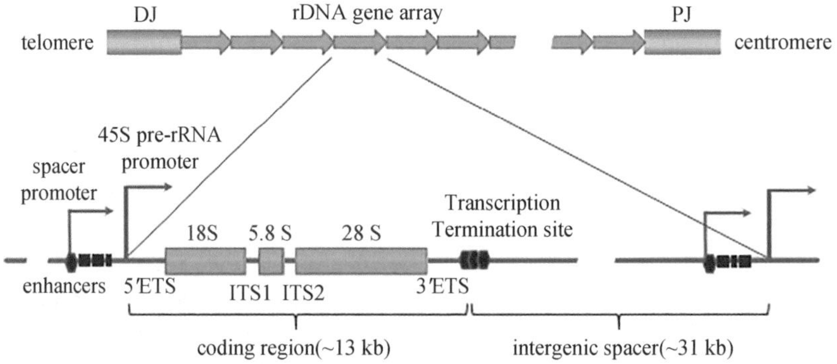

Figure 8-23 Schematic representation of human rDNA genes. In the human genome, rDNA repeats consist of a coding region and intergenic spacer. Two promoters are denoted by arrows. The region distal to spacer promoter contains enhancer elements indicated by bars. Abbreviations: PJ & DJ: proximal and distal junctions (From: Chromosome Res. 2019)

Mammalian rDNA transcription units are 43~45 kb in length. Each rDNA repeat unit consists of the coding region producing a precursor transcript (pre-rRNA), often referred to as 45S (or 47S) that is processed into three mature rRNA molecules: 18S, 5.8S, and 28S (Figure 8-23). Boundaries of the coding region contain external transcribed spacers (5′ETS and 3′ ETS), and coding parts of the 45S sequence are separated by internal transcribed spacers (ITS1 and ITS2). Sequences encoding pre-rRNAs are separated by long intergenic spacers (IGSs) approximately 30 kb in length. These repeat units are flanked at their ends by one or more terminator sequences that stop the transcript elongation by RNAPI. The rRNA coding sequences are highly conserved in animals, plants, and fungi. However, the IGS sequence is less conserved and contains regulatory elements including promoters and repetitive enhancer elements. There are two promoters within the vertebrate rDNA repeat unit: the gene promoter and the spacer promoter. Transcription of the 45S pre-rRNA is controlled by the gene promoter that consists of two parts — a core promoter and an upstream control element. The spacer promoter, located within the IGS, can give rise to the non-coding RNA that is thought to participate in gene silencing.

8.4.3 Transcription and processing of pre-rRNA

Transcription of rDNA genes requires a specialized transcription machinery. Key components of this machinery are transcription factor UBF, RNAPI, and topoisomerases. These repeated rDNA genes are very actively transcribed by RNAPI, allowing their transcription to be readily visualized by electron microscopy. In such electron micrographs, each of the tandemly arrayed rRNA genes is surrounded by densely packed growing RNA chains, forming a structure that looks like a Christmas tree. The high density of growing RNA chains reflects that of RNA polymerase molecules, which are present at a maximal density of approximately one polymerase per hundred base pairs of template DNA.

The primary transcript of the rRNA genes is the large 47S pre-rRNA, which contains the 18S, 5.8S, and 28S rRNA as well as transcribed spacer regions. Two ETSs are present at both the 5′ and 3′ ends of the pre-rRNAs, and two ITSs lie between the 18S, 5.8S, and 28S rRNA sequences. In human cells, the order of the processing events was chiefly determined in the early 1990s. The initial steps include cleavage of the primary transcript close to 5′ end (position +414) and removal of 3′ ETS. The resulting 45S rRNA may be cleaved at the 5′ end of 18S RNA gene (the main pathway) or within the first internal transcribed spacer (relatively rare alternative pathway). The pre-rRNA is processed via a series of cleavages to yield the mature rRNA species (Figure 8-24).

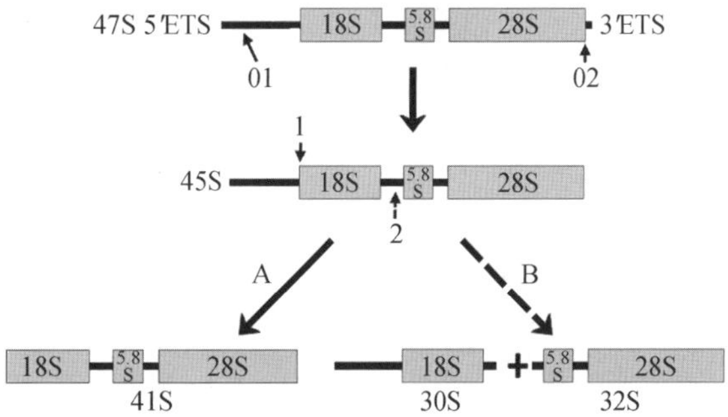

Figure 8-24 First steps of human pre-rRNA processing. (A) represents the main pathway. (B) is a relatively rare alternative. 01, Cleavage of 5′ETS at position +414; 02, elimination of 3′ETS; 1, cleavage at the 5′ end of 18S sequence, which is followed by degradation of 5′ETS remnant; 2, alternative cleavage within the first ITS. Sites A′ and A0 within 5′ETS is not shown (From: Nucleus. 2013)

The processing of pre-rRNA requires the action of both proteins and RNA that are localized to the nucleolus. Nucleoli contain a large number of small nucleolar RNA (snoRNA) that functions in pre-rRNA processing. Like the spliceosomal small nuclear RNA (snRNA), the snoRNA are complexed with proteins, forming snoRNPs. Individual snoRNPs consist of single snoRNA associated with eight to ten proteins. The snoRNPs then assemble on the pre-rRNA to form processing complexes in a manner analogous to the formation of spliceosomes on pre-mRNA. Some snoRNA are responsible for the cleavages of pre-rRNA into 18S, 5.8S, and 28S products.

The majority of snoRNA, however, function to direct the specific base modifications of pre-rRNA, including the methylation of specific ribose residues and the formation of pseudouridines. Most of the snoRNA contain short sequences of approximately 15 nucleotides that are complementary to 18S or 28S rRNA. Importantly, these regions of complementarity include the sites of base modification in the rRNA. By base pairing with specific regions of the pre-rRNA, the snoRNA act as guide RNAs that target the enzymes responsible for ribose methylation or pseudouridylation to the correct site on the pre-rRNA molecule.

8.4.4 Ribosome assembly

The formation of ribosomes involves the assembly of the pre-rRNA with both ribosomal

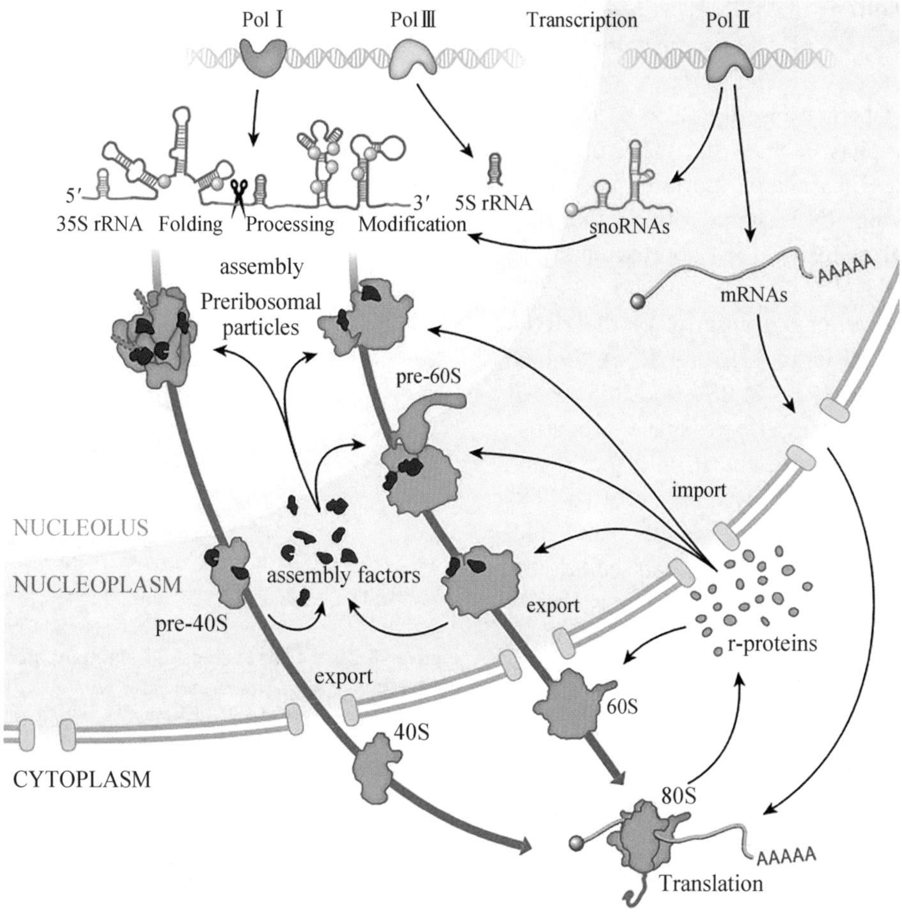

Figure 8-25 Overview of the eukaryotic ribosome assembly pathway. Ribosome assembly requires all three RNA Pols: Pol Ⅰ synthesizing the 35S pre-rRNA, Pol Ⅲ synthesizing the 5S rRNA, and Pol Ⅱ synthesizing snoRNAs and mRNAs, which also encode the r-proteins. During transcription, the pre-rRNA is folded, nucleolytically processed, and modified (methylated or pseudouridinylated). Already at this early stage, ribosome assembly factors and r-proteins assemble with the nascent pre-rRNA to form preribosomal particles, which subsequently go through further maturation steps, thereby moving from the nucleolus and through the nucleoplasm into the cytoplasm, where mature 60S and 40S subunits join to form the 80S ribosomes (From: Annu. Rev. Biochem. 2019)

proteins (r-proteins) and 5S rRNA. The genes that encode ribosomal proteins are transcribed outside of the nucleolus by RNA Pol Ⅱ, yielding mRNA that are translated on cytoplasmic ribosomes. The ribosomal proteins are then transported from the cytoplasm to the nucleolus, where they are assembled with rRNA to form preribosomal particles. Although the genes for 5S rRNA are also transcribed outside of the nucleolus, in this case by RNA Pol Ⅲ, 5S rRNA similarly is assembled into preribosomal particles within the nucleolus (Figure 8-25).

During the complicated biogenesis of eukaryotic ribosomes, the rRNA is folded, modified, processed, and assembled with r-proteins to form the two ribosomal subunits. This process takes place largely in a specialized nuclear compartment, the nucleolus, and can be divided into five major activities: synthesis of the pre-rRNA and r-proteins, base modification of the pre-rRNA, folding of the pre-rRNA, assembly of the pre-rRNA with the r-proteins, and endo- and exonucleolytic processing of the pre-rRNA to remove external and internal transcribed spacer rRNAs (Figure 8-24). Most of these assembly and processing steps are tightly coupled and occur within preribosomal particles that travel, as folding and maturation progresses, from the nucleolus, through the nucleoplasm, and across NPCs into the cytoplasm, where they ultimately mature into translation-competent ribosomal subunits. Ribosome assembly is the most energy-consuming process in a growing cell, requiring extensive regulation and coordination with other cellular pathways. Perhaps, then, it is no surprise that a growing number of diseases (termed ribosomopathies) are associated with defects in ribosome synthesis. Moreover, strong upregulation of ribosome assembly is an important molecular alteration of rapidly dividing cancer cells, owing to the high demand for ribosomes. All these findings make the entire ribosome synthesis machinery an attractive target for the treatment of cancer.

8.4.5 Nucleolar dynamics

Nucleoli are dynamic structures that differ, both in size and appearance, from one cell type to another depending on transcriptional activity. These membrane-less organelles comprise a distinct structural compartment in which major steps of ribosome biogenesis take place. As mentioned previously, transcription of rRNA genes by RNA Pol Ⅰ occurs at the interface between FCs and DFCs, while processing of pre-rRNA and assembly with r-proteins takes place in the DFC and in the GC. This tripartite nucleolar architecture, a characteristic of most amniotes, gets lost if rDNA transcription is perturbed. Inhibition of Pol Ⅰ transcription by drugs, commonly by low doses of actinomycin D, leads to rearrangement of nucleolar components, culminating in complete nucleolar disintegration and marked alterations in the nucleolar proteome. These dramatic changes in nucleolar morphology, in turn inhibit major nuclear pathways that affect transcription, replication and DNA repair. The molecular mechanisms that govern the shuttling dynamics of nucleolar proteins and nucleolar architecture are complex and intertwined, supporting that the nucleolus is a key player that regulates cell growth, survival, senescence, and stress responses (Figure 8-26).

The nucleolus is also responsive to DNA damage, and RNA Pol Ⅰ transcription is inhibited in response to double-strand breaks both in rDNA and outside the nucleolus. This disruption of the transcription process leads to substantial internal dynamics and modification of nucleolar structures in which FC and DFC go to the GC periphery, together with DNA repair machineries and numerous DNA damage response proteins to form the nucleolar cap (NC), replacing FCs and DFCs, and the rDNA is dislocated to the periphery of the

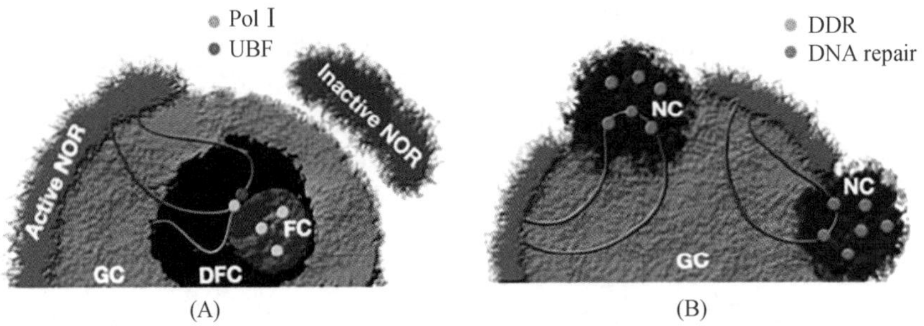

Figure 8-26 Nucleolar organization and reorganization upon DNA damage. (A) Genomic regions that surround the nucleolus-associated domains (NADs) preferentially acquire a silenced condensed state. The rDNA in NORs may invaginate into the internal layers of the nucleolus, forming the intranucleolar stretches that contain both inactive and silenced genes. In normal conditions, active NORs form the FCs and DFCs, inside the GC. (B) Nucleolar reorganization upon DNA damage. Abbreviations: DDR: DNA damage response (From: Trends Genet. 2019)

nucleolus where the DNA repair machinery can access and repair the rDNA (Figure 8-26).

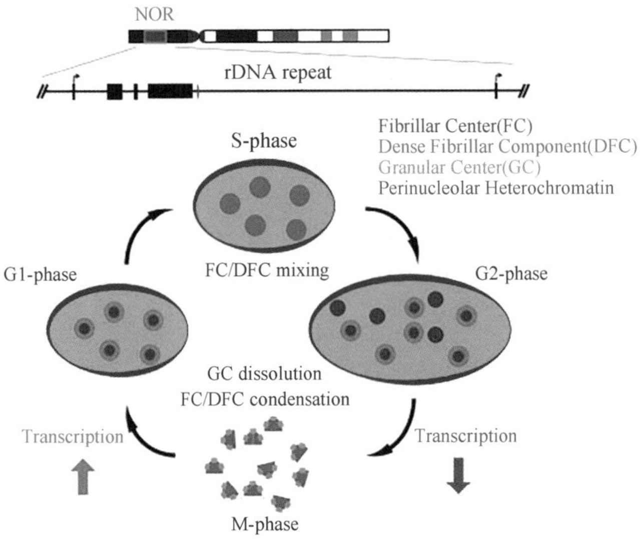

Figure 8-27 Cell cycle-dependent nucleolar dynamics. An ideogram of a NOR-bearing human acrocentric chromosome is shown above. The cartoon below depicts characteristic changes of nucleolar architecture in different phases of the cell cycle. The main structural components are indicated. In S-phase, RNA Pol Ⅰ may transiently leave active rRNA genes which leads to disappearance of the FC (dark green). Large, fused nucleoli in the G_2-phase may contain poised/inactive rDNA units, presented as dark green FCs without surrounding DFC. In the M-phase, the FC/DFC colored semi-circles represent active NORs which remain associated with UBF and part of the transcription and processing machineries (From: Current Opinion in Cell Biology. 2018)

The remarkable plasticity of the mammalian nucleolus is recognized during each cell cycle. As cells approach mitotic metaphase, the nucleolus disassembles and many nucleolar proteins associate with the surface of metaphase chromosomes. UBF remains associated with active rDNA repeats, serving as mitotic bookmark and binding platform for the RNA Pol Ⅰ machinery. At the exit from mitosis, nucleolar proteins coalesce to form **prenucleolar bodies** (PNBs), dynamic structures in which processing complexes are concentrated. PNB disassembly and nucleolus assembly require translocation of the processing complexes to the NORs, involving dynamic long-distance interactions between spatially separated genomic regions. Resumption of rDNA transcription at telophase/early G_1-phase correlates with fusion of NOR-bearing chromosomes, which is attributed to the liquid droplet-like behavior of nucleolar structures. During progression through G_1-phase, FC/DFC/GC structures evolve around active transcription units and nucleolar fusion occurs, resulting in the formation of a few large nucleoli that contain all active NORs. During early S-phase RNA Pol Ⅰ is transiently released from rDNA to prevent collision with the replication machinery. The disrupted FC/DFC structure is restored in mid Sphase, coupling nucleolar architecture to cell cycle-dependent functions. Cell cycle-dependent changes in nucleolar structure and function are mostly brought about by reversible chemical modifications of components of the Pol Ⅰ transcription machinery and factors required for ribosome biogenesis (Figure 8-27).

Nucleolar/rDNA changes through the cell cycle and during DNA repair are likely to substantially alter the nuclear landscape. All in all, dynamic changes to the nucleolus during the cell cycle, DNA repair, or tissue differentiation are expected to influence the architecture of the entire nucleus and play a role modulating the epigenetic, transcriptional, metabolic, and other functional states of the cell.

8.5 The nuclear matrix

Descriptions of an insoluble proteinaceous nuclear substructure, in some ways analogous to the cytoskeleton, have existed for several decade years. However, the difficulties associated with studying this nuclear fraction mean that there are still many unanswered questions about structure and function, and even some residual controversy about its very existence. The nuclear substructure has been termed the **nuclear matrix** (NM), nuclear scaffold or nuclear skeleton (or nucleoskeleton) on the basis of the technique used to reveal it. Depending on extraction technique, the residual protein fraction varies slightly in composition but the core components revealed by several large-scale screens are similar. These include lamins, matrins, hnRNPs, other structural proteins, and various proteins involved in DNA metabolism, many of which are listed and categorized in a database of NM Proteins, NMPdb.

8.5.1 What is the nuclear matrix?

When isolated nuclei are treated with a mild nonionic detergent and high salt (e.g., 2 mol/L NaCl), which remove lipids and nearly all of the histone and nonhistone proteins of the chromatin, the DNA is seen as a halo surrounding a residual nuclear core. It has been known for many years that when mammalian cells are treated with DNase I, a fibrillar network of protein and RNA remains in the region of the nucleus. This highly organized view of the nucleus implies that there is an underlying nuclear substructure, the NM. It is a non-chromatin, proteinaceous scaffold-like network that is resistant to high ionic strength buffers, nonionic detergents, and nucleolytic enzymes. This protein network is composed of actin and numerous other protein components that have not been fully characterized, including components of the chromosome scaffold that rearranges and condenses to form metaphase chromosomes during mitosis. Although the role of actin in the NM remains mysterious, there had been hints of an interaction between nuclear actin and nuclear RNAs. However, snRNPs remain associated with the NM prepared from detergent-extracted, DNase I-treated cells. Moreover, when the NM is prepared with a low concentration of salt, pre-mRNAs associated with the matrix undergo splicing when ATP is added. These results suggest that the RNA-processing foci observed microscopically may be associated with specific regions of the NM.

Although the concept of the NM has been disputed for several decades, the NM is still being hypothesized as a platform for three-dimensional genome organization. In fact, the NM is a proteinaceous entity that is thought to act as a structural and functional framework of the nucleus. It fulfills a structural role in eukaryotic cells and is responsible for maintaining the shape of the nucleus and the spatial organization of chromatin. Moreover, the NM participates in several cellular processes, such as DNA replication/repair, splicing, chromatin remodeling, cell signaling and differentiation, cell cycle regulation, apoptosis and carcinogenesis.

8.5.2 Features of the scaffold/matrix attachment regions

In eukaryotes, the genome is compartmentalized into chromatin domains by the attachment of chromatin to a supporting structure, the NM. The interactions between chromatin and the NM occur via specific AT-rich DNA sequences called nuclear **scaffold/matrix attachment regions** (S/MARs). S/MARs are quite ubiquitous in the genome and generally are located at distances that are smaller than the size of DNA loops. They interact with DNA topoisomerase Ⅱ (Topo Ⅱ) molecules and some specific proteins, including special AT-binding protein. As specific DNA fragments, S/MARs can fix the chromatin to the NM and separate chromatin into prominent discrete DNA loops with an open chromatin conformation in both metaphase and interphase cells.

A surprising feature in S/MAR fragments is the lack of conservation of sequence. They are usually ~70% AT-rich sequences, but otherwise lack any consensus sequences. However, other interesting sequences often are in the DNA stretch containing the S/MAR. The recognition site for Topo Ⅱ is usually present in the S/MAR. It is therefore possible that an S/MAR serves more than one function, providing a site for attachment to the NM, but also containing other sites at which topological changes in DNA are effected.

So far, it is not understood how the folded 30-nm chromatin fiber is anchored to the linear chromosome axis, but evidence suggests that in mitotic chromosomes, the bases of chromosomal loops are enriched both in condensins and in DNA Topo enzymes, which allow DNA to swivel when anchored (Figure 8-28). These two proteins may form much of the axis at metaphase. By means of such chromosomal attachment sites, the NM might help to organize chromosomes, localize genes, and regulate gene expression and DNA replication. For instance, if cells are incubated with fluorescently or radioactively labeled RNA or DNA precursors for a brief period, nearly all of the newly synthesized nucleic acid is found to be associated with the fibrils of the NM.

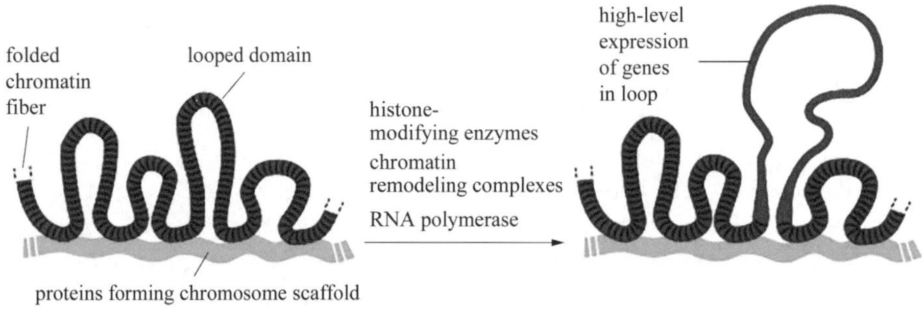

Figure 8-28 A model for the organization of an interphase chromosome. A section of an interphase chromosome is shown folded into a series of looped domains, each containing perhaps 50,000~200,000 or more nucleotide pairs of DNA condensed into a chromatin fiber. The chromatin in each individual loop is further condensed through poorly understood folding processes that are reversed when the cell requires direct access to the DNA packaged in the loop. Neither the composition of the postulated chromosomal axis nor how the folded chromatin fiber is anchored to it is clear (From: Alberts B, et al. 2015)

For cell biologists, the critical questions remaining to be answered include what is the relationship between the chromosome scaffold of dividing cells and the NM of interphase cells and are the same DNA sequences attached to both structures? In several cases, the same DNA fragments that are found with the NM *in vivo* can be retrieved from the metaphase scaffold. And fragments that contain S/MAR sequences can bind to a metaphase scaffold. It therefore seems likely that DNA contains a single type of attachment site, which in interphase cells is connected to the NM, and in mitotic cells is connected to the chromosome scaffold. All in all, the NM and chromosome scaffold consist of different proteins, although there are some common components. Topo Ⅱ is a prominent component of the chromosome scaffold, and is a constituent of the NM, suggesting that the control of topology is important in both cases.

Summary

The nucleus is the largest structure and the most prominent organelle in the eukaryotic cell. It consists of nuclear envelope (including NPC), chromatin, nucleolus and nuclear matrix. The NE and the NPC may be single most important feature that distinguishes eukaryotes from prokaryotes. The NE provides a unique molecular and biochemical environmental protective mechanism against potentially damaging cytoplasmic enzymatic activities. It functions like a selectively permeable barrier allowing macromolecules to move between the nucleus and the cytoplasm via a gatekeeper, the NPC. The NPC is a huge protein complex that fuses the internal and external nuclear membrane to form an aqueous channel. The nuclear transport is mediated by several distinct types of transport signals that are recognized by specific transport receptors or via a variety of adaptor proteins. These transport receptors can then interact with components of the NPC and with the RanGTPase protein. The vast majority (typically 99.5%) of DNA in most eukaryotes is found in the nucleus. The DNA in eukaryotes is tightly bound to an equal mass of histones, which form repeated arrays of DNA-protein particles called nucleosomes. The nucleosome is composed of an octameric core of histone proteins around which the DNA double helix is wrapped. Nucleosomes are spaced at intervals of about 200 nucleotide pairs, and they are usually packed together into quasi-regular arrays to form a 30-nm chromatin fiber. Elucidating just how a nucleosomal array can be compacted into higher-order chromatin structures is central to understanding the dynamics of chromatin structure. Although the existence of the 30 nm fiber *in vivo* remains controversial, the cryo-EM structure of nucleosomal arrays revealed how nucleosomal arrays stack with each other. Especially, the crystal structure of tetranucleosomes showed that nucleosomes form a fiber in a zigzag manner and the tetranucleosomes serve as a unit to form a high-order structure. The chromatin is divided into euchromatin and heterochromatin. During interphase, the general mass of chromatin is in the form of euchromatin, which is slightly less tightly packed than mitotic chromosomes. Regions of heterochromatin remain densely packed throughout interphase. During cell division, chromatin fibers coil and condense into a number of linear structures called chromosomes. Three functional elements are required for replication and stable inheritance of chromosome: the centromere is essential for segregation during cell division; telomeres are specialized structures, comprising DNA and protein, which cap the ends of eukaryotic chromosomes and play critical roles in chromosome replication and maintenance; eukaryotic replication origins are not only the sites of replication initiation, but also control the timing of DNA replication. The nucleolus is a nonmembrane-bound nuclear organelle that is dedicated to the task of rRNA transcription and ribosome assembly, and that occupies a substantial and disproportionate volume of the cell nucleus. The nuclear matrix is not a concept, it is a readily observed cellular structure, i. e. , a non-chromatin, proteinaceous scaffold-like network. It has been proposed as a platform for nuclear compartmentalization and as an anchor for DNA structure, the site of transcription, DNA replication/repair, splicing, epigenetic remodeling.

Questions

1. Please explain the following terms: nucleosome, nucleolus, constitutive/facultative heterochromatin, chromatin remodeling, nuclear envelope, NLS, nuclear pore complex.
2. What role do the following cellular components play in the storage, expression, or transmission of genetic information: (a) chromatin, (b) nucleolus, (c) ribosome, (d) centromere?
3. Multiple-choice questions
   (1) Concerning the nucleus:
      (a) The diameter is generally about 5 μm
      (b) Euchromatin is not actively expressed
      (c) Nucleoli are very active in the synthesis of mRNA
      (d) All cells in the human body have nuclei
   (2) Within the nucleus:
      (e) Chromatin consists of DNA and RNA only
      (f) Histone proteins are negatively charged
      (g) The nucleosome core protein consists of eight histone subunits
      (h) H1 histone is not found in the core protein
4. Short-answer questions:
   What is a nucleosome and what are the proteins present in this structure?
5. Define the following terms and their relationships to one another:
   (a) Interphase chromosome; (b) Mitotic chromosome; (c) Chromatin; (d) Heterochromatin; (e) Histones; (f) Nucleosome.
6. If a human nucleus is 10 μm in diameter, and it must hold as much as 2 m of DNA, which is complexed into nucleosomes that during full extension are 11 nm in diameter, what percentage of the volume of the nucleus is occupied by the genetic material?
7. The DNA in a cell associates with proteins to form chromatin. What is a nucleosome? What role do histones play in nucleosomes? How are nucleosomes arranged in condensed 30-nm fibers?
8. Replication and segregation of eukaryotic chromosomes require three functional elements: replication origins, a centromere, and telomeres. How would a chromosome be affected if it lacked (a) replication origins or (b) a centromere?
9. Describe the problem that occurs during DNA replication at the end of chromosomes. How are telomeres related to this problem?

(曲志才)

Chapter 9

Cell cycle and cell division

Cell division from an already existing cell is the only way to make a new cell. A cell reproduces by its contents and then dividing in two. In unicellular organisms, such as bacteria and yeasts, each cell division produces a complete new life, whereas a new functioning organism is produced in multicellular cells. Even in our adults, cell division also occurs to replace those dead cells.

Most eukaryotic cells proceed through an ordered series of events, constituting the **cell cycle**, during which their chromosomes are duplicated and division to each of two daughter cells. Regulation of the cell cycle is critical for the normal development of multicellular organisms. Loss of control ultimately leads to some diseases such as cancer. In the late 1980s, it became clear that the molecular processes regulating the main events in the cell cycle are fundamentally similar in all eukaryotic cells. Because of this similarity, research with diverse organisms, each with its own particular experimental advantages, has contributed to a growing understanding of how these events are coordinated and controlled. In this chapter we focus on the major stages in detail of the cell division and how they are regulated in eukaryotes.

9.1 An overview of the cell cycle

The division cycle of most eukaryotic cells consists of four coordinated processes: cell growth, DNA replication, distribution of the duplicated chromosomes to daughter cells, and cell division (Figure 9-1). In bacteria, these four events occur throughout most of the cell cycle. In eukaryotes, however, the cell cycle is more complex and consists of four discrete phases. Although cell growth is a continuous process, DNA is synthesized during only one phase of the cell cycle, and the replicated chromosomes are then distributed to daughter nuclei by a complex series of events preceding cell division. Progression between these stages of the cell cycle is controlled by a conserved regulatory apparatus, which not only

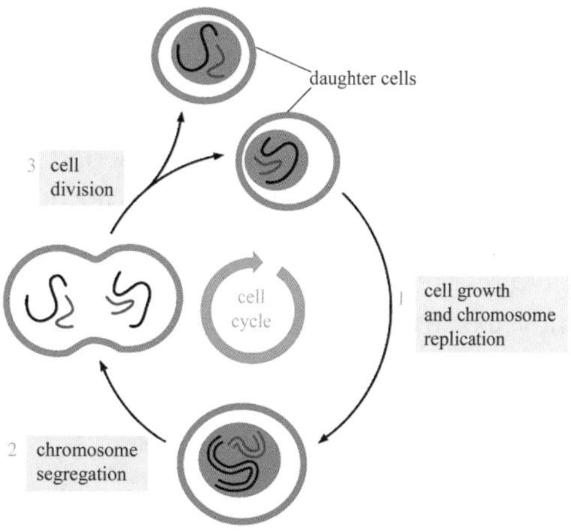

Figure 9-1 Four processes of cell cycle in eukaryotic cell (From: Alberts B, et al. 2012)

coordinates the different events of the cell cycle, but also links the cell cycle with extracellular signals that control cell proliferation.

A cell reproduces by carrying out an orderly sequence of events in which it duplicates its contents and then divides into two. This cycle of duplication and division is the essential mechanism by which all living things reproduce. The cell cycle can be regarded as the life cycle of an individual cell, which can be divided into two major phases, **interphase** and **M phase** (Figure 9-2). Interphase, the initial stage of the cycle, is a time for preparing cell divison events. During this stage, a cell engages in diverse metabolic activities for cell growth and then DNA replication. M phase, an easily visible process with a light microscope, have two partially overlapping stages named as nuclear division (mitosis) and cytokinesis. During M phase, the duplicated chromosomes are separated into two nuclei and finally the cell splits in two.

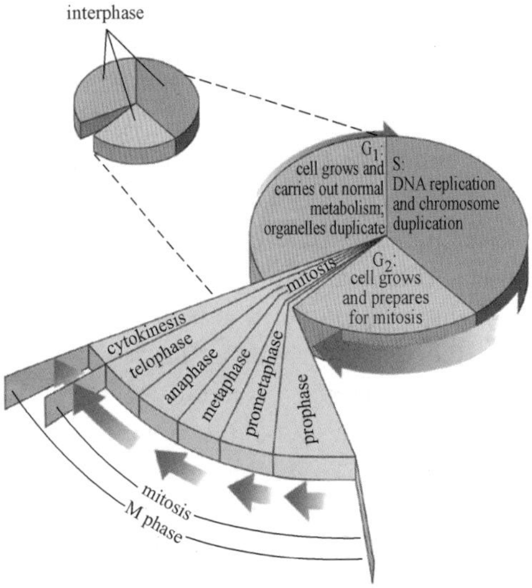

Figure 9-2 An overview of the eukaryotic cell cycle. This diagram of the cell cycle indicates the stages through which a cell passes from one division to the next. The cell cycle can be divided into two stages: interphase and M phase (From: Karp G. 2005)

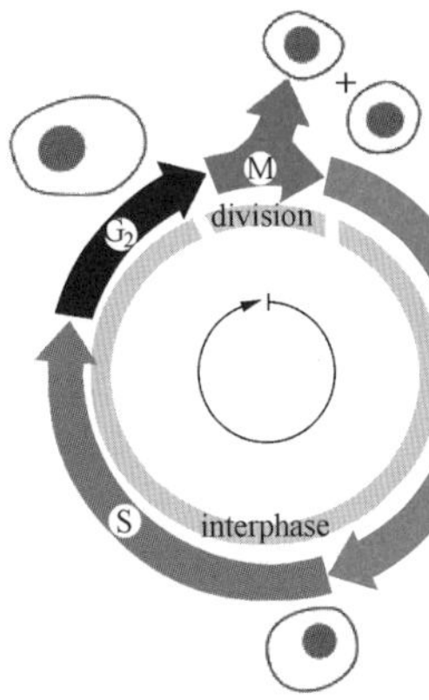

Figure 9-3 The eukaryotic cell cycle is comprised of four successive phases. Nuclear division and then cytoplasmic division occur in M phase. Interphase is divided into three phases: G_1, S and G_2 phases. The cell grows continuously during interphase but stops growing during M phase. (From: Alberts B, et al. 2002)

Although M phase is the period when the contents of a cell are actually divided, numerous preparations for an upcoming mitosis occur during interphase, including replication of the DNA of each chromosome. Studies in the early 1950s on asynchronous cultures showed that DNA replication occurs during a defined period of the cell cycle. And this specific period is called S phase (DNA synthesis phase). Investigations of this nature show two another periods during interphase, one before and one after DNA synthesis occurs, which are designated G_1 (gap I) and G_2 (gap Ⅱ), respectively. During both of these intervals, as well as during S, intensive metabolic activity, cell growth, and cell differentiation. By the end of G_2, the volume of the cell has roughly doubled, DNA has been replicated, and mitosis (M phase) is initiated. Following mitosis, continuously dividing cells then repeat this cycle (G_1, S, G_2, M) over and over (Figure 9-3).

Analysis of a large variety of cells has revealed great variability in the duration of the cell cycle. While M phase usually lasts only an hour or so in mammalian cells, interphase may extend for days, weeks and even longer, which depend on the cell type and the environmental conditions. For example, cell cycles can range from as short as 30 min in a cleaving frog embryo, whose cell cycles lack both G_1 and G_2 phases, and the time of the cycle occupies several months in slowly growing tissues, such as the mammalian liver. G_1 is the most variable among four stages of cell cycle. While the total time of S, G_2 and M phases is relatively fixed, with a few notable exceptions. For example, cell that has stopped dividing, whether temporarily or permanently, whether in the body or in culture, are present in a stage preceding the initiation of DNA synthesis.

9.2 Regulation of the cell cycle

The events of the cell cycle must occur in a particular sequence, and this sequence must be preserved even if one of the steps takes longer than usual. To coordinate these activities, eukaryotic cells have evolved a complex network of regulatory proteins, known as the cell-cycle control system. All of the nuclear DNA, for example, must be replicated before the nucleus begins to divide. If DNA synthesis is slowed down or stalled, mitosis and cell division will be delayed. The same events also occur at other stages of the cell cycle. It is the cell-cycle control system that ensures correct progression through the cell cycle by regulating the cell-cycle machinery.

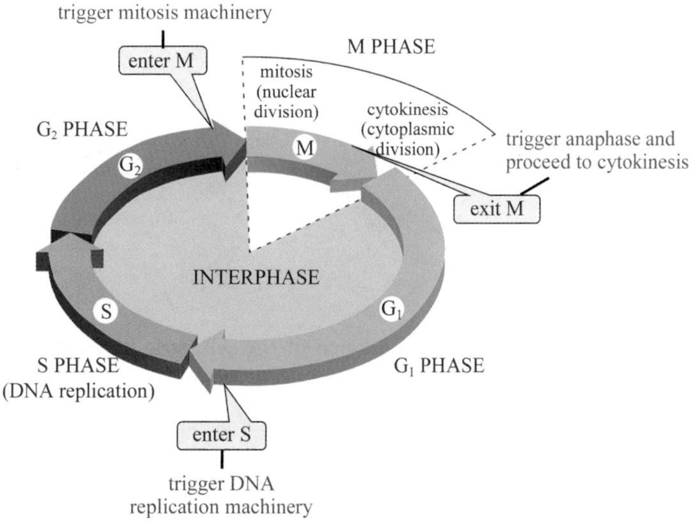

Figure 9-4 The control of the cell cycle. The essential processes of the cell cycle — such as DNA replication, mitosis, and cytokinesis are triggered by a cell-cycle control system. By analogy with a washing machine, the cell-cycle control system is shown here as a central arm — the controller — that rotates clockwise, triggering essential processes when it reaches specific points on the outer dial(From: Alberts B, et al. 2012)

For many years, it was not even clear whether there was a separate control system, or whether the processes of DNA synthesis, mitosis, and cytokinesis somehow controlled themselves. A major breakthrough came in the late 1980s with the identification of the key proteins of the control system, along with the realization that they are distinct from some proteins that perform the processes of DNA replication, chromosome segregation, and so on. In this case, a central controller triggers each process in a set sequence (Figure 9-4).

9.2.1 Cell cycle control by cell growth and extracellular signals

The progression of cells through the division cycle is regulated by **extracellular signals** from the environment, as well as by internal signals that monitor and coordinate the various processes that take place during different cell cycle phases. An example of cell cycle regulation by extracellular signals is provided by the effect of growth factors on animal cell proliferation. In addition, different cellular processes, such as cell growth, DNA replication, and mitosis, all must be coordinated during cell cycle progression. This is accomplished by a series of control points that regulate progression through various phases of the cell cycle.

A major cell cycle regulatory point in many types of cells occurs late in G_1 and controls progression from G_1 to S. This regulatory point was first defined by studies of budding yeast *S. cerevisiae*, where it is known as **START** (Figure 9-5). Once cells have passed START, they are committed to entering S phase and undergoing one cell division cycle. However, passage through START is a highly regulated event in the yeast cell cycle, where it is controlled by external signals, such as the availability of nutrients and cell size.

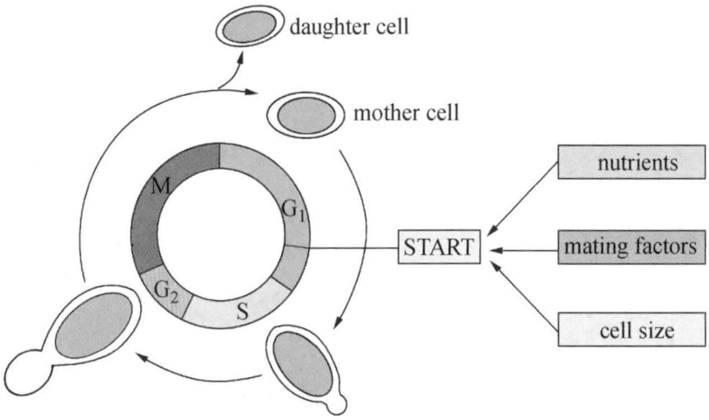

Figure 9-5 Regulation of the cell cycle of budding yeast. Buds form just after START and continue growing until they separate from the mother cell after mitosis. The daughter cell formed from the bud is smaller than the mother cell and therefore requires more time to grow during the G_1 phase of the next cell cycle. Although G_1 and S phases occur normally, the mitotic spindle begins to form during S phase, so the cell cycle of budding yeast lacks a distinct G_2 phase (From: Alberts B, et al. 2012)

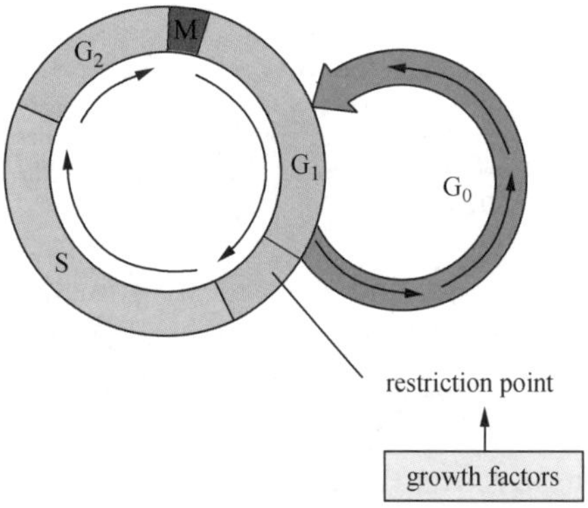

Figure 9-6 Regulation of animal cell cycles by growth factors. The availability of growth factors controls the animal cell cycle at the restriction point. If growth factors are not available during G_1, the cells enter a quiescent stage of the cycle called G_0 (From: Alberts B, et al. 2002)

The importance of START regulation is particularly evident in budding yeasts, in which cell division produces progeny cells of very different sizes: a large mother cell and a small daughter cell. In order for yeast cells to maintain a constant size, the small daughter cell must grow more than the large mother cell does before they divide again. Thus, cell size must be monitored in order to coordinate cell growth with other cell cycle events. This regulation is accomplished by a control mechanism that requires each cell to reach a minimum size before it can pass START. Consequently, the small daughter cell spends a longer time in G_1 and grows more than the mother cell.

The proliferation of most animal cells is similarly regulated in the G_1 phase. In particular, a decision point in late G_1, called the **restriction point** in animal cells, functions analogously to START in yeasts (Figure 9-6). In contrast to yeasts, however, the passage of animal cells through the cell cycle is regulated primarily by the extracellular growth factors that signal cell proliferation, rather than by the availability of nutrients. In the presence of the appropriate growth factors, cells pass the restriction point and enter S phase. Once it has passed through the restriction point, the cell is committed to proceed through S phase and the rest of the cell cycle, even in the absence of further growth factor stimulation. If appropriate growth factors are not available in G_1, progression through the cell cycle stops at the restriction point. Such arrested cells then enter a quiescent stage of the cell cycle called G_0(G zero), in which they remain for long periods of time without proliferating. G_0 cells are metabolically active, although they cease growth and have reduced rates of protein synthesis. Once activated by appropriate growth factors or suitable extracellular condition and signals, many G_0 cells can return to cell cycle and continue through mitosis. For example, skin fibroblasts are arrested in G_0 until they are stimulated to divide as required to repair damage resulting from a wound. The proliferation of these cells can be triggered by platelet-derived growth factor, which is released from blood platelets during clotting and signals the proliferation of fibroblasts in the vicinity of the injured tissue.

Although the proliferation of most cells is regulated primarily in G_1, some cell cycles are instead controlled principally in G_2. For example, vertebrate oocytes can remain arrested in G_2 for long periods of time (several decades in humans) until their progression to M phase is triggered by hormonal stimulation. Extracellular signals can thus control cell proliferation by regulating progression from the G_2 to M as well as the G_1 to S phases of the cell cycle.

9.2.2 Cell cycle checkpoints

The controls discussed in the previous section regulate cell cycle progression in response to cell size and extracellular signals, such as nutrients and growth factors. In addition, the events that take place during different stages of the cell cycle must be coordinated with one another so that they occur in the appropriate order. For example, it is critically important that the cell not begin mitosis until replication of the genome has been completed. The alternative would be a catastrophic cell division, in which the daughter cells failed to inherit complete copies of the genetic material. In most cells, this coordination between different phases of the cell cycle is dependent on a system of **checkpoints** and feedback controls that prevent entry into the next phase of the cell cycle until the events of the preceding phase have been completed.

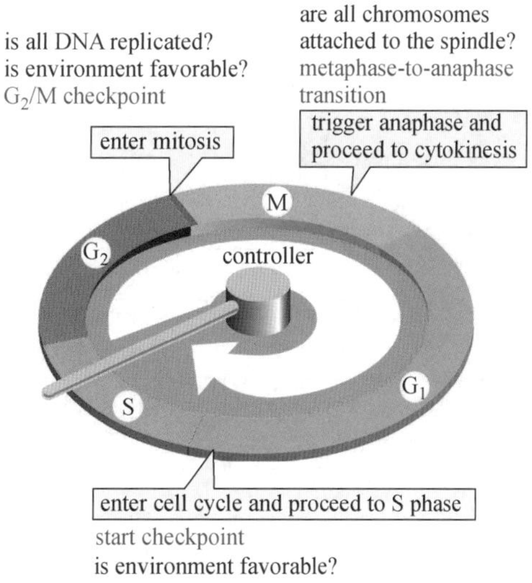

Figure 9-7 Checkpoints in the cell-cycle control system. Information about the completion of cell-cycle events, as well as signals from the environment, can cause the control system to arrest the cycle at specific checkpoints. The most prominent checkpoints occur at locations marked with yellow boxes (From: Alberts B, et al. 2012)

At least three major checkpoints exist within the cell cycle, where the cell is monitored or "checked" before it can proceed to the next stage of the cycle to ensure that incomplete or damaged chromosomes are not replicated and passed on to daughter cells (Figure 9-7).

The first is G_1/S checkpoint, which monitors the cell size achieved following the previous mitosis, and whether the DNA has been damaged. If the cell has not achieved an adequate size, or if the DNA has been damaged, further progress through the cycle is arrested until these conditions are "corrected", so to speak. If both conditions are initially "normal", then the checkpoint is traversed and the cell proceeds to the S phase of the cycle.

An interesting related finding involves the protein product of the *p*53 gene in humans and its involvement during scrutiny at the G_1/S checkpoint. This protein functions during the regulation of apoptosis, the genetic process whereby programmed cell death occurs. In mammalian cells, arrest at the G_1/S checkpoint is mediated by the action of the *p*53 gene protein, which is rapidly induced in response to damaged DNA (Figure 9-8). Interestingly, the gene encoding *p*53 is frequently mutated in human cancers. Loss of *p*53 function as a result of these mutations prevents G_1 arrest in response to DNA damage, so the damaged DNA is replicated and passed on to daughter cells instead of being repaired.

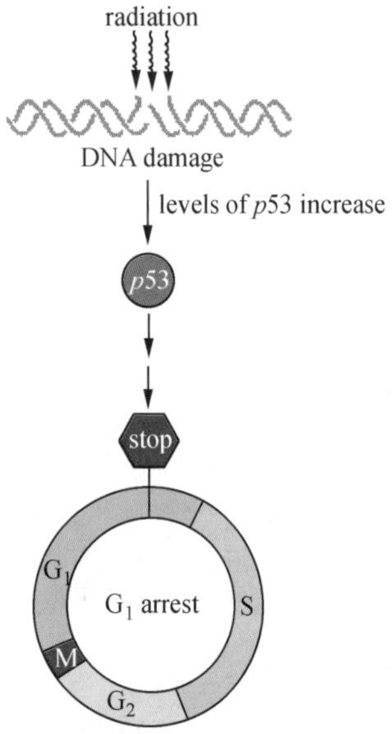

Figure 9-8 Role of *p*53 in G_1 arrest induced by DNA damage. DNA damage, such as that resulting from irradiation, leads to rapid increases in *p*53 levels. The protein *p*53 then signals cell cycle arrest at the G_1 checkpoint.

The second checkpoint is the G_2/M checkpoint, where physiological conditions in the cell are monitored prior to entering mitosis. It prevents the initiation of mitosis until DNA replication is completed. This checkpoint also checks whether centrosome has finished duplication. This G_2/M checkpoint senses unreplicated DNA and centrosomes, which generates a signal that leads to cell cycle arrest. Operation of the G_2/M checkpoint therefore prevents the initiation of M phase before completion of S phase, so cells remain in G_2 until the genome has been completely replicated. Progression through the cell cycle is also arrested at the G_2/M checkpoint in response to DNA damage after replication, such as that resulting from irradiation. This arrest allows time for the damage to be repaired, rather than being passed on to daughter cells.

The final checkpoint occurs during mitosis and is called the M checkpoint. It maintains the integrity of the genome occurs toward the end of mitosis. This checkpoint monitors the correct alignment of chromosomes on the mitotic spindle, thus ensuring that a complete set of chromosomes is distributed accurately to the daughter cells. For example, the failure of one or more chromosomes to align properly on the spindle causes mitosis to arrest at metaphase, prior to the segregation of the newly replicated chromosomes to daughter nuclei. As a result of this checkpoint, the chromosomes do not separate until a complete complement of chromosomes has been organized for distribution to each daughter cell.

9.2.3 Coupling of S phase to M phase

The G_2/M checkpoint prevents the initiation of mitosis prior to the completion of S phase, thereby ensuring that incompletely replicated DNA is not distributed to daughter cells.

Thus, once DNA has been replicated, control mechanisms must exist to prevent initiation of a new S phase prior to mitosis. These controls prevent cells in G_2 from reentering S phase and block the initiation of another round of DNA replication until after mitosis, at which point the cell has entered the G_1 phase of the next cell cycle.

Initial insights into this dependence of S phase on M phase came from cell fusion experiments carried out by Potu Rao and Robert Johnson of the University of Colorado in 1970 (Figure 9-9). These investigators isolated cells in different phases of the cycle and then fused these cells to each other to form cell hybrids. When interphase cells fused with M phase cells, different shapes of chromosomes condensation have found in the fused cells, which called premature chromosome condensation (PCC). It thus appeared that there was a kind of factor which can induce chromosome to condense. This factor was named as maturation promoting factor (MPF, mitosis promoting factor).

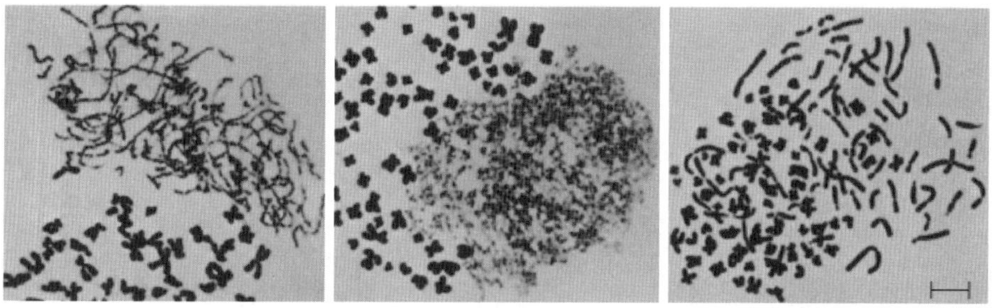

(A) M-G_1 fusion (B) M-S fusion (C) M-G_2 fusion

Figure 9-9 Cell fusion experiments demonstrating the maturation promoting factor in M phase. When G_1 cells were fused with M phase cells, the G_1 nucleus condensed as single line chromosomes (A) In contrast, when S cells were fused with M phase cells, the chromosomes looked like powder (B) The fusion of G_2 phase cells with M phase cells, the G_2 nucleus condensed to double-line chromosomes. All these premature chromosomes condensations were induced by maturation promoting factor (MPF) in M phase cells.

9.2.4 The role of cyclins

While the cell fusion experiments revealed the existence of some factors that regulated the whole cell cycle, they provided no information about the biochemical properties of these factors. Insights into the nature of the agents that trigger DNA replication and promote a cell into **mitosis** (or **meiosis**) were first gained in a series of experiments on the oocytes and early embryos of frogs and invertebrates. Results of these experiments showed that entry of a cell into M phase is initiated by a protein kinase called MPF (maturation-promoting factor). It consists of two subunits: ①a subunit with kinase activity that transfers phosphate group from ATP to specific Ser and Thr residues of specific protein substrates; ② a regulatory subunit called cyclin (Figure 9-10A). The term "cyclin" was coined because the concentration of this regulatory protein rises and falls in a predictable pattern with each cell cycle progresses (Figure 9-11A). The cyclin that helps drive cells into M phase is called M-cyclin. Synthesis of M-cyclin starts immediately after cell division continues steadily throughout interphase and carries out their regulatory functions in M phase.

Then more kinds of cyclins whose concentrations raised in different phases were found gradually (Figure 9-11B). Cyclins are proteins that accumulate continuously throughout the cell cycle and are then destroyed by subiquitation during mitosis. When the cyclin concentration rises, the kinase that cyclins bound is activated, promoting cell cycle going forword. These results suggested that: ①progression of cell cycle depends on an enzyme whose sole activity is to phosphorylate other proteins; ②the activity of this enzyme is controlled by a subunit whose concentration varies from one stage of the cell cycle to another. The cyclin that becomes active toward the end of G_1 phase and is responsible for driving the cell into S phase is called G_1-cyclin, for example, cyclin C and cyclin D. Whereas, the cyclin driving the cell into M phase is called M-cyclin, such as cyclin A and cyclin B.

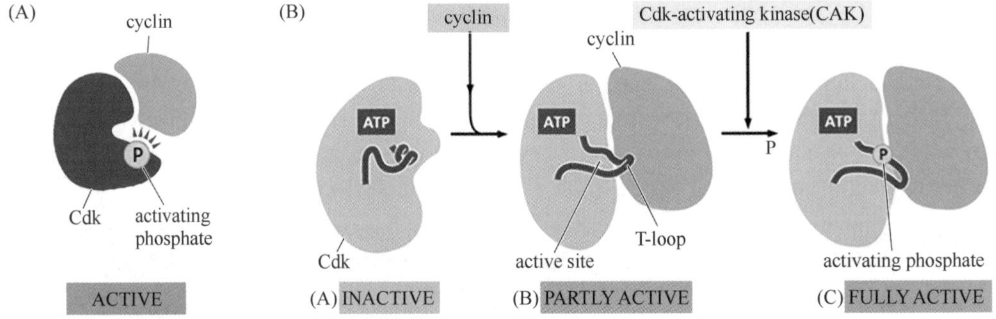

Figure 9-10 The structural basis of the cyclin-dependent protein kinases. (A) Cdk is a complex formed by a regulated subunit called cyclin and a catalytic subunit having kinase activity. Without cyclin, Cdk is inactive. (B) The active site of Cdk2 is blocked by a region of the protein called the T-loop without cyclin bound. When cyclin binds to the Cdk2, the T-loop move out of the active site, causing the partial activation of the Cdk2. After the phosphorylation of Cdk2 by CAK(Cdk-activating kinase) at a threonine residue in the T-loop, the enzyme changes its shape and the fully activation was aroused(From: Alberts B, et al. 2012)

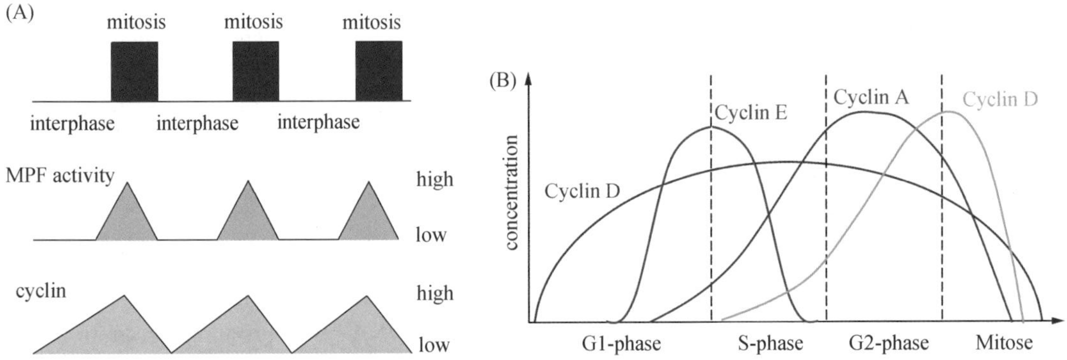

Figure 9-11 Fluctuation of cyclin and MPF levels during the cell cycle. This drawing depicts the cyclical changes that occur during early frog development when mitotic divisions occur synchronously in all cells of the embryo. The top tracing shows the alternation between periods of mitosis and interphase, the middle tracing shows the cyclical changes in the concentrations of cyclins that control the relative activity of the MPF kinase(From: Karp G. 2005) (B) The concentration curves of different cyclins during cell cycle.

9.2.5 Cyclin-dependent kinase

Over the past two decades, a large number of laboratories focused on MPF-like enzymes, called **cyclin-dependent kinases** (**Cdks**) which are composed of cyclins and cell-cycle kinases. It has been found that Cdks are not only involved in M phase, but alse are the key agents that orchestrate activities throughout the cell cycle. Cdks have no enzymatic activity until they bind to the cyclins and phosphorylated themselves. The cyclical changes in cyclin levels result in the cyclin assembly and activation of the cyclin-Cdk complexes; this activation of these complexes in turn triggers various cell-cycle events, such as entry into S phase and M phase(Figure 9-12).

9.2.6 Progression regulatory of cell cycle

Different cyclin-Cdk complexes trigger different steps of the cell cycle. Research on the genetic control of the cell cycle in yeast began in the 1970s in two laboratories, Leland Hartwell at University of Washington and Paul Nurse at the University of Oxford. Both laboratories discovered a gene family that, when mutated, would cause the cycle of cells at elevated temperature to stop at certain points. In fission yeast, the first found product of this gene was called *cdc*2 or Cdk1. Its sequence was eventually found to be homologous to that of the catalytic subunit of MPF. The *cdc* is an abbreviation for " cell division cycle ". Subsequent studies on yeast and many different mammalian cells have supported the concept that the progression of a eukaryotic cell through its cell cycle is regulated at distinct stages. The cell cycle control system achieves each step by means of molecular brakes that can stop the cycle at various checkpoints.

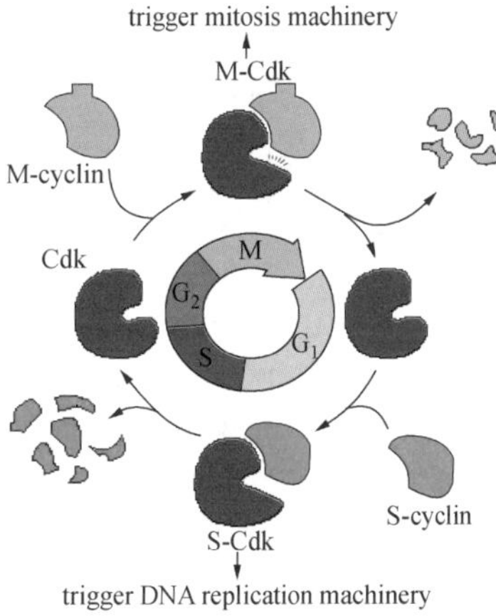

Figure 9-12 A simplified view of the core of the cell cycle control system. Cdk associates successively with different cyclins to trigger the different events of the cycle. Cdk activity is usually terminated by cyclin degradation. For simplicity, only the cyclins that act in S phase (S-cyclin) and M phase (M-cyclin) are shown, and they interact with a single Cdk; as indicated, the resulting cyclin-Cdk complexes are referred to as S-Cdk and M-Cdk, respectively (From: Alberts B, et al. 2002)

The first transition point, i. e. START in fission yeast, occurs before the end of G_1. Passage through START requires the activation of *cdc*2 by one or more G_1-cyclins, whose levels rise during late G_1 (Figure 9-13). Activation of *cdc*2 by these cyclins leads to the initiation of replication at sites where prereplication complexes had previously assembled. The transition from G_2 to mitosis requires activation of *cdc*2 in fission yeast. Cdks containing an M-cyclin phosphorylate substrates that are required for the cell to enter mitosis. The concentration of each type of cyclin rises gradually, and this helps to activate the appropriate Cdk partner, while the rapid falls returns the Cdk to its inactive state. Exit from mitosis and entry into G_1 depends on a rapid decrease in Cdk activity that results from a plunge in concentration of the M-cyclins (Figure 9-13).

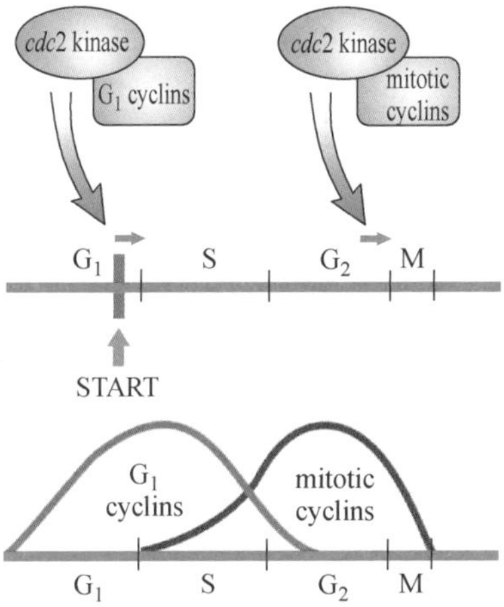

Figure 9-13 A model for cell cycle regulation in fission yeast. The cell cycle is controlled primarily at two points, START and the G_2 to M transition. In fission yeast, cyclins can be divided into two groups, G_1 cyclins and M cyclins. Passage of a cell through these two critical junctures requires the activation of the same *cdc*2 kinase by a different kind of cyclin (From: Karp G. 2005)

Cyclin-dependent kinases are often described as the "engines" that drive the cell cycle through its various stages. The activities of these enzymes are regulated by a variety of " brakes" and " accelerators" that operate in combination with one another. These include:

(1) *Cyclin binding* A rise and fall of cyclin levels plays an important part in regulating Cdk activity during the cell cycle. When a cyclin is present in the cell, it binds to the catalytic subunit of the Cdk partner. The binding causes a major change in the conformation of the catalytic subunit and the movement of a flexible loop of

the Cdk polypeptide chain away from the opening to the enzyme's active site, allowing the Cdk to phosphorylate its protein substrates (Figure 9-10B).

(2) *Cdk phosphorylation state* Activity of Cdk is also regulated by phosphorylation and dephosphorylation. The cyclin-Cdk complex to be active it has to be phosphorylated at one or more sites by a specific protein kinase, and be phosphorylated at other sites by a specific protein phosphatase. For example, CDK1 is activated as following steps: In step 1, one of the kinases, called CAK (Cdk-activating kinase), phosphorylates a critical Thr residue (Thr 161 in Figure 9-14). Phosphorylation of this residue is necessary, but not sufficient, for the Cdk to be active. A second protein kinase shown in step 1, called Weel, phosphorylates a Thr residue and a Tyr residue in the enzyme (Thr14 and Tyr 15 in Figure 9-14A). If these two residues are phosphorylated, the enzyme is inactive, regardless of the phosphorylation state of any other residue. In other words, the effect of Weel overrides the effect of CAK, keeping the Cdk in an inactive state. Line 2 of figure 9-14B shows the phenotype of the cells with a mutant *wee*1 gene. These mutants cannot maintain the Cdk in an inactive state to divide at an early stage in the cell cycle producing smaller cells, hence the name "wee." In normal cells, Weel keeps the Cdk inactive until the end of G_2. Then, at the end of G_2, the inhibitors of phosphate at Thr14 and Tyr15 are removed by the third enzyme, a phosphatase named Cdc25 (step 2, Figure 9-14A). Removal of this phosphate switches the stored cyclin-Cdk molecules into the active state, driving the yeast cell into mitosis. Line 3 of figure 9-14B shows the phenotype of cells with a mutant *Cdc*25 gene. These mutants cannot remove the inhibitory phosphate from the Cdk and cannot enter mitosis. The balance between Weel kinase and Cdc25 phosphatase activities, which normally determines whether the cell will remain in G_2 or progress into mitosis, is regulated by still other kinases and phosphatases.

(3) *Cdk inhibitors* Cdk activity can be blocked by a variety of inhibitors. In budding yeast, for example, a protein called Sic1 acts as a Cdk inhibitor during G_1. The degradation of Sic1 allows the cyclin-Cdk that is present in the cell to initiate DNA replication. In mammal cells, P27 and P21 proteins can suppress G_1/S-Cdk and S-Cdk activities when cells get into differentiation and DNA damage respectively. The protein P16 is also a kind of inhibitor for G_1-Cdk, which can prevent cells over-dividing. So P16 is frequently inactivated in cancer cells.

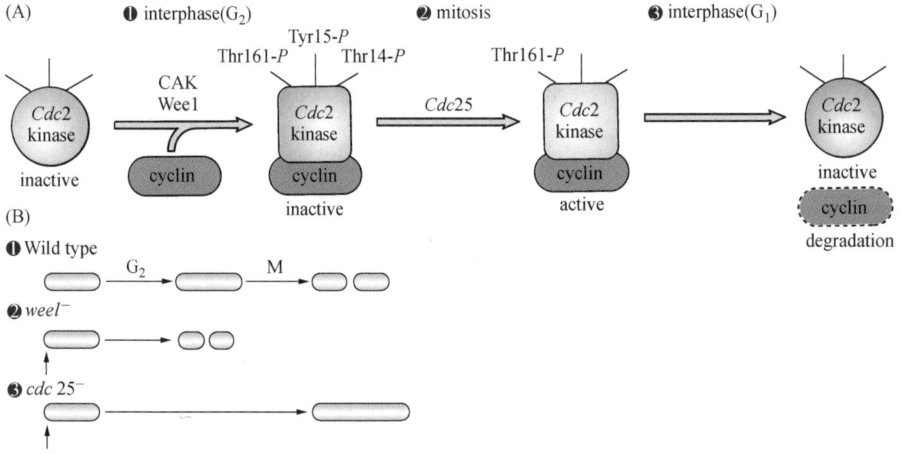

Figure 9-14 Progression through the fission yeast cell cycle requires the phosphorylation and dephosphorylation of critical Cdc2 residues (From: Karp G. 2005)

(4) *Controlled proteolysis* From figures 9-11 and 9-13 we can see that the variety of the cyclin concentration leads to change in the activity of Cdk. Cells regulate the concentration of cyclins and other key cell cycle proteins by adjusting both the rate of synthesis and destruction at different points of the cell cycle. Degradation is accomplished by means of the ubiquitin-proteasome pathway. Regulation of the cell cycle requires two important classes of multisubunit complexes, SCF and APC complexes. These complexes recognize proteins to be degraded and link these proteins to a polyubiquitin chain, which ensures their destruction in a proteasome. The SCF complex is active from late G_1 through early mitosis and mediates the destruction of G_1 cyclins, Cdk inhibitors, *et al* (Figure 9-15A).

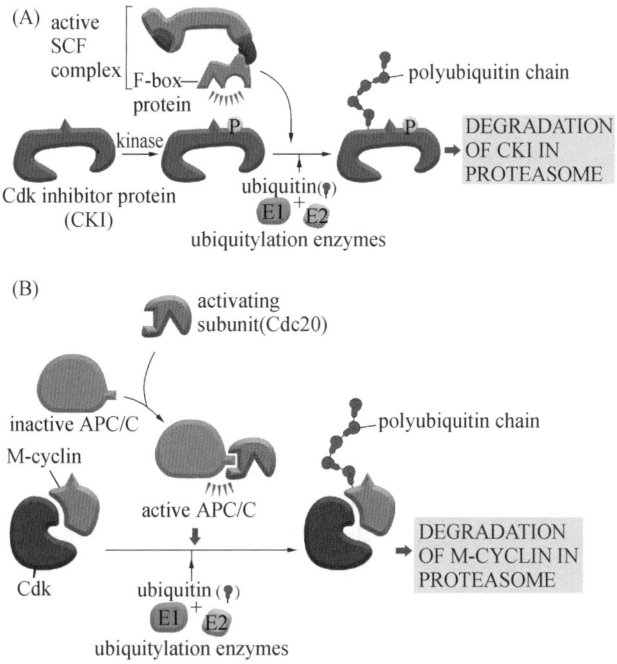

Figure 9-15 Control of proteolysis by SCF and APC complex during the cell cycle (From: Alberts B, et al. 2008)

These proteins become targets for an SCF after they are phosphorylated by the protein kinases. The subunits called F-box protein help SCF recognize the target proteins and then SCF catalyzes ubiquitylation of the targets. For example, SCF ubiquitylates a Cdk inhibitor protein (CKI) in late G_1 stage to help control DNA replication and cells entering into S phage. APC is an abbreviation for "anaphase promoting complex". The activity of APC acts in mitosis and degrades a number of key mitotic proteins, including the mitotic cyclins. When M-Cdk accumulates in M phage, it will stimulate Cdc20 activity. The Cdc20 protein, a subunit of APC complex triggers initial activation of APC to degrade M-Cdk kinases. Destruction of the mitotic cyclins allows a cell to exit mitosis and enter a new cell cycle (Figure 9-15B).

As noted above, the proteins and processes that control the cell cycle are remarkably conserved among eukaryotes. As in yeast, successive waves of synthesis and degradation of different cyclin play a key role in driving mammalian cells from one stage to the next. The mammalian cells have several different versions of this protein kinase. The pairing between individual cyclin and Cdks is highly specific, and only certain combinations are found (Figure 9-16). For example, the activity of a cyclin E-Cdk2 complex triggers the cell into S

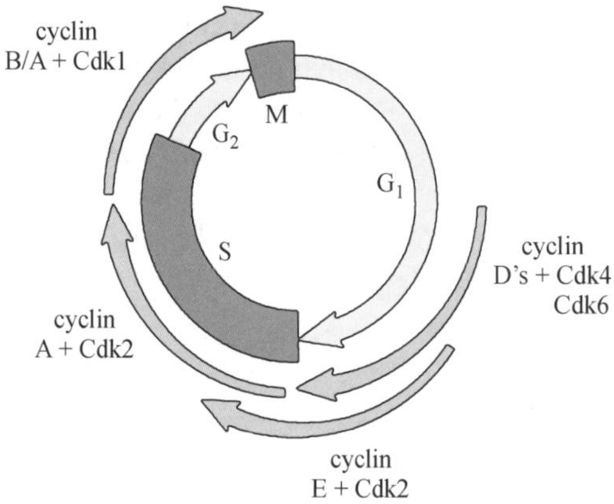

Figure 9-16 Combinations between various cyclins and cyclin-dependent kinases at different stages in the mammalian cell cycle (From: Karp G. 2005)

phase, whereas activity of a cyclin B-Cdk1 complex drives the cell into mitosis. Cdks do not always stimulate activities, but can also inhibit inappropriate events. For example, cyclin B-Cdk1 activity during G_2 prevents a cell from re-replicating DNA that has already been replicated earlier in the cell cycle. This helps ensuring that each region of the genome is replicated once and only once per cell cycle.

The regulation system for cyclin/Cdk complexes during the whole cell cycle is very important to cell division. It was wide known that an uncontrolled cell cycle is an obvious maker of tumor cell. This dysregulated cell cycle which can be caused by misregulated CDKs promise in the treatment of diverse cancers. CDK4/6, which play a crucial role in the G_1-S phase transition, have been found be over-activated in cancers. Several inhibitors of CDK4/6 have recently been approved as a new therapeutic strategy to treat cancers such as breast cancer.

9.3 Cell division

According to the third tenet of the cell theory, new cells originate only from other living cells. The process by which this occurs is called **cell division**. For a multicellular organism, such as a human or an oak tree, countless divisions of a single-cell zygote produce an organism of astonishing cellular complexity and organization. Cell division does not stop with the formation of the mature organism but continues in certain tissues throughout life. Millions of cells residing within the marrow of your bones or the lining of your intestinal tract are undergoing division at this very moment. This enormous output of cells is needed to replace cells that have aged or died.

Although cell division occurs in all organisms, it takes place very differently in prokaryotes and eukaryotes. We will restrict discussion to the eukaryotic version here. Mitosis and meiosis, the two distinct types of eukaryotic cell division will be discussed. Mitosis leads to production of cells that are genetically identical to their parent, whereas meiosis leads to production of cells with half the genetic content of the parent. Mitosis serves as the basis for producing new cells, meiosis as the basis for producing sexually reproducing organisms. Together, these two types of cell division form the links in the chain between parents and their offspring and, in a broader sense, between living species and the earliest eukaryotic life forms present on Earth.

9.3.1 Mitosis

Mitosis is a process of nuclear division in which replicated DNA molecules of each chromosome are faithfully partitioned into two nuclei. The cell division that follows is called **cytokinesis**, the separation of the two cells by division of the plasma membrane. These two stages compose the M phage in the cell cycle (Figure 9-17A). The two daughter cells resulting mitosis and cytokinesis possess a genetic content identical to each other and to the mother cell from which they arose. Therefore, mitosis maintains a constant amount of genetic material from cell generation to cell generation. Mitosis can take place in either haploid or diploid cells. Although mitosis proceeds as a continuous sequence of events, it is generally divided into six stages: prophase, prometaphase, metaphase, anaphase, and telophase, each characterized by a particular series of events (Figure 9-17B). A cell can pause in metaphase, but once it passes this stage, the cell will carry on to the end of mitosis and go into cytokinesis. Notably, we must keep in mind that the division of mitosis into arbitrary phases is done only for the sake of discussion and experimentation.

(A)

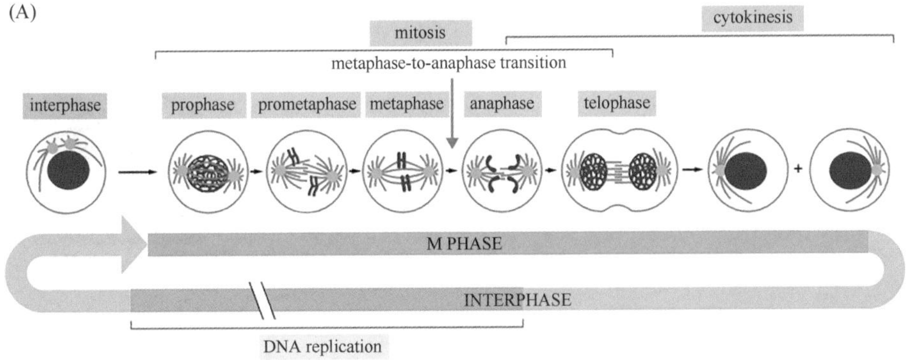

(B)

(1) The replicated chromosomal material condenses to form compact chromosomes, consisting of two associated sister chromatids (2) Cytoskeleton is disassembled and mitotic spindle is assembled (3) Golgi complex and ER fragment (4) Nuclear envelope disperses	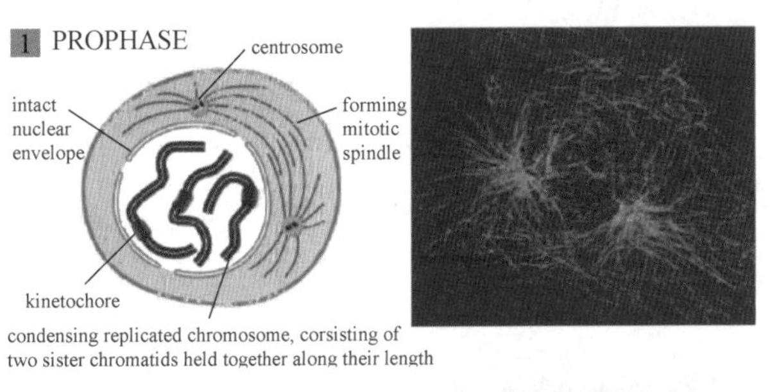
(1) Chromosomal microtubules attach to kinetochores of chromosomes (2) Chromosomes are moved to spindle equator	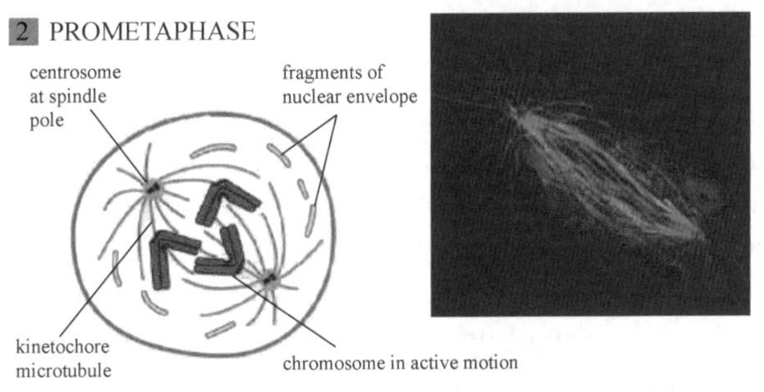
Chromosomes are aligned at the equator of the spindle, attached to opposite of the spindle with kinetochores	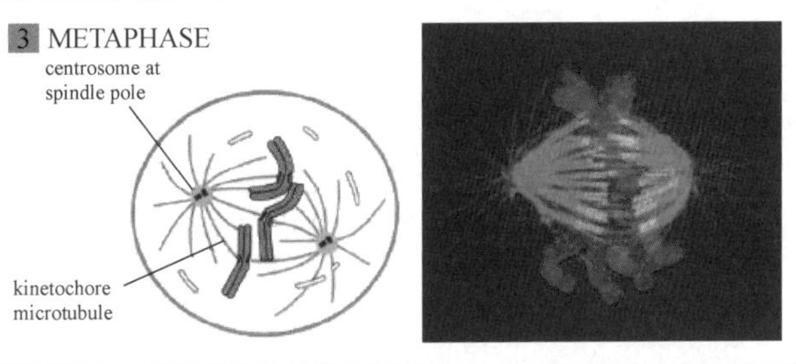

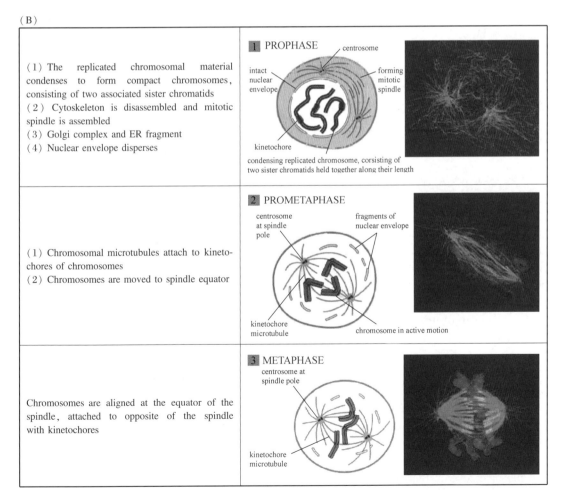

(1) Centromeres split, and sister chromatids separate to form two daughter chromosomes
(2) The separated chromosomes move to opposite spindle poles

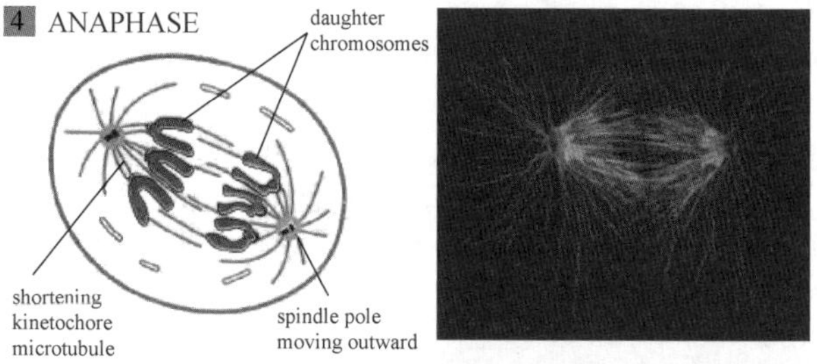

(1) The two sets of daughter chromosomes cluster at opposite spindle poles and decondense
(2) Nuclear envelope reassembles around each sets of chromosomes cluster
(3) Golgi complex and ER reforms

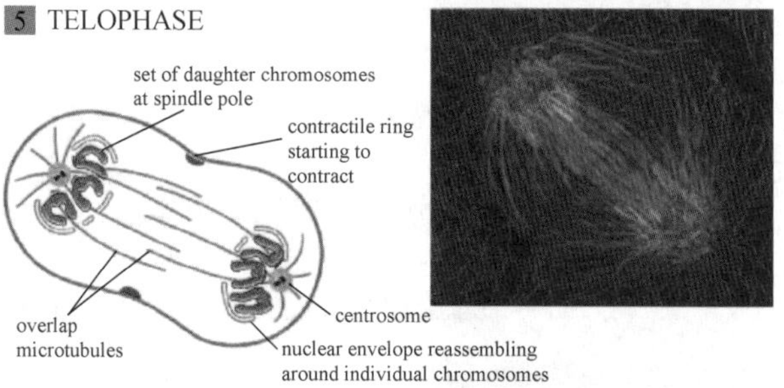

The cytoplasm is divided at equator plate to create two daughters, each with one nucleus

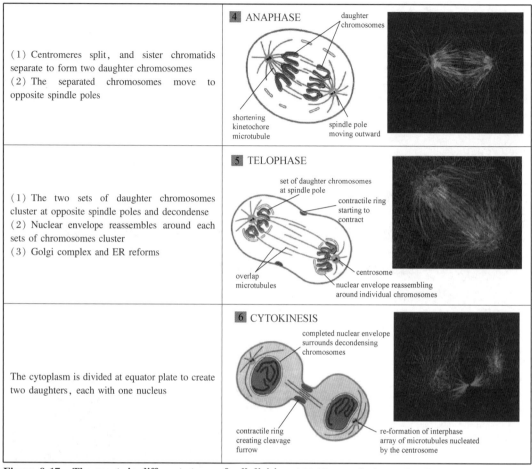

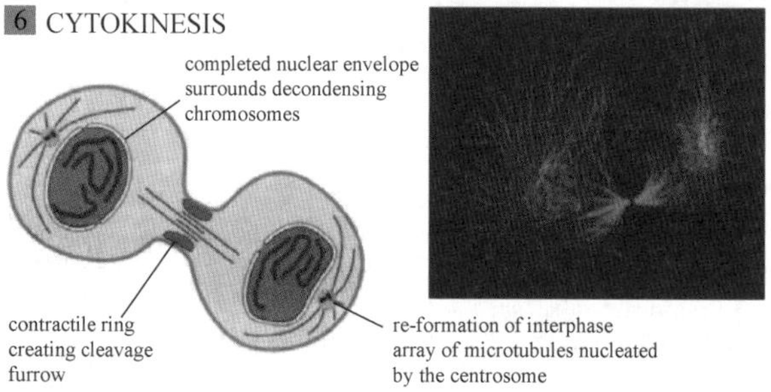

Figure 9-17 The events in different stages of cell division. (A) The overall of stages in cell cycle. The processes of M phase include nuclear division (mitosis) and cell division (cytokinesis) occupies a small fraction of the cell cycle. (B) The principal stage of M phase (Mitosis and Cytokinesis) in an animal cell. In the fluorescence micrograph, chromosomes are stained red and microtubules are green (From: Albert B, et al. 2012)

1. Stages of Mitosis

(1) Prophase During the first stage of mitosis, **prophase**, the duplicated chromosomes are prepared for segregation, and the mitotic machinery is assembled.

1) **Formation of the mitotic chromosome**: The extended state of interphase replicated begin to condense and to be converted into much shorter, thicker structures by a remarkable process of **chromosome compaction** (or **chromosome condensation**), and they are easily visible in the light microscope as chromosomes (Figure 9-17).

As the result of compaction, the chromosomes of mitotic cell appear as distinct, rod-like structures. Each duplicated chromosome is seen as a pair of sister chromatids joined by the duplicated but unseparated centromere (Figure 9-18). Sister chromatids are a result of replication in the previous interphase.

2) **Centromeres and kinetochores**: The most notable landmark on a mitotic chromosome is an indentation or primary constriction, which marks the position of the **centromere** (Figure 9-18). The centromere is the residence of highly repeated DNA sequences (satellite DNA) that serve as the binding sites for specific proteins. Examination of sections through a mitotic chromosome reveals the presence of a proteinaceous, button-like structure, called **kineto-**

chore, at the outer surface of the centromere of each sister-chromatid (Figure 9-19A, B). The kinetochore assembles on the centromere during prophase, and they form the centromere-kinetochore complex. As will be apparent shortly, the kinetochore functions as: ①the site of attachment of the chromosome to the dynamic microtubules of the mitotic spindle (Figure 9-19A, B); ②the residence of several motor proteins involved in chromosome motility (Figure 9-19C, D); ③a key component in the signaling pathway of an important checkpoint.

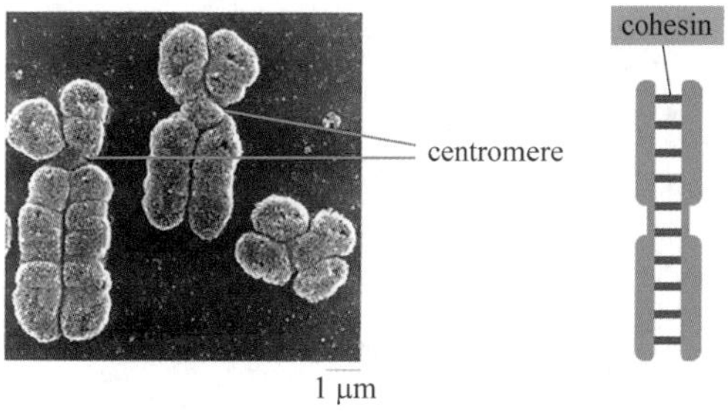

Figure 9-18 Scanning electron micrograph of mitotic chromosomes. Each mitotic chromosome is comprised of a pair of sister chromatids connected to one another by the protein complex **cohesin** (From: Karp G. 2005 and Albert B, et al. 2012)

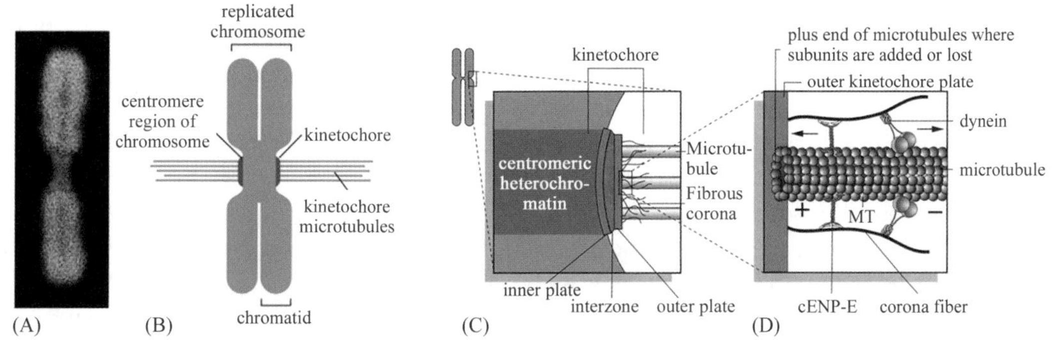

Figure 9-19 The kinetochore. (A) Fluorescence micrograph of a metaphase chromosome. The two kinetochores associated with each sister chromatid are stained red using specific kinetochore protein antibody. (B) A drawing of a metaphase chromosome. Its two sister chromatids attached to the ends of kinetochore microtubules. (C) A drawing of a section through the kinetochore of a mammalian metaphase chromosome, showing its three-layered structure (inner plate, interzone and outer plate). The inner plate is thought to consist of a specialized layer of chromatin that is attached to the centromeric heterochromatin of the chromosome, whereas the outer plate contains proteins that are thought to the plus (+) end of microtubules. (D) A schematic model of the disposition of motor proteins at the outer surface of the kinetochore. These motors play a role in tethering the microtubule to the kinetochore. (From: Karp G. 2005 and Albert B, et al. 2012)

3) **Formation of the mitotic spindle**: As discussed earlier, we knew how microtubule assembly in animal cells is initiated by a special microtubule-organizing structure, the centrosome. As a cell progresses past G_2 and into mitosis, the microtubules of the cytoskeleton undergo sweeping disassembly in preparation for their reassembly as components of a complex, microtubule-containing "machine" called the **mitotic spindle.** The rapid disassembly of the interphase cytoskeleton is thought to be accomplished by the inactivation of proteins that stabilize microtubules (e. g., **microtubule-associated proteins**, or **MAPs**) and the activation of proteins that destabilize these polymers.

As DNA replication begins in the nucleus at the onset of S phase, each centriole duplicates its centrosome to produce two daughter centrosomes, which initially remain together at one side of the nucleus. At the beginning of prophase, the two daughter centrosomes separate (Figure 9-20). They organize their own array of microtubules and begin to move to opposite poles of the cell and gradually form the framework of the mitotic spindle. The two centrosomes that give rise to these microtubules are called **spindle poles**. The correct equipment of mitotic spindle is very important for mitosis, especially for average distribution of chromosomes into two daughter cells.

4) **Dissolution of nuclear envelope and the partitioning of cytoplasmic organelles**: The nuclear envelope disappears at the end of prophase. Meanwhile, this signals the beginning of

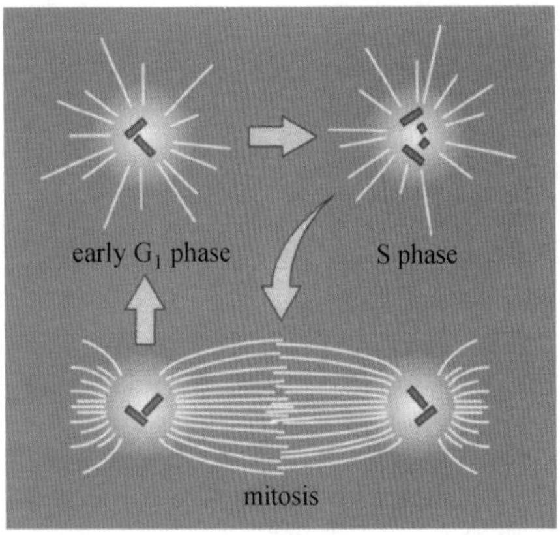

Figure 9-20 The centrosome cycle of an animal cell (From: Karp G. 2005)

the next stage of cell cycle called prometaphase. The disassembly of the nuclear lamina, which constitutes the inner layer of the nuclear envelope, is promoted by phosphorylation of the lamina molecules by the mitotic Cdk kinase (M-Cdk). The phosphorylation phenomenon leads to disassembly of the nuclear lamina and the nuclear envelope is fragmented into a population of small vesicles (nuclear lamina A) or soluble monomer proteins (nuclear lamina B) that disperse throughout the mitotic cell. During the process of mitosis, the Golgi membranes also become fragmented to form a distinct population of small vesicles that are divided between daughter cells. It is generally held that the endoplasmic reticulum undergoes fragmentation during prophase to give rise to large numbers of vesicles that become dispersed throughout the cytoplasm.

(2) Prometaphase Upon nuclear envelope breakdown, the cell enters the second phase of mitosis, **prometaphase**, during which mitotic spindle assembly is formed and the chromosomes attach to microtubules in the spindle via their kinetochores and move into the center of the cell (Figure 9-17). At the beginning of prometaphase, compacted chromosomes are scattered throughout the space that was the nuclear region. As the microtubules of the spindle penetrate into the central region of the cell, the free (plus) ends of the microtubules are seen to grow and shrink in a highly dynamic fashion, searching for a chromosome. Once attached, the chromosomes start to align along the metaphase plate in the center of the spindle.

The spindle microtubules end up attached to the chromosomes through specialized protein complexes called kinetochores, which assemble on each centromere of chromosome. Once the nuclear envelope has broken down, the microtubule encounters a chromosome and captures it. As kinetochores on sister chromatids face in opposite directions, they tend to attach to microtubules from opposite poles of the spindle. So, the two sister chromatids of each mitotic chromosome ultimately become connected by their kinetochores to spindles (microtubules) that extend to form opposite poles. With the chromosomes moving toward the center of the mitotic spindle, the longer microtubules are shortened, while the shorter microtubules are elongated. Shortening and elongation of microtubules carry out primarily by loss or gain of subunits at the plus end of the microtubule (Figure 9-21). Remarkably, this dynamic activity occurs while the plus end of each microtubule remains attached to a kinetochore. The motor proteins (dyneins) located at the kinetochore (Figure 9-19D) may play a key role in tethering the microtubule to the chromosome while it loses and gains subunits.

Eventually, each chromosome aligns at the center of the spindle, so that microtubules from each pole are equivalent in length. It marks the end of prometaphase and the beginning of metaphase.

(3) Metaphase Metaphase is characterized by the lining up of the chromosomes along the equator of the spindle, midway between the spindle poles (Figure 9-17, 9-22). The plane of alignment of the chromosomes at metaphase is called the **metaphase plate**. The paired kinetochore microtubules on each chromosome attach to opposite poles of the spindle. The

mitotic spindle of the metaphase cell contains a highly organized array of microtubules that is ideally suited for separating the duplicated chromatids positioned at the center of the cell.

In animal cells, the microtubules of the metaphase spindle can be functionally divided into three types (Figure 9-22). The interacting microtubules are called **interpolar microtubules**, the microtubules with free plus ends are called **astral microtubules**, and the microtubules which catch chromosomes are called **kinetochore microtubules**:

1) **Astral microtubules**: That radiate outward from the **centrosome** into the region outside the body of the spindle. They contact the cell cortex and help position the spindle apparatus in the cell.

2) **Chromosomal (or kinetochore) microtubules**: That stretch from the centrosome to the kinetochores of the chromosomes. During metaphase, the chromosomal microtubules exert a pulling force on the kinetochore. As a result, the chromosomes are maintained in the equatorial plane by a "tug-of-war" between balanced pulling forces exerted by chromosomal spindle fibers from opposite poles. During anaphase, chromosomal microtubules are required for the movement of the chromosomes toward the poles.

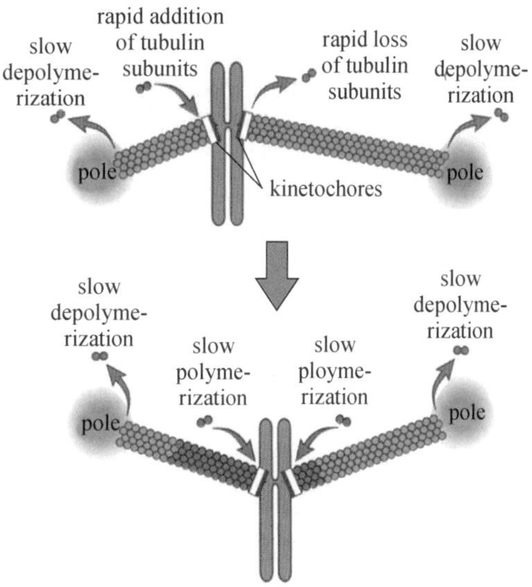

Figure 9-21 Microtubule behavior during formation of the metaphase plate. Initially, the chromosome is connected to microtubules from opposite poles that are very different in length. As prometaphase continues, this imbalance is corrected as the result of the shortening of microtubules from one pole, due to the rapid loss of tubulin subunits at the kinetochore, and the lengthening of microtubules from the opposite pole, due to the rapid addition of tubulin subunits at the kinetochore (From: Karp G. 2005)

3) **polar microtubules**: That extend from the centrosome past the chromosomes. Inter polar microtubules from one centrosome overlap with their counterparts from the opposite centrosome. The inter polar microtubules form a structural basket that maintains the mechanical integrity of the spindle.

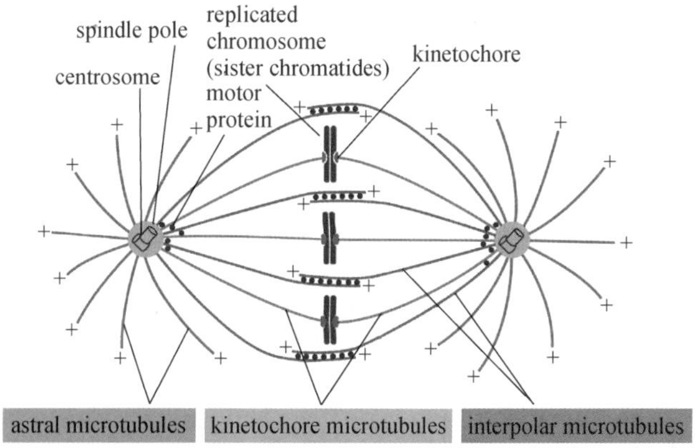

Figure 9-22 Three classes of microtubules make up the mitotic spindle. Schematic drawing of a spindle with chromosomes attached, showing the three types of spindle microtubules: astral microtubules, kinetochore microtubules, and inter polar microtubules. Each spindle pole contains a pair of centrioles surrounded by amorphous pericentriolar material at which the microtubules are nucleated. All of the spindle microtubules have their minus ends at centrosome. (From: Alberts B, et al. 2008)

(4) Anaphase The paired chromatids synchronously separate to form two daughter chromosomes at the onset of anaphase, and the chromatids begin their poleward migration (Figure 9-17). The movement of each chromosome toward a pole results from the shortening of the chromosomal microtubules attached to its kinetochore. As the chromosome moves during anaphase, its centromere is seen at

its leading edge with the arms of the chromosome trailing behind. Anaphase can be divided into two continuous stages, anaphase A and anaphase B. In anaphase A, the kinetochore microtubules are shortened by depolymerization, and the attached chromosomes move poleward. The driving force for the movements of each chromatid is provided mainly by the action of **dynein**, a kind of microtubule motor protein. This process uses the energy of ATP hydrolysis to remove tubulin subunits from the microtubule.

In anaphase B, after the sister chromatids have separated, the polar microtubules are gradually elongated, and the two spindle poles move farther apart. At the same time, the astral microtubules at each spindle pole pull the poles away from each other (Figure 9-23). The driving forces for pushing the spindle poles and the sister chromatids farther apart are thought to be from two sets of motor proteins-members of the **kinesin** and **dynein** families. The dyneins which interact with kinetochore microtubules and astral microtubules walk towards the centrosome to help the sister chromatids separate each other and fix the centrosomes at the two poles of the cell. The elongation of the interpolar microtubules during anaphase B is accompanied by the net addition of tubulin subunits to the plus ends of the polar microtubules. Thus, the kinesins connecting two interpolar spindles move to the plus ends of the microtubules and push the poles apart each other (Figure 9-24).

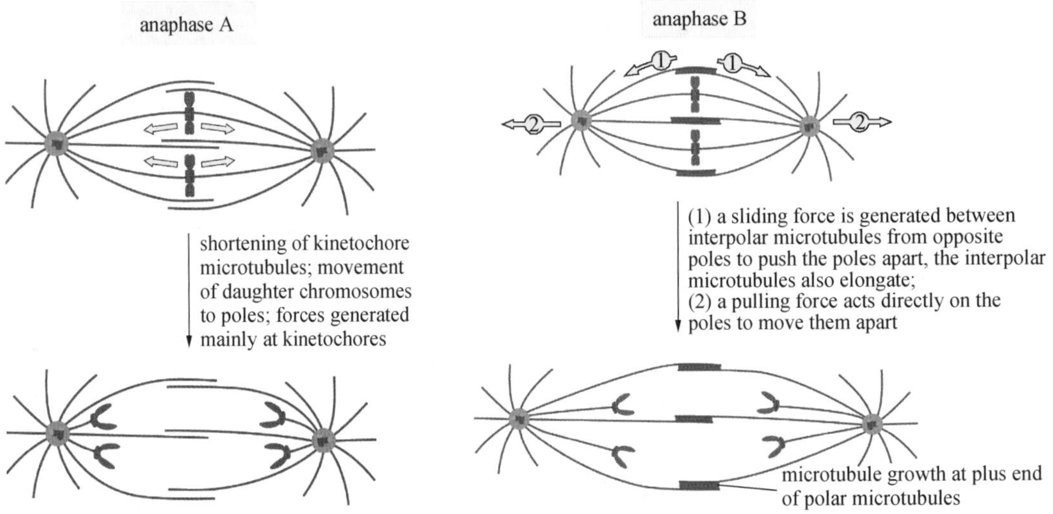
anaphase A
anaphase B
shortening of kinetochore microtubules; movement of daughter chromosomes to poles; forces generated mainly at kinetochores
(1) a sliding force is generated between interpolar microtubules from opposite poles to push the poles apart, the interpolar microtubules also elongate;
(2) a pulling force acts directly on the poles to move them apart
microtubule growth at plus end of polar microtubules

Figure 9-23 Two processes separate sister chromatids at anaphase (From: Alberts B, et al. 2008)

(5) Telophase During telophase, chromatids arrive at opposite poles of cell, and new membranes form around the daughter nuclei. The nuclear envelope reforms and the chromosomes become disperse and are no longer visible under the light microscope (Figure 9-25). The division of the cytoplasm begins with the assembly of the contractile ring (Figures 9-17, 9-25). Meanwhile, the spindle fibers disperse and the kinetochore microtubules begin to disintegrate at this stage.

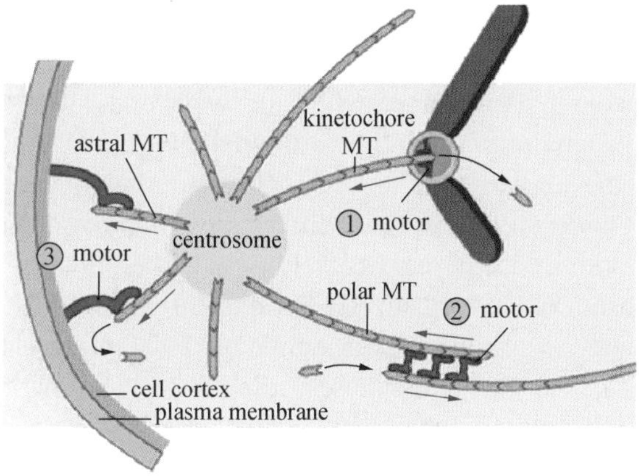

Figure 9-24 Microtubules and motors in the spindle. In anaphase A, ①motors (dyneins) move to the minus ends of kinetochore microtubules and help the sister chromatids separate apart. In anaphase B. ② motors (kinesins) walk to the plus ends of interpolar microtubules and push the ploes to be far away each other. ③motors (dyneins) which is on the astral MT will fix two centrosomes at the poles of the cells.

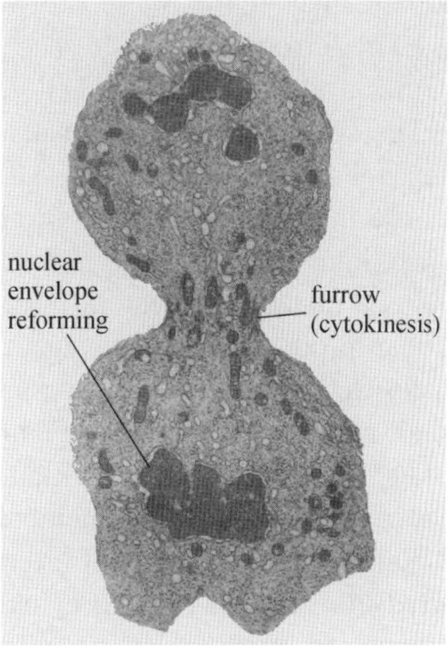

Figure 9-25 Telophase. Elcectron micrograph of a section through an ovarian granulose cell in telophase (From: Karp G. 2005)

2. Cytokinesis

(1) Cytokinesis in animal cells The final stage in the process of cell division is known as **cytokinesis**, which usually begins in anaphase but is not completed until after the two daughter nuclei have formed. During cytokinesis, the cytoplasm divided by a process termed cleavage, driven by the tightening of a **contractile ring** composed of actin and myosin protein subunits (Figure 9-26A). The first visible sign of cytokinesis is when the cell begins to pucker in, a process called **furrowing** (Figure 9-26B, C). As the ring of cytoskeletal proteins contracts, a cleavage furrow is formed perpendicular to the mitotic spindle and gradually splits the cytoplasm and its contents into two daughter cells. At the beginning of cytokinesis, a narrow band forms around the cell surface. As time progresses, the indentation deepens to form a furrow completely encircling the cell. The plane of the furrow lies in the same plane previously occupied by the chromosome of the metaphase

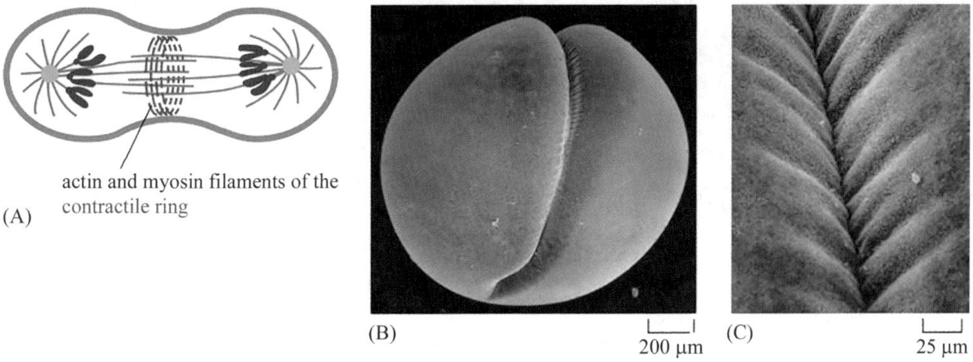

Figure 9-26 Formation and operation of the contractile ring during cytokinesis. (A) The actin-myosin bundles of the contractile ring. (B) The cleavage furrow of a cleaving frog egg in the low-magnification scanning electron. (C) The surface of the furrow at higher magnification (From: Albert B, et al. 2012)

plate, so that the two sets of chromosomes are ultimately partitioned into different cells. The furrow continues to deepen until opposing surfaces make contact with one another and fuse in the center of the cell, thus splitting the cell in two.

The contractile ring is finally dispensed with altogether when cleavage ends, as the plasma membrane of the cleavage furrow narrows to form the **midbody** (Figure 9-27A, B). The midbody contains the remains of the central spindle and acts as a tether between the two daughter cells. The midbody can remain on the inside of the plasma membrane and serve as a mark on the cortex to orient the spindle in the next cell division. The contractile ring that produces cell cleavage is composed of an organized cytoskeletal network that includes actin and bipolar myosin-II filaments working together in a sliding action that mimics muscle contraction (Figure 9-27C). In early telophase, after sister chromatids separate, actin and myosin filaments which release when cell enters mitosis begin to polymerize rapidly on the membrane to form the contractile ring. Once assembled, the contractile ring is capable of exerting a force strong enough to tend a fine glass needle inserted into the cell prior to cytokinesis. The force-generating mechanism operating during cytokinesis is thought to be similar to the actinomyosin-based contraction of muscle cells. Whereas the sliding of actin filaments of a muscle cell brings about the shortening of the muscle fiber, sliding of the filaments of the contractile ring pulls the cortex and attached plasma membrane toward the center of the cell. GTPase RhoA has been found in regulating of the contraction of actin-myosin ring in some research. As a result, the contractile ring constricts the equatorial region of the cell, which much likes pulling on a purse string narrows the opening of a purse.

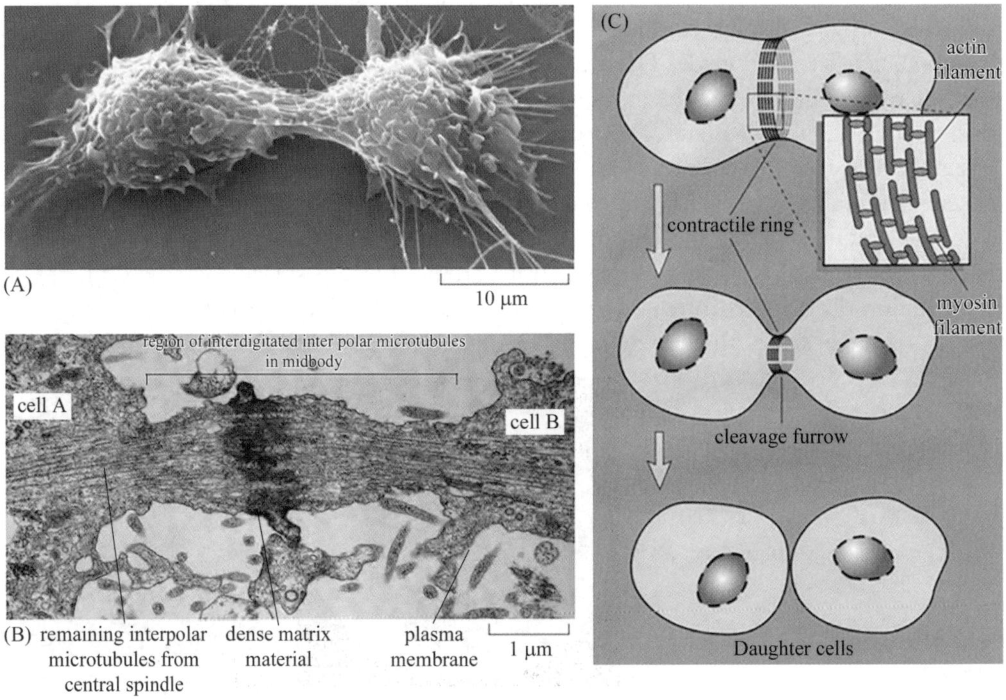

Figure 9-27 The contractile ring divides the cell in two. (A) Scanning electron micrograph of an animal cell in culture in the last stages of dividing. The midbody joins the two daughter cells. (B) A conventional electron micrograph of the midbody in an animal cell. (From: Albert B, et al. 2012) (C) Actin filaments become assembled in a ring at the cell equator. Contraction of the ring, which requires the action of myosin, causes the formation of a furrow that splits the cell in two (From: Karp G. 2005)

(2) Cytokinesis in plant cells In higher plant cells, cytokinesis is regulated by the cell wall and occurs by a different mechanism. Unlike animal cells, plant cells must construct an extracellular wall inside a living cell. Wall formation starts in the center of the cell and grows outward to meet the existing lateral walls. The formation of a new wall begins with the construction of a simple precursor, which is called the **cell plate**.

The new cell wall starts to assemble in the cytoplasm between the two sets of segregated chromosomes in late anaphase. The assembly process is guided by a structure called the **phragmoplast**, which is formed by the remains of the interpolar microtubules at the equator of the old mitotic spindle (Figure 9-28B). After formation of the phragmoplast, small Golgi-derived secretory vesicles move into the region, probably transported along the microtubules, and become aligned along a plane between the daughter nuclei. These vesicles are filled with polysaccharide and glycoproteins which are pivotal elements of the new cell wall. As more vesicles go to the equator, the cell plate expands until it bumps into the cell membrane and fuses with it (Figure 9-28C). Thus, the plant cell divides into two with new cell wall between them. Later, cellulose microfibrils are laid down within the matrix to complete the construction of the new cell wall (Figure 9-28D).

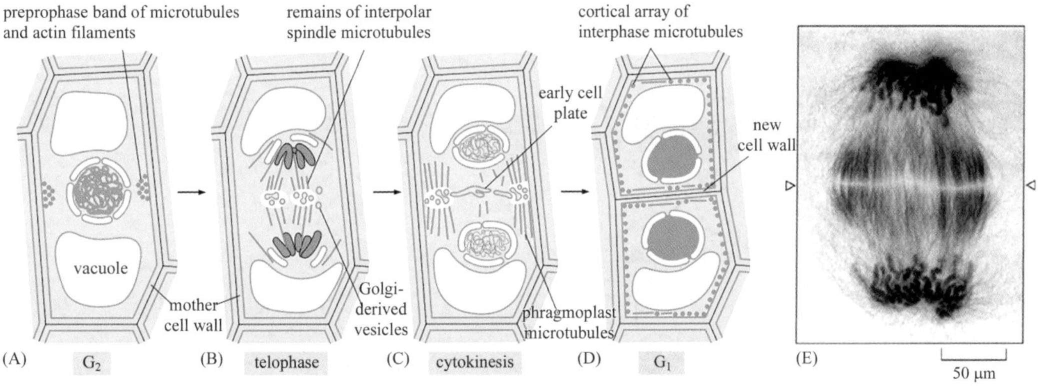

Figure 9-28 The features of cytokinesis in a dividing higher-plant cell. The membrane-enclosed vesicles, derived from the Golgi apparatus and filled with cell wall material, fuse to form the growing new cell wall, which grows outward to reach the plasma membrane and original cell wall. The plasma membrane and the membrane surrounding the new cell wall fuse, completely separating the two daughter cells (From: Albert B, et al. 2012)

9.3.2 Meiosis

Meiosis is a special kind of cell division, which only occurs at some stage of the sexual cell reproduction. The production of offspring by sexual reproduction includes the union of two cells, each with a complete haploid set of chromosomes (Figure 9-29). The doubling of the chromosomes at fertilization is compensated by an equivalent reduction in chromosome number at meiosis. Meiosis ensures production of a haploid phase in the life cycle, and fertilization ensures a diploid phase. Without meiosis, the chromosome number would double with each generation, and sexual reproduction would not be possible. Meiosis involves a single round of DNA replication that duplicates the chromosomes, followed by two successive cell divisions, called meiosis I (the first meiotic division) and meiosis II (the second meiotic division), without further DNA replication. Meiosis I, which results in reduction in the number of chromosomes, can be factitiously divided into five stages: prophase I, prometaphase I, metaphase I, anaphase I and telophase I. Meiosis II is very similar to a mitotic division. As is the case in any cell division, the interphase of meiosis includes G_1 phase, S phase and G_2 phase. In order to distinguish from mitosis, the interphase of meiosis is called premeiotic interphase. The premeiotic S phase has two marked characteristics, one is that it takes several times longer than the premeiotic S phase; the other is that partial DNA (from 99.7% to 99.9%) is replicated.

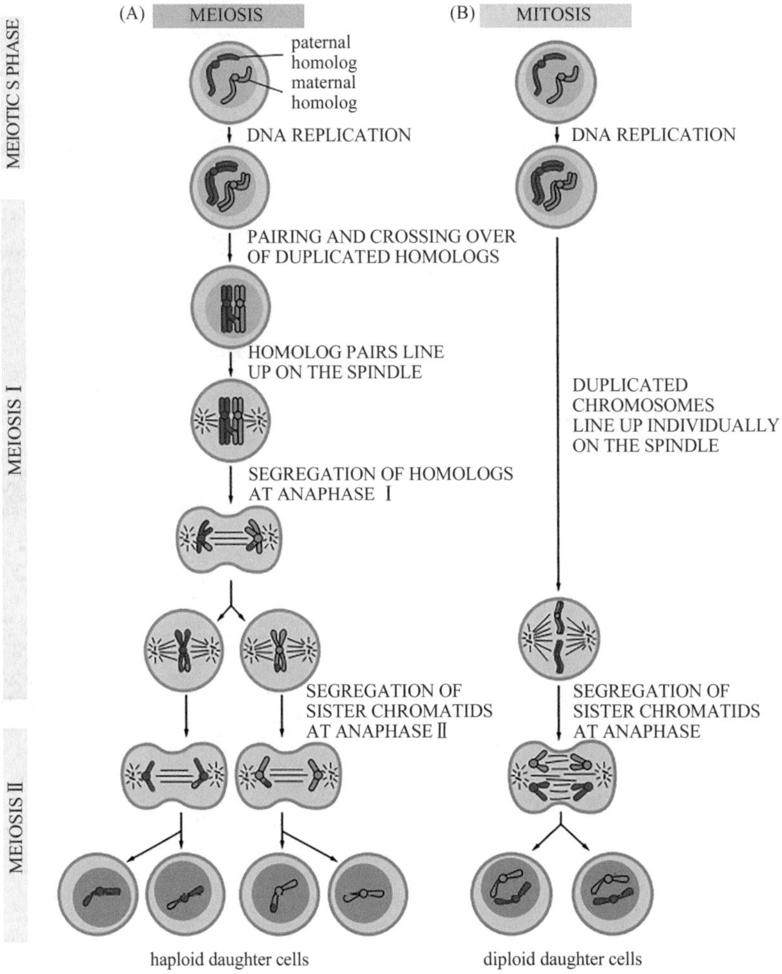

Figure 9-29 Overall flowchart of meiosis and compared to mitosis. (A) In Meiosis, there is only a single round of chromosome duplication and followed by two rounds of chromosome segregation. The duplicated homologs, each consisting of tightly bound sister chromatids, pair up and are segregated into different daughter nuclei in meiosis Ⅰ; the sister chromatids are segregated in meiosis Ⅱ. After meiosis, each diploid cell produces four genetically different haploid nuclei, which differentiate into gametes. (B) In mitosis, homologs do not pair up and the sister chromatids are segregated during the single division (From: Albert B, et al. 2012)

(1) The first meiotic division A series of complex events occurs during the long prophase of meiotic division Ⅰ: duplicated homologous chromosomes pair, genetic recombination is initiated between nonsister chromatids, and each pair of duplicated homologs assembles into an elaborate structure called the **synaptonemal complex** (SC). In some organisms, genetic recombination begins before the synaptonemal complex assembles and is required for the complex to form; in others, the complex can form in the absence of recombination. In all organisms, however, the recombination process is completed while the DNA is held in the synaptonemal complex, which serves to space out the crossover events along each chromosome.

1) **Prophase** I : Prophase I is traditionally divided into five sequential stages: **leptotene**, **zygotene**, **pachytene**, **diplotene**, and **diakinesis** — defined by the morphological changes associated with the assembly (synapsis) and disassembly (desynapsis) of the synaptonemal complex (Figure 9-30).

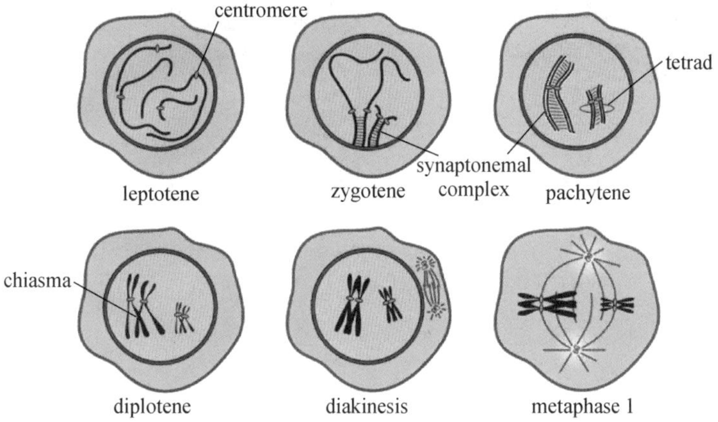

Figure 9-30 Stages of prophase I (From: Karp G. 2005)

Leptotene: The first stage of prophase I is **leptotene** (leptonema), during which the chromosomes condense and become gradually visible in the light microscope. Although the chromosomes have replicated at an earlier stage, there is no indication that each chromosome is actually composed of a pair of identical chromatids. In the electron microscope, however, the chromosomes are revealed to be composed of paired chromatids. The telomeres of maize leptotene chromosomes are distributed throughout the nucleus. Then, near the end of leptotene, there is a dramatic reorganization so that the telomeres become localized at the inner surface of the nuclear envelope at one side of the nucleus. Clustering of telomeres at one end of the nuclear envelope occurs in a wide variety of eukaryotic cells and causes the chromosomes to resemble a bouquet of flowers. And the clustering of telomeres at the nuclear envelope is thought to facilitate the alignment of homologues in preparation for synapsis.

Zygotene: The second stage of prophase I, which is called zygotene (pairing stage), is marked by the visible association of homologues with one another. This process of association is called synapsis and the complex formed by a pair of synapsed homologous chromosomes is called a bivalent or tetrad. The former term reflects the fact that the complex contains two homologues, whereas the latter term calls attention to the presence of four chromatids. It had been assumed for years that interaction between homologous chromosomes first begins as chromosomes initiate synapsis. Studies on yeast cells demonstrated that homologous regions of DNA from homologous chromosomes are already in contact with one another during leptotene. Chromosome compaction and synapsis during zygotene simply make this arrangement visible under the microscope. The first step in genetic recombination is the occurrence of double-stranded breaks in aligned DNA molecules. Studies in both yeast and mice suggest that the DNA breaks occur in leptotene, well before the chromosomes are visibly paired.

Electron micrographs indicate that chromosome synapsis is accompanied by the formation of a complex structure called the synaptonemal complex. The SC is a ladder-like structure with transverse protein filaments connecting the two lateral elements. The chromatin of

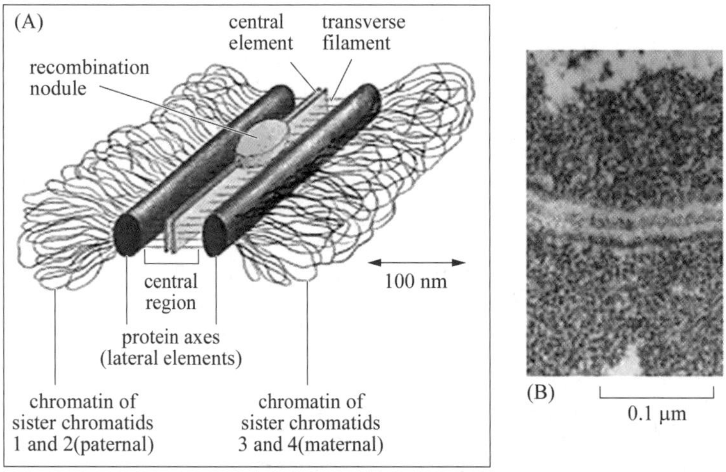

Figure 9-31 A mature synaptonemal complex. (A) Only a short section of the long ladder-like complex is shown. A similar synaptonemal complex is present in organisms as diverse as yeasts and humans (From: Karp G. 2005) (B) An electron micrograph of a synaptonemal complex at pachytene in a meiotic lily flower cell (From: Albert B, et al. 2012)

each homologue is organized into loops that extend from one of the lateral elements of the SC. The lateral elements are composed primarily of cohesin, which presumably binds together the chromatin of the sister chromatids. The SC is not required for genetic recombination but functions mainly as a scaffold to allow interacting chromatids to complete their crossover activities. In this stage, the rest 0. 1% ~ 0. 3% DNA, called zygDNA, finish replicating. The zygDNA happen transcription actively, which probably make homologous chromosomes pair each other.

Pachytene: The end of synapsis marks the end of zygotene and the beginning of the next stage of prophase I, called **pachytene** (pachynema), which is characterized by a fully formed SC. During pachytene, the homologues are held closely together along their length by the SC. The DNA of sister chromatids is extended into parallel loops (Figure 9-31). Under the electron microscope, a number of electron-dense bodies about 100 nm in diameter are seen within the center of the SC. These structures have been named recombination nodules because they correspond to the sites where crossing over is taking contain the enzymatic machinery that facilitates genetic recombination, which is completed by the end of pachytene.

Diplotene: Diplotene follows the pachytene, in which each chromosome continues shortening and thickening and the two chromosomes in each bivalent begin to repel each other and a split occurs between the chromosomes (Figure 9-32). The separation is not accomplished, but the homologous chromosomes stick together at certain points, the **chiasmata** (singular chiasma). Chiasmata are formed by covalent junctions between a chromatid from one homologue

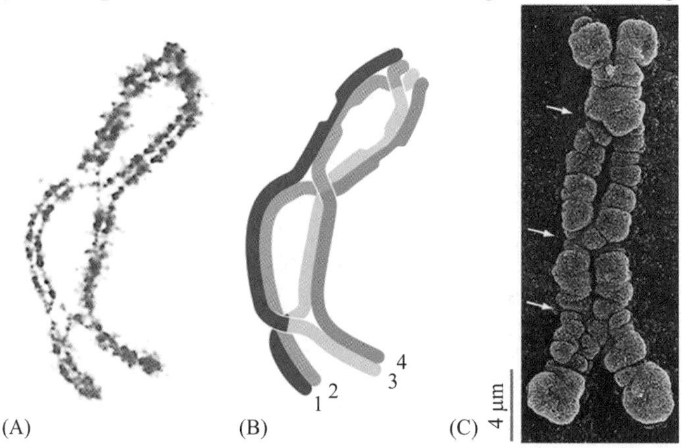

Figure 9-32 Visible evidence of chiasmata resulting from crossover events. place. Recombination nodules (A) Crossovers in a grasshopper bivalent in light micrograph. (B) Pairs of diplotene bivalents from the grasshopper showing the chiasmata formed between chromatids of each homologous chromosome. The accompanying inset indicates the crossover that has presumably occurred within the bivalent. Chromatid 1 exchanges with chromatid 3; chromatid 2 exchanges with chromatids 3 and 4. The chromatids of each diplotene chromosome are closely apposed except at the chiasmata (From: Karp G. 2005 and Albert B, et al. 2012)

and nonsister chromatid from extent of genetic recombination, which play a crucial part in holding the compact homologs together. The chiasmata are made more visible by a tendency for the homologues to separate from one another at the diplotene stage.

Diakinesis: During the final stage of meiotic prophase I, called **diakinesis**, the meiotic spindle is assembled, and the chromosomes are prepared for separation. The chromosomes condense and become more compact. Diakinesis ends with the disappearance of the nucleolus, the breakdown of the nuclear envelope, and the movement of the tetrads to the metaphase plate. In vertebrate oocytes, these events are triggered by an increase in the level of the protein kinase activity of MPF.

In some organism, just before or during diakinesis the chiasmata move to the ends of the chromosome arms. While in most eukaryotic species, chiasmata can still be seen in homologous chromosomes aligned at the metaphase plate of meiosis I. In humans and other vertebrates, every pair of homologues typically contains at least one chiasma, and the longer chromosomes tend to have two or three of them. It is thought that some mechanism exists to ensure that even the smallest chromosomes form a chiasma. If a chiasma does not occur between a pair of homologous chromosomes, the chromosomes of that bivalent tend to separate from one another after dissolution of the SC. This premature separation of homologues often results in the formation of nuclei with an abnormal number of chromosomes.

2) **Metaphase I and anaphase I**: In metaphase I, the bivalent become aligned in the center and their own homologous chromosomes are connected to the chromosomal fibers from opposite poles (Figure 9-33). In contrast, the kinetochores of sister chromatids are connected as a unit to microtubules from the same spindle pole. The orientation of the maternal and paternal chromosomes of each bivalent on the metaphase I plate is random; the maternal member of a particular bivalent has an equal likelihood of facing either pole. During anaphase I, the homologous chromosomes separate and move to the opposite pole, while the sister chromatids remain together. Thus, anaphase I is the cytological event that corresponds to Mendel's law of independent assortment. As a result of independent assortment, organisms are capable of generating a nearly unlimited variety of gametes.

Separation of homologous chromosomes at anaphase Ⅰ require the dissolution of the chiasmata that hold the bivalent together. The chiasmata are maintained by cohesin between sister chromatids in regions that flank these sites recombination (Figure 9-32A). The chiasmata disappear at the onset of anaphase Ⅰ, as the arms of the chromatids of each bivalent lose cohesin. Loss of cohesin between the arms is accomplished by proteolytic cleavage of the cohesin molecules in those regions of the chromosome. In contrast, cohesion between the joined centromeres of sister chromatids remains strong, because the cohesin situated there is protected from proteolytic attack. As a result, sister chromatids remain firmly attached to one another as they move together toward a spindle pole together during anaphase Ⅰ.

3) **Telophase I**: Telophase I of meiosis is very similar to the telophase of mitosis. During telophase Ⅰ, the homologous chromosomes pairs complete their migration to the two poles as a result of the action of the spindle. A haploid set of chromosomes is at each pole, with each chromosome still having two chromatids. The nuclear envelope may or may not reform around each chromosome set, the spindle disappears, and cytokinese follows. Meiotic division Ⅰ is far more complex and requires much more time than either mitosis or meiotic division Ⅱ. Even the preparatory DNA replication during meiotic division Ⅰ tends to take much longer than an ordinary S phase, and cells can then spend days, months, or even

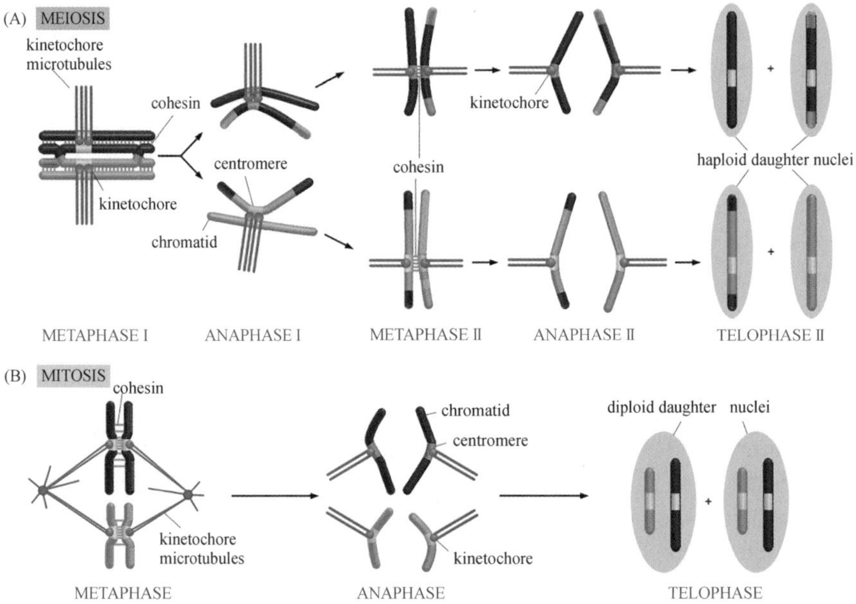

Figure 9-33 Separation of homologous chromosomes during meiosis Ⅰ and separation of chromatids during meiosis Ⅱ. (A) In meiosis Ⅰ, the two sets of kinetochore microtubules are located side-by-side on each homolog at the sister centromeres. The proteolytic destruction of the cohesion complexes along the sister chromatid arms unglues the arms and destroy the crossovers, which allow the duplicated homologs to separate at anaphase Ⅰ, yet the residual cohesin complexes at the centromeres keep the sisters together. The proteolytic destruction of the residual cohesin complexes at the centromeres allows the sister chromatids to separate at anaphase Ⅱ. (B) In mitosis, the two kinetochores which attach to microtubules from different spindle poles, come apart at the start of anaphase and segregate into two daughter cells (From: Albert B, et al. 2012)

years in prophase Ⅰ, which depend on the species and on the gamete being formed.

(2) The second meiotic division When meiotic division Ⅰ endcs, nuclear membranes reform around the two daughter nuclei and the short-lived interphase of meiotic division Ⅱ begins. During this period, the chromosomes may decondense somewhat, but usually they soon recondense and prophase Ⅱ begins (because there is no DNA synthesis during this interval, in some organisms the chromosomes seem to pass almost directly from one division phase into the next). Prophase Ⅱ is brief: If the nuclear envelope had reformed in telophase Ⅰ, it is broken down again. The chromosomes become recompacted and line up at the metaphase plate. Unlike metaphase Ⅰ, the kinetochores of sister chromatids of metaphase Ⅱ face opposite poles and become attached to opposing sets of chromosomal spindle fibers (Figure 9-33). Anaphase Ⅱ begins with the synchronous splitting of the centromeres, which had held the sister chromatids together, allowing them to move toward opposite poles of the cell. Meiosis Ⅱ ends with telophase Ⅱ, in which the chromosomes are once again enclosed by a nuclear envelope. The products of meiosis are haploid cells with a 1C amount of nuclear DNA.

In female animals, to end meiosis, the germ cell only produces one mature egg (oocyte), the rest three haploid cells called polar body will disintegrate. However, in male animals, after meiosis, the germ cell will produce four haploid products called sperm.

Summary

Cell division is one of the important directions of cell. The division cycle of most cells consists of four coordinated processes: cell growth, DNA replication, distribution of the duplicated chromosomes, and cell division. The cell cycle can be regarded as the life cycle of an individual cell. It can be divided into two major phases based on cellular activities readily visible with a light microscope, M phase and interphase. Interphase is the interval between divisions during which the cell undergoes its functions and prepares for mitosis. The events of the cell cycle must also occur in a particular sequence, and this sequence must be preserved even if one of the steps takes longer than usual. To coordinate these activities, eukaryotic cells have evolved a complex network of regulatory proteins, known as the cell-cycle control system. The cell cycle is tightly regulated at three checkpoints: G_1/S, G_2/M and M phases. Cyclins control the cell cycle by interacting with cyclin-dependent kinases (CDKs) complexes. These in turn are activated and stimulate cell cycle progression by phosphorylation of specific target in the cell. Dysregulation of cell division will cause diseases such as cancer and degenerative disease.

Mitosis is a process of nuclear division in which replicated DNA molecules of each chromosome are faithfully partitioned into two nuclei. Mitosis is conventionally divided into six stages: prophase, prometaphase, metaphase, anaphase, telophase and cytokines, and each characterized by a particular series of events.

Meiosis is a special kind of cell division, which only occurs at some stage of the sexual cell reproduction. Meiosis ensures production of a haploid phase in the life cycle, and fertilization ensures a diploid phase. Without meiosis, the chromosome number would double with each generation, and sexual reproduction would not be possible. Meiosis involves a single round of DNA replication that duplicates the chromosomes, followed by two successive cell divisions, called meiosis Ⅰ and meiosis Ⅱ. Meiosis Ⅰ, which results in reduction in the number of chromosomes, can be factitiously divided into five stages, prophase Ⅰ, prometaphase Ⅰ, metaphase Ⅰ, anaphase Ⅰ, and telophase Ⅰ. Prophase Ⅰ is traditionally divided into five sequential stages: leptotene, zygotene, pachytene, diplotene, and diakinesis, which defined by the morphological changes associated with the assembly (synapsis) and disassembly (desynapsis) of the synaptonemal complex. Meiosis Ⅱ is very similar to a mitotic division.

Questions

1. Give short definitions of the following terms: bivalent and tetrad, Cdk-activiting kinase (CAK), cell cycle and cell cycle control, cell cycle checkpoint, centromere and kinetochore, chiasma (chiasmata) and crossing over, cyclin and cyclin-dependent kinase (Cdk), homologous chromosome and sister chromatid, maturation-promoting factor (MPF), SCF and APC complexes, microtubule-associated protein (MAP), mitosis and meiosis, phragmoplast, restriction point, START, synapsis and synaptonemal complex (SC).
2. Short-answer questions
   (1) Describe the chief events of mitosis and show how these differ from meiosis.
   (2) Describe the phases of the cell cycle and the events that characterize each phase.
   (3) What "checkpoints" occur in the cell cycle? What is the role of each?
   (4) How is MPF (cyclin/Cdk complex) activated and what are functions of this complex in regulating cell cycle?
3. Essay questions:
   (1) Describe the function mechanism of cyclin in the regulation of the cell cycle?
   (2) What types of force-generating mechanisms might be responsible for chromosome movement during anaphase?

(3) Contrast the events that take place during mitosis and meiosis, and their roles in the lives of a plant or animal?
(4) What is the main formation of the spindle, and their function respectively?
(5) What is the cell cycle? What are the stages of the cell cycle in mitosis? How does the cell cycle vary among different types of cells?
(6) During meiosis, what's the meaning for recombination events among homologous chromosomes?
(7) Why mitosis is a kind of faithful cell division? Meanwhile, in meiosis exchange between homologous chromosome is a frequent event?

(李秀兰,曲志中,任　华修改)

Chapter 10

Cell differentiation

10.1 Introduction

Cell differentiation is the process by which cells acquire the specialized functions that allow them to play their roles in the tissues and organs of the adult animal. This process occurs through changes in gene expression. It is from the proteins expressed in a cell that a cell acquires its properties, and it is through the regulation of gene expression that different cells express different proteins and thus become different from each other.

10.1.1 The biology of cell differentiation

Molecular biology of cell differentiation specifically emphasize on mammalian systems in development, which focuses on biological systems, *in vivo* and in cell culture, and examines molecular mechanisms of gene expression during differentiation.

(1) Cell determination and differentiation With few exceptions, all of the cells in a multicellular organism contain the same DNA. This is not surprising, given that organisms develop from mitotic divisions of one original cell, called a **zygote**. However, it is clear that there are many different cell types in the bodies of multicellular organisms. How do these cells "mature" to take on specific roles for an organism? The processes by which cells mature, commonly referred to as cell determination and differentiation.

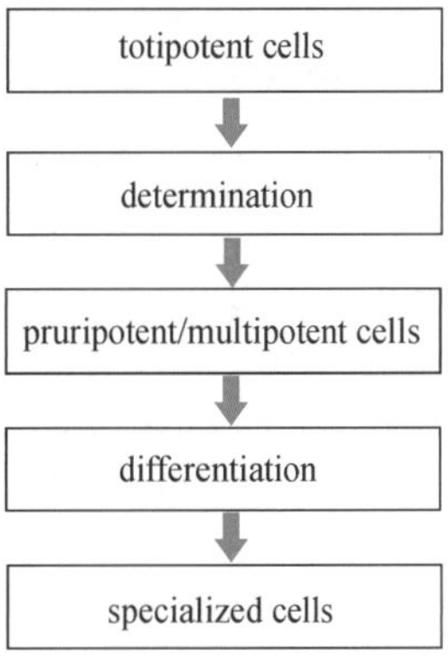

During early embryogenesis, cells divide and gradually become committed to specific patterns of gene activity through a process called cell **determination**. Specific genes are associated with the determination event. Because the daughter cells of each "determined" cell have the same limited potential as their parent cell, determination is considered heritable. Determination is permanent under normal conditions but it is possible to reverse the process experimentally.

In fact, cell determination is the process by which portions of the genome are selected for expression in different embryonic cells. This involves developmental decisions that gradually restrict cell fate. Cells can progress from totipotent to pluripotent to determined. Regulated gene expression underlies cell differentiation while cell differentiation is stable but can occasionally be reversed. Currently, epigenetic mechanisms account for the stability of determination and differentiation.

(2) Cell differentiation plays important roles in development Cell differentiation also describes the progressive restriction of the developmental potential and increasing specialization of function which takes place during the development of the embryo and leads to the formation of specialized cells, tissues, and organs. The individual cells become more and more restricted in their developmental potential.

Cell differentiation leads the continuous loss of physiological and cytological characters of young cells, resulting in getting the characters of adult cells. The unspecialized cells become modified and specialized for the performance of specific functions. Differentiation results from the controlled activation and de-activation of genes. When young cells develop as a result of cell division, they become differentiated to suit the function they were made for. This is done by the genetic information present in the cell, which will have genes "switched on" and "switched off" to determine the function of the cell.

(3) Cell differentiation occurs through changes in gene expression A cell expresses a fraction of its genes, and the different types of cells in multicellular organisms arise because different sets of genes are expressed during differentiation. It is from the proteins expressed in a cell that a cell acquires its properties, and it is through the regulation of gene expression that different cells express different proteins and thus become different from each other. A wide variety of proteins are capable of affecting a cell's differentiated state. Many of the key proteins are **transcription factors**, but intercellular signaling plays a critical role in influencing the presence and activity of the transcription factors in a cell. Cells can also change the pattern of genes they express in response to the changes in their environment. Ultimately, before differentiation can be considered to be complete, the new state of gene expression must be stabilized.

10.1.2 The majority of organisms consist of many types of cells

Living organisms are made up of cells that are usually too small to see with the naked eye. Many cells, for example most animal cells, plant and bacterial cells, have a fixed shape. However, large organisms contain many cells, and the human body contains billions of cells of many different types.

(1) Multicellular organisms have many specialized cells The single-celled procaryotic and eukaryotic cells lacked the specialization and organization we see in many plants and animals today. Single-celled organisms need to perform every life function within the confines of a single cell. Multicellular organisms, like ourselves, can have cells specialized for a particular function. Thus, a liver cell doesn't have to "worry" about all the chemical reactions and interactions with other cells that a retinal cell in the back of your eye has to worry about.

(2) This specialization allows for more efficiency Progressive restriction of the developmental potential and increasing specialization of function which takes place during the development of the embryo and leads to the formation of specialized cells, tissues, and organs. Some cells provide protection; some give structural support or assist in locomotion; others offer a means of transporting nutrients. All cells develop and function as part of the organized system — the organism — they make up.

10.2 Cells with potency of differentiation

Multicellular organisms are formed from a single **totipotent** stem cell — a single fertilized egg or zygote. As this cell and its progeny undergo cell divisions, the potential of the cells becomes restricted, and they specialize to generate cells of a certain lineage. In the majority of organisms, there are more than one type of cell. Indeed, about 220 different types of cells — many highly specialized — make up the tissues and organs of the human body.

10.2.1 Cells with different potency of differentiation

According to the potency of differentiation, cells could be classified as **totipotent cells**; **pluripotent cells**; **multipotent cells**; terminal differentiated cell, or specialized cell.

Except of totipotent cells, the pluripotent cells and multipotent cells are real stem cells, which are further divided as embryonic stem cells and adult stem cells, therefore they are also named as pluripotent stem cells and multipotent stem cells. Multipotent stem cells finally produce terminal differentiated cells, such as red blood cells, white blood cells(Figure 10-1).

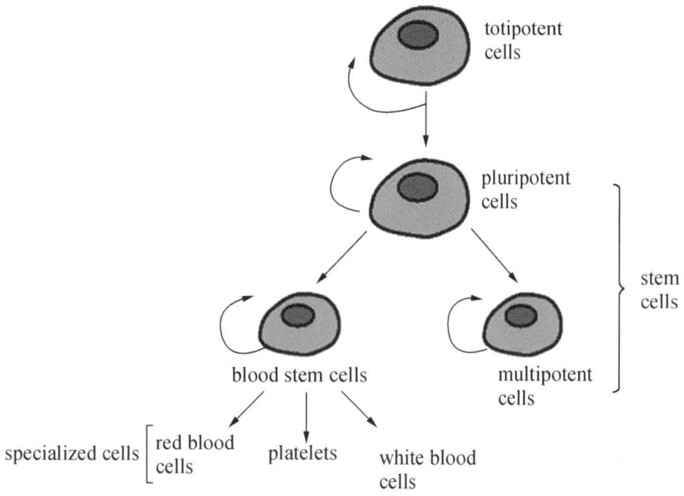

Figure 10-1 Totipotent cells, pluripotent cells, multipotent cells

10.2.2 Human body's development

In order to throughly understand totipotent, pluripotent, multipotent and terminal differentiated cells, it is important to view them in the context of human development (Figure 10-2).

At the time of fertilization, the one zygote cell is capable of forming an entire organism. These cells are classified as totipotent, which means that the potential of the cell is unlimited. For a short time, zygote division creates identical totipotent cells. Any of these cells formed during the first division after fertilization could be placed in a woman's uterus and develop into a fetus.

By the fourth day, the cells begin to form a **blastocyte**, or bundle of cells. The outer layer of the blastocyte forms the placenta and other necessary tissues in the uterus required for the fetus to develop. The inner cluster of cells will continue to develop into nearly all of the tissues of the human body. Although those inner cells will form virtually every type of tissue in the body, they cannot give rise to the placenta or other supporting tissues for the uterus. Thus, they are unable to form an organism on their own if placed in a woman's uterus, and are therefore referred to as pluripotent. As the pluripotent cells continue to specialize, they become cells that only lead to the development of specific tissues. Some will lead to bone marrow, while others will lead to blood or skin. The stem cells that carry this extra specialization are considered multipotent.

It is well known that multipotent stem cells play a vital role in fetal development, however, multipotent cells can still be found during the course of a person's adult life.

Although usually found in very small quantities, adult stem cells play a critical role in sustaining life. For example, red blood cells are continuously replaced, and the production of new red blood cells is initiated by multipotent blood stem cells. Virtually any body function that requires growth involves stem cells.

The difference between totipotent, pluripotent, and multipotent cells is essential in understanding cell differentiation as well as stem cells.

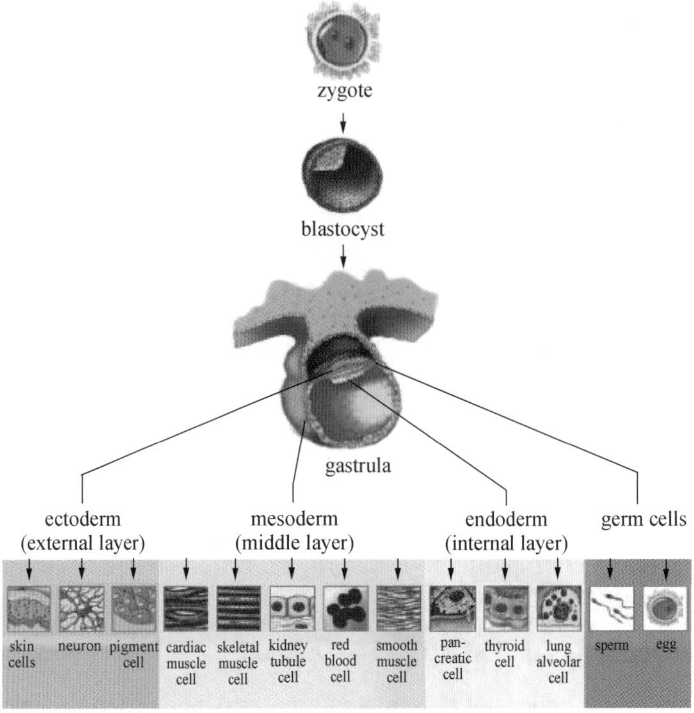

Figure 10-2 Differentiation of human tissues

10.2.3 Totipotent cells

A cell that has the capacity to independently develop into a complete organism is called totipotent. The only totipotent cells are the fertilized egg and the first 4 or so cells produced by its cleavage (as shown by the ability of mammals to produce identical twins, triplets, etc.). An example is a fertilized egg in the uterus. A totipotent cell by itself can give rise to an entire embryo.

Progressively the developing cells lose, in fact, the ability to use this information. No information is lost only the ability to use it is lost.

10.2.4 pluripotent cells

Pluripotent cells have the potential to make any differentiated cell in the body, but cannot contribute to making the extraembryonic membranes (which are derived from the trophoblast). A pluripotent cell cannot give rise to an embryo, but it can give rise to all the different types of cells found in the organs of the body.

In contrast to totipotent cells, pluripotent cells have a more restricted capacity to differentiate. By introducing pluripotent cells in different tissues they may be induced to act like the surrounding cells. However, a single pluripotent cell can not develop into an independent organism. Pluripotent cells can be re-transformed into totipotent cells by means of transplantation of nuclei. Pluripotent stem cells are most widely used by scientists in stem cell research.

10.2.5 multipotent cells

Multipotent cells can only differentiate into a limited number of cell types. They are found in adult animals; almost all organs in the body (e. g., brain, liver, bone marrow, skin, etc.) contain them where they replace dead or damaged cells. A multipotent cell can give rise to more than one type of cell but not all the different cells in all the organs. For example, hematopoietic stem cell can only produce mature blood cells.

10.2.6 terminal differentiated cells

All cells carry out basic functions, such as respiration, growth, division, synthesis. Most cells have specialized capabilities, and traditionally, cells in their final, differentiated state usually have very unique characteristics, reflected in their morphology and function. These cells are terminal differentiated cells. They may be characterized by: ①no proliferative capacity; ② never seem to divide; ③irreplaceable; ④ usually have a long life span. Scientists now find exceptions that some terminal differentiated cells could undergo: non-constantly renewing; return to active cycling (cell cycle/division) in response to critical cell depletion; renewal by simple duplication of existing differentiated cells division gives rise to daughter cells of the same type.

For example, proliferation in the liver, which is normally characterized by very low cell turnover, can be dramatically stimulated in the partial hepatectomy model. After surgical removal of 2/3 of the liver, the remaining hepatocytes exit G_0 and synchronously enter the cell cycle in order to restore liver mass.

10.3 Stem cells

There are stem cells in almost all animals, from mouse to humans. There are types of stem cells: **Embryonic stem cells**, **adult stem cells**, and **induced pluripotent stem (iPS) cells**.

Stem cells are powerful. They are able to transform into any one of the 220 cell types in the human body. One stem cell can also divide to produce millions more stem cells. The potential for stem cells to renew themselves is almost infinite. In the 3-to 5-day-old embryo, called a blastocyst, stem cells in developing tissues give rise to the multiple specialized cell types that make up the heart, lung, skin, and other tissues. In some adult tissues, such as bone marrow, muscle, and brain, discrete populations of adult stem cells generate replacements for cells that are lost through normal wear and tear, injury, or disease.

10.3.1 The Unique Properties of All Stem Cells

Stem cells have two important characteristics that distinguish them from other types of cells. First, they are unspecialized cells that renew themselves for long periods through cell division. The second is that under certain physiological or experimental conditions, they can be induced to become cells with specific functions such as the beating cells of the heart muscle or the insulin-producing cells of the pancreas.

All stem cells—regardless of their source—have three general properties: stem cells are capable of dividing and renewing themselves for long periods; stem cells are unspecialized; stem cells give rise to specialized cells.

(1) **Stem cells must self-renew** One of the most important issues in stem cell biology is to understand the mechanisms that regulate self-renewal. Self-renewal is crucial to stem cell function, because it is required by many types of stem cells to persist for the lifetime of the animal. Moreover, whereas stem cells from different organs may vary in their developmental potential, all stem cells must self-renew and regulate the relative balance between self-renewal and differentiation. Understanding the regulation of normal stem cell self-renewal is also fundamental to understanding the regulation of cancer cell proliferation, because cancer can be considered to be a disease of unregulated self-renewal.

(2) **Stem cells are unspecialized** One of the fundamental properties of a stem cell is that it does not have any tissue-specific structures that allow it to perform specialized functions. A stem cell cannot work with its neighbors to pump blood through the body (like a heart muscle cell); it cannot carry molecules of oxygen through the bloodstream (like a red blood cell); and it cannot fire electrochemical signals to other cells that allow the body to move or speak (like a nerve cell). However, unspecialized stem cells can give rise to specialized cells, including heart muscle cells, blood cells, or nerve cells.

(3) **Stem cells are capable of dividing and renewing themselves for long periods** Unlike muscle cells, blood cells, or nerve cells—which do not normally replicate themselves—stem cells may replicate many times. When cells replicate themselves many times over it is called proliferation. A starting population of stem cells that proliferates for many months in the laboratory can yield millions of cells. If the resulting cells continue to be unspecialized, like the parent stem cells, the cells are said to be capable of long-term self-renewal.

(4) **Stem cells can produce specialized cells** When unspecialized stem cells give rise to specialized cells, the process is called differentiation. Stem cell differentiation is triggered by the signals inside and outside cells. The internal signals are controlled by a cell's genes, which are interspersed across long strands of DNA, and carry coded instructions for all the structures and functions of a cell. The external signals for cell differentiation include chemicals secreted by other cells, physical contact with neighboring cells, and certain molecules in the microenvironment.

10.3.2 Embryonic stem cells

Embryonic stem cells (ES cells) are from the inner cell mass, which is part of the early (4~5 day old) embryo called the blastocyst. Once removed, the cells of the inner cell mass can be cultured into embryonic stem cells.

In 1998 we saw the publication of two papers describing the growth *in vitro* of human embryonic stem (ES) cells derived either from the inner cell mass (ICM) of the early blastocyst or the primitive gonadal regions of early aborted fetuses. Work on murine ES cells over many years had already established the amazing flexibility of ES cells, essentially able to differentiate into almost all cells that arise from the three germ layers. The realization of such pluripotentiality has, of course, resulted in the field of stem cell research going into overdrive (Figure 10-3).

(1) **The generation of embryonic stem cells** Embryonic stem cells, as their name suggests, are derived from embryos. The embryos from which human embryonic stem cells are derived are typically four or five days old and are a hollow microscopic ball of cells called the blastocyst. The blastocyst is constructed with three structures: the trophoblast,

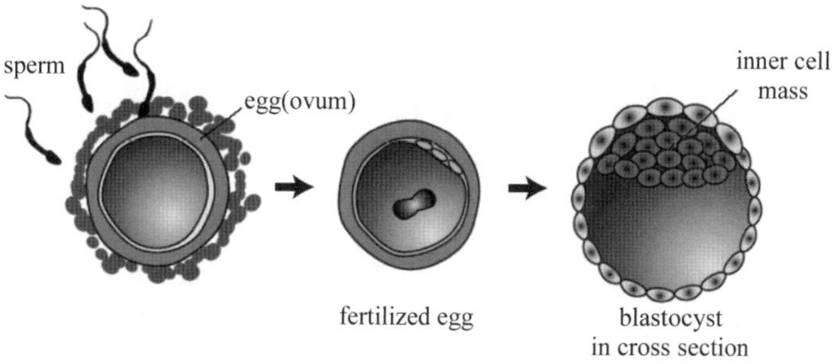

Figure 10-3 The blastocyst. Blastocyst is the stage of development at which the embryo's inner cell mass forms. It is the inner cell mass that harbors embryonic stem cells. Stem cells exist only fleetingly at this stage of development (From: artist Darryl Leja)

which is the layer of cells that surrounds the blastocyst; the blastocoel, which is the hollow cavity inside the blastocyst; and the inner cell mass, which is a group of approximately 30 cells at one end of the blastocoel.

(2) ES differentiation The capacity of embryonic stem cells for virtually unlimited self-renewal and differentiation capacity has opened up the prospect of widespread applications in biomedical research and regenerative medicine. For the latter, the cells provide hope that it will be possible to overcome the problems of donor tissue shortage and also, by making the cells immunocompatible with the recipient.

Stem cells begin their transformation into the different types of cells that make up the organisms during cell differentiation in the development. In vertebrates, differentiation begins during a stage called gastrulation, when distinct tissue layers first form. Like most other developmental processes, differentiation is controlled by genes, the genetic instructions encoded in the DNA of every cell. Genes instruct each cell to build the proteins that allow it to create the structures, and ultimately perform the functions, specific to its type of cell.

(3) Embryonic stem cells grown *in vitro* Growing cells in the laboratory is known as cell culture. Human embryonic stem cells are isolated by transferring the inner cell mass into a plastic laboratory culture dish that contains a nutrient broth known as culture medium. The cells spread over the surface of the dish and divide. The inner surface of the culture dish is typically coated with mouse embryonic skin cells that have been treated so they will not divide. This coating layer of cells is called a feeder layer.

Over the course of several days, the cells of the inner cell mass proliferate and begin to cover the culture dish. When this occurs, they are removed gently and plated into several fresh culture dishes. The process of replating the cells is called subculturing, and repeated many times and for many months. Each cycle of subculturing the cells is referred to as a passage. After six months or more, the original 30 cells of the inner cell mass yield millions of embryonic stem cells.

(4) Induced differentiation of embryonic stem cells As long as the embryonic stem cells in culture are grown under certain conditions, they can remain undifferentiated (unspecialized). To generate cultures of specific types of differentiated cells — muscle cells, blood cells, or nerve cells, the specific conditions should be applied to control the differentiation of embryonic stem cells. These methods are: change the chemical composition of the culture medium; alter the surface of the culture dish; modify the cells by inserting specific genes; 3-D Dynamic culture.

10.3.3 Adult stem cells

An adult stem cell is an undifferentiated cell found among differentiated cells in a matured tissue or organ, can renew itself, and can differentiate to yield the major specialized cell types of the tissue or organ(Figure 10-4).

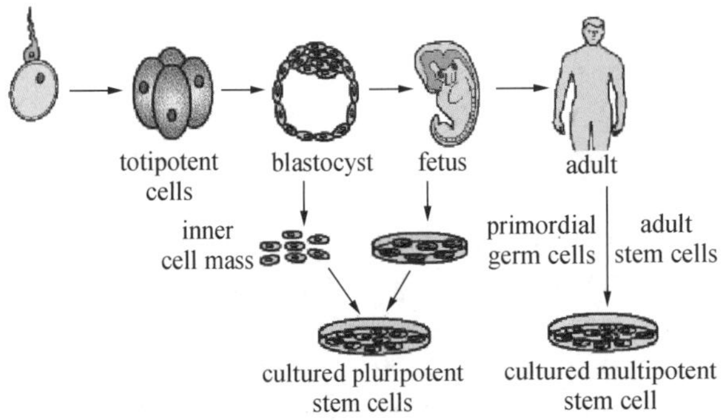

Figure 10-4 This picture illustrates three sources of stem cells. Embryonic stem cells can be isolated from the blastocyst, embryonic germ cells from the fetus, and adult stem cells from the tissues of adults (From: www. nih. gov)

(1) The history of adult stem cell researches The history of research on adult stem cells began about 50 years ago. In the 1960s, researchers discovered that the bone marrow contains at least two kinds of stem cells. One population, called hematopoietic stem cells, forms all the types of blood cells in the body. A second population, called bone marrow stromal cells, was discovered a few years later. Stromal cells are a mixed cell population that generates bone, cartilage, fat, and fibrous connective tissue.

Also in the 1960s, scientists who were studying rats discovered two regions of the brain that contained dividing cells, which become nerve cells. Despite these reports, most scientists believed that new nerve cells could not be generated in the adult brain. It was not until the 1990s that scientists agreed that the adult brain does contain stem cells that are able to generate the brain's three major cell types — astrocytes and oligodendrocytes, which are non-neuronal cells, and neurons, or nerve cells. Therefore, it was found that adult stem cells in many more tissues than they once thought possible.

(2) Location of the adult stem cells Adult stem cells have been identified in many organs and tissues. One important point to understand about adult stem cells is that there are a very small number of stem cells in each tissue. Stem cells are thought to reside in a special stem niches of certain tissues where they may remain quiescent (non-dividing) for many years until they are activated by disease or tissue injury. The adult tissues reported to contain stem cells include brain, gut epithelium, bone marrow, peripheral blood, blood vessels, skeletal muscle, skin, pancreas, renal, connective tissue, liver and so on.

(3) The importance of adult stem cell The primary roles of adult stem cells in a living organism are to maintain and repair the tissue in which they are found. Some scientists now use the term somatic stem cell instead of adult stem cell. Unlike embryonic stem cells, which are defined by their origin (the inner cell mass of the blastocyst), the origin of adult stem cells in mature tissues is unknown.

Adult stem cells originate from mature adults. These can also be referred to as multipotent stem cells, as the number of cell types which they can differentiate into is limited. Adult stem cells serve as a fresh source of cells in living organisms. They replace cells that need to be replaced on a regular basis in a living organism, such as blood (which has a 120 day lifespan) and other connective tissues.

Scientists in many laboratories are trying to find ways to grow adult stem cells in cell culture and manipulate them to generate specific cell types so they can be used to treat injury or disease. Some examples of potential treatments include replacing the dopamine-producing cells in the brains of Parkinson's patients, developing insulin-producing cells for type Ⅰ diabetes and repairing damaged heart muscle following a heart attack with cardiac muscle cells.

(4) The plasticity and transdifferentiatin of adult stem cell As reported, adult stem cells occur in many tissues and that they enter normal differentiation pathways to form the specialized cell types of the tissue in which they reside. Adult stem cells may also exhibit the ability to form specialized cell types of other tissues, which is known as transdifferentiation or plasticity.

In the adult, organ formation and regeneration were thought to occur through the action of organ- or tissue-restricted stem cells (i. e., hematopoietic stem cells giving rise to all the cells of the blood, neural stem cells making neurons, astrocytes, and oligodendrocytes). However, it is now believed that stem cells from one organ system, for example the haematopoietic compartment can develop into the differentiated cells within another organ system, such as the liver, brain or kidney. Thus, certain adult stem cells may turn out to be as malleable as ES cells and so also be useful in regenerative medicine.

The following list offers examples of adult stem cell plasticity that have been reported during the past few years.

- Hematopoietic stem cells may differentiate into: three major types of brain cells (neurons, oligodendrocytes, and astrocytes), skeletal muscle cells, cardiac muscle cells, and liver cells.
- Bone marrow stromal cells may differentiate into: cardiac muscle cells and skeletal muscle cells.
- Brain stem cells may differentiate into: blood cells and skeletal muscle cells.
- Muscle precursors can give rise to hematopoietic cells, osteogenic and adipogenic differentiation potential that is normally retrieved in mesoderm-derived stromal cells.

Current research is aimed at determining the mechanisms that underlie adult stem cell plasticity. If such mechanisms can be identified and controlled, existing stem cells from a healthy tissue might be induced to repopulate and repair a diseased tissue.

10.3.4 Induced pluripotent stem cells (iPS cells)

In the past decade, researchers have successfully make stem cells from regular mature, differentiated cells. These cells are called induced pluripotent stem cells because researchers force them to become pluripotent even after they have reached a differentiated state. By turning up the expression of just a few genes, scientists can force a mature cell to retrace its developmental pathway backwards all the way to a flexible pluripotent state.

In 2006, Shinya Yamanaka and his colleagues from Kyoto University turned mouse skin cells — specifically, fibroblasts — into stem cells. This is the first time that induced pluripotent stem cells are reported. They selected 24 genes that they thought might contribute to a stem cell's pluripotent nature. They turned on these genes in the mouse fibroblasts, in many different combinations. In the end, they found that a mix of four genes Oct3/4, Sox2, c-Myc and Klf4 that encode transcription factors — molecules that turn genes on and off — could turn fibroblasts into iPS cells.

Like embryonic stem cells, iPS cells can divide indefinitely and differentiate into the three germ layers which grants them a central role in modeling human disorders and in the field of regenerative medicine. Unlike embryonic stem cells, however, iPS cells used to be differentiated cells. Recent findings suggest that they retain some genetic or cellular "memory" of their previous life, in the form of epigenetic markers on the DNA. These markers are methyl, acetyl and other chemical groups that attach to the DNA, turning some genes off and others on. Cells acquire epigenetic markers as they differentiate, and they maintain some of them when they dedifferentiate into iPS cells.

10.4 Controls of cell differentiation

The progressive restriction of differentiation potential from pluripotent embryonic stem cells, via multipotent progenitor cells to terminally differentiated, mature somatic cells, involves step-wise changes in transcription patterns that are tightly controlled by the coordinated action of key transcription factors and changes in epigenetic modifications.

10.4.1 Differentiation involves gene expression

Every cell from different tissue or organ has the same set of genes in a same individual. A heart cell nucleus contains skin cell genes, as well as the genes that instruct stomach cells how to absorb nutrients. This suggests that in order for cells to differentiate — to become different from one another — certain genes must somehow be activated, while others remain inactive. That is that the differential gene expression leads to cell differentiation, because all adult cells have the genes necessary to become any type of cells of individual.

(1) Differentiation is controlled by genes The formation and function of tissues requires the acquisition of cellular phenotype during differentiation. Differentiation involves expression of genes that specify cellular fate and down-regulation of genes that hold cells in an original state. Once a tissue forms, changes in gene expression are also important as cells within a tissue adapt to environmental changes through a pattern of new gene expression. These cellular adaptations allow cells within a tissue to survive and maintain function. Thus, the control of gene expression is important for the formation, function, and maintenance of cells.

(2) Different cells dependent on different proteins Indeed, about 220 different types of cells — many highly specialized — make up the tissues and organs of the human body. In the human body, virtually every cell contains the same genetic material, the same DNA. Although humans have many different types of cell, for example liver cells or nerve cells, each of them contain all the information needed to make all human proteins. Even if a cell does not actually make a protein, it has the genetic information to do so. The different cell types of an organism are distinguishable both by the amounts and types of proteins they produced. Some proteins may be unique to liver cells, some to nerve cells etc. At the same time, most proteins are the same from one cell type to another; for example, human actins in one cell type is much the same as in another.

10.4.2 Transcription factors associated with cell differentiation

DNA binding transcription factors (TFs) play critical roles in initiating global changes in gene activity, such as activation or repression. The binding of TFs to target genes leads to the recruitment of co-factors which include chromatin remodeling and modification complexes. The regulation of gene target networks by specific TFs at different stages constitutes the transcriptional programs that direct cellular differentiation.

A wide variety of proteins are capable of affecting a cell's differentiated state. Many of the key proteins are transcription factors, but intercellular signaling plays a critical role in influencing the presence and activity of the transcription factors in a cell. Before differentiation can be considered to be complete, the new state of gene expression must be stabilized.

(1) Pluripotent transcription factors ES cells have the ability to self-renew or differentiate into cells of all three germ layers (endoderm, ectoderm, and mesoderm). In

ES cells, OCT4, SOX2, and NANOG form a core transcriptional network influencing the stem cell self-renewal machinery. Several hundred target genes co-occupied by OCT4, SOX2, and NANOG can be classified into two groups of downstream genes exerting opposing functions. One group includes actively transcribed genes associated with proliferation and transcription factors necessary to maintain the ES cell state. The other group includes transcriptionally silent genes encoding developmental regulators that are only activated as cells differentiate and commit to particular lineages. Establishment of **induced pluripotent stem cells** has shown the transcriptional activity of these key pluripotency factors.

Recent evidence highlights the important role of post-translational modifications, including ubiquitination, sumoylation, phosphorylation, methylation and acetylation, in regulating the levels and activity of pluripotency factors to achieve a balance between pluripotency and differentiation(Figure 10-5).

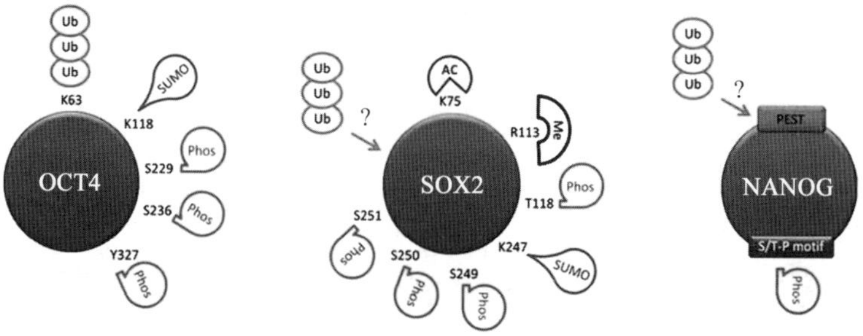

Figure 10-5 Post-translational modifications of pluripotency-associated transcription factors (Ub: ubiquitination; SUMO: sumoylation; Me: methylation; AC: acetylation; Phos: phosphorylation; K118, S229, etc: indicate the amino acid residues modified)

(2) Lineage-specific transcription factors The identification of transcriptional regulatory networks, which control lineage-specific development and function, is of central importance to the understanding of cell biology.

Lineage commitment and differentiation are likely to be coordinated by the combined effects of multiple tTFs acting on numerous different target genes. The examples of lineage-specific TFs include GATA-1, GATA2, PU1 in erythroid differentiation, HNF-4α in hepatocytic differentiation, Pax6 in neural differentiation in the developing cerebral cortex, etc.

The interplay between lineage-specific TFs and chromatin modifying or remodeling complexes allows chromatin modifications at specific lineage gene loci and promotes transcriptionally prone conformations. Molecular and biochemical studies have demonstrated that lineagespecific TFs, even when expressed at basal levels in progeny cells can bind to gene regulatory regions and gene promoters of variety loci. From the very early steps, lineage-specific TFs can recruit chromatin modifying and remodeling complexes to gene regulatory regions, hence favoring the recruitment of general transcription factors(GTFs) and preinitiation complex(PIC) assembly, necessary for the transcription of protein-coding genes in eukaryotes. Therefore, by potentiating lineagespecific gene expression, lineage-specific TFs can influence the identity of progeny cells. Among various differentiations, hematopoiesis represents one of the best-understood models of lineage commitment and cell differentiation(Figure 10-6).

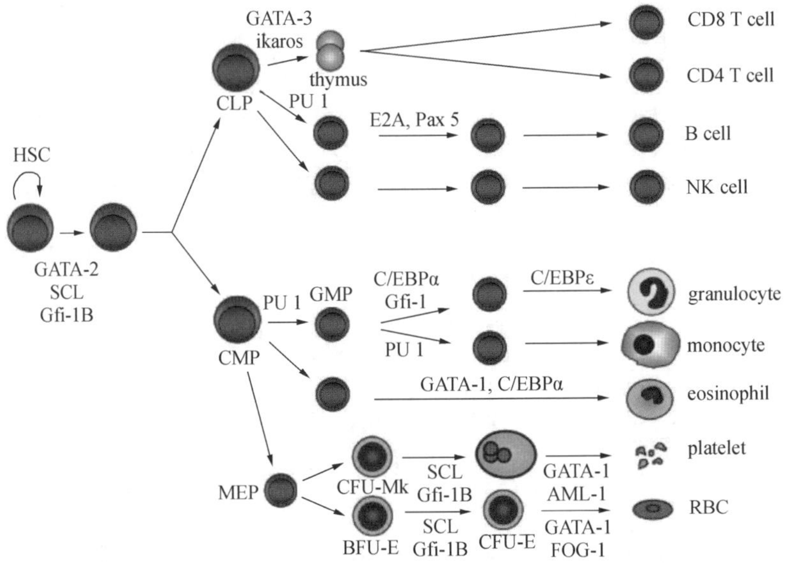

Figure 10-6 Transcription factors regulating hematopoietic differentiation. Transcription factors regulating lineage specification at various differentiation branch points or facilitating differentiation along specific lineage pathways are indicated(From: Keio J Med, 2011, 60: 47-55)

10.4.3 Epigenetic view of cell differentiation

Since a stem cell, progenitor cells, or mature cells, regardless of cell type, possess same genome of that species, determination of cell type must occur at the level of gene expression. As it turns out, epigenetic processes play a crucial role in the control of differentiation. During differentiation, specialized cells take in genetic information from stem cell or progenitor cell, but they exhibit different phenotype (structure and function). Scientists demonstrated that it is the alternation of genes expression pattern, not the change of DNA sequence that results in the different phenotype.

(1) What is epigenetics Epigenetics study the heritable changes in gene function that occur without a change in the DNA sequence. Epigenetics describe the alteration of the chromatin architecture to either permit or deny access of transcription factor machinery, and the basis of this control is brought about by **DNA methylation** of cytosine at specific CpG islands or by covalent modification of the N-terminal tails of the histones that form the structure of the nucleosome.

Current researches about epigenetics produce following conclusions: ①The study of stable alterations in gene expression that arises during development and cell proliferation. ②Epigenetic phenomena do NOT change the actual, primary genetic sequence. ③Epigenetic phenomena are important because, together with promotor sequences and transcription factors, they modulate when and at what level genes are expressed. ④The protein context of a cell can be understood as epigenetic phenomenon. ⑤Examples include: DNA methylation, histone acetylation, Chromatin modifications, X-inactivation and genetic imprinting.

(2) Epigenetic regulation of cell differentiation Epigenetic mechanisms such as DNA methylation, histone acetylation and methylation, chromatin remodeling, and RNA interference, and their effects in gene activation and inactivation, are increasingly understood to be more than "bit players" in phenotype transmission and differentiation,

representing a fundamental regulatory mechanism with consequences for cell differentiation and development.

DNA methylation and histone modifications, are closely related and involved chromatin structure. Successful epigenetic control of gene expression often requires the cooperation and interaction of both mechanisms, and disruption of either of those processes leads to aberrant gene expression.

1) **DNA methylation**: DNA methylation turns off gene expression and its profile is reprogrammed during cell differentiation. Establishing a silent chromatin state is achieved through the addition of a methyl-group to (predominately) the cytosine residue of a CpG dinucleotide. The addition of the methyl-group causes a physical change in the state of the chromatin that inhibits the expression of any genes in the methylated region. And, interestingly, this inhibitory chromatin state is passed on to daughter cells during cell division.

During the development of fertilized egg, dynamic changes in both histone modification and DNA methylation can be seen, generating cells with a broad developmental potential. The first true differentiation step in early development is the distinction between the inner cell masses, which go on to form the embryo, and trophectodermal cells, which form extraembryonic tissues.

2) **Histone acetylation**: The nucleus contains the chromosomes of the cell. Each chromosome consists of a single molecule of DNA complexed with an equal mass of proteins. Collectively, the DNA of the nucleus with its associated proteins is called chromatin.

Each nucleosome consists of eight histone molecules (two of each histones are H2A, H2B, H3, and H4) associated with 146 nucleotide pairs of DNA and a stretch of linker DNA about 60 nucleotide pairs in length. The diameter of the nucleosome "bead," or core particle, is about 10 nm. Histone H1 is thought to bind to the linker DNA and facilitate the packing of nucleosomes into 30 nm fibers (Figure 10-7).

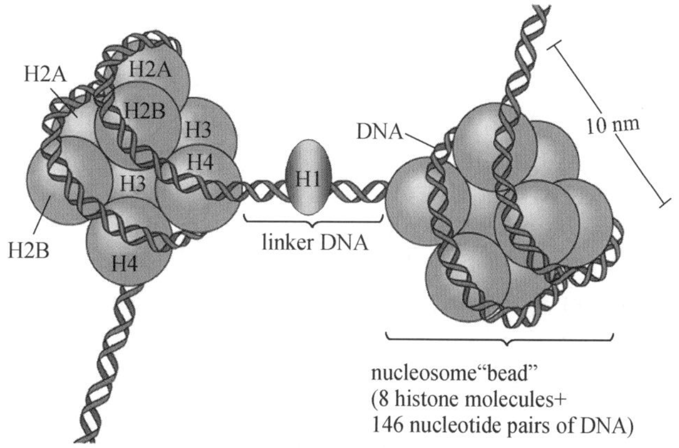

Figure 10-7 Nucleosome. Subunit of chromatin composed of a short length of DNA wrapped around a core of histone proteins (From: Longman AW. 1999)

Histone acetylation and deacetylation are chromatin-modifying processes that have fundamental importance for transcriptional regulation. Transcriptionally active chromatin regions show a high degree of histone acetylation, whereas deacetylation events are generally linked to transcriptional silencing. Acetylation and deacetylation of histone tails are catalyzed by histone acetyltransferases and deacetylases (HDACs), respectively. Histone acetyltransferases have been shown to increase the activity of several transcription factors, including nuclear hormone receptors, by inducing histone acetylation, which facilitates promoter access to the transcriptional machinery. Conversely, HDACs reduce levels of histone acetylation and are associated with transcriptional repression.

3) **Histone methylation**: Post-translational addition of methyl groups to the aminoterminal

tails of histone proteins was discovered more than four decades ago. Only now, however, is the biological significance of lysine and arginine methylation of histone tails being elucidated. Recent findings indicate that methylation of certain core histones is catalyzed by a family of conserved proteins known as the histone methyltransferases (HMTs).

The basic amino acid side chains of arginine and lysine residues on histones can be methylated by HMTs. Most modified residues are located in the N-terminal tails of histone H3 and H4 (Figure 10-7).

Arginines can either be mono or dimethylated(symmetric or asymmetric methylation), while lysines can be mono or ditrimethylated. Different modification levels of multiple residues give high combinatorial complexity, enabling histone methylation to be involved in different DNA-templated events.

4) **Chromatin remodeling**: Chromatin remodeling refers to numerous *in vitro* ATP-dependent changes in a chromatin substrate, including disruption of histone-DNA contacts within nucleosomes, movement of histone octamers *in cis* and *in trans*, loss of negative supercoils from circular minichromosomes, and increased accessibility of nucleosomal DNA to transcription factors and restriction endonucleases.

Histone modification induces chromatin remodeling by regulating the way in which nucleosomes are packaged into: ① Open chromatin conformations which permit gene expression; ②Inaccessible chromatin conformations in which genes are repressed. Specific modifications define the activity state of a gene.

During cell differentiation, DNA methylation is important in the control of gene transcription and chromatin structure. Emerging evidence suggests that chromatin remodeling enzymes and histone methylation are essential for proper DNA methylation patterns. Other histone modifications, such as acetylation and phosphorylation, in turn, affect histone methylation and histone methylation also appears to be highly reliant on chromatin remodeling enzymes. Finally, the genes for survival and genes associated with specialized structure and functions are activated, while other genes are suppressed.

5) **MicroRNA**: MicroRNAs (miRNAs) are small, often phylogenetically conserved, non-protein coding RNAs that mediate posttranscriptional gene repression by inhibiting protein translation or by destabilizing target transcripts. miRNAs provide an additional way of gene regulation by functioning at the posttranscriptional level.

Studies demonstrated the requirement of mature miRNAs for the maintenance of ES cell self-renewal and pluripotency. Additionally, several independent studies have identified distinct miRNAs expressed in ES cells and their differentiated counterparts, reinforcing the role of ES cell-specific miRNAs in regulating ES cell identity(Figure 10-8).

10.4.4 Signal pathways for differentiation

The induction and patterning of tissues and organs in multicellular organisms are regulated by extracellular signals derived from neighboring cells and tissues that control the proliferation, differentiation and death of the cells.

The signals, either soluble or membrane-bound, interact with receptors or receptor complexes that stimulate intracellular signaling pathways, which in most cases lead to the reactions in nucleus, resulting in changes in gene expression. A limited number of different families of extracellular signals appear to be sufficient to induce and pattern most tissues and organs. Each family of signals and receptors consists, however, of multiple members that interact with different specificity.

Most of these signals and pathways were initially identified and analyzed genetically in Drosophila and Caenorhabditis elegans, but have subsequently been shown to perform the

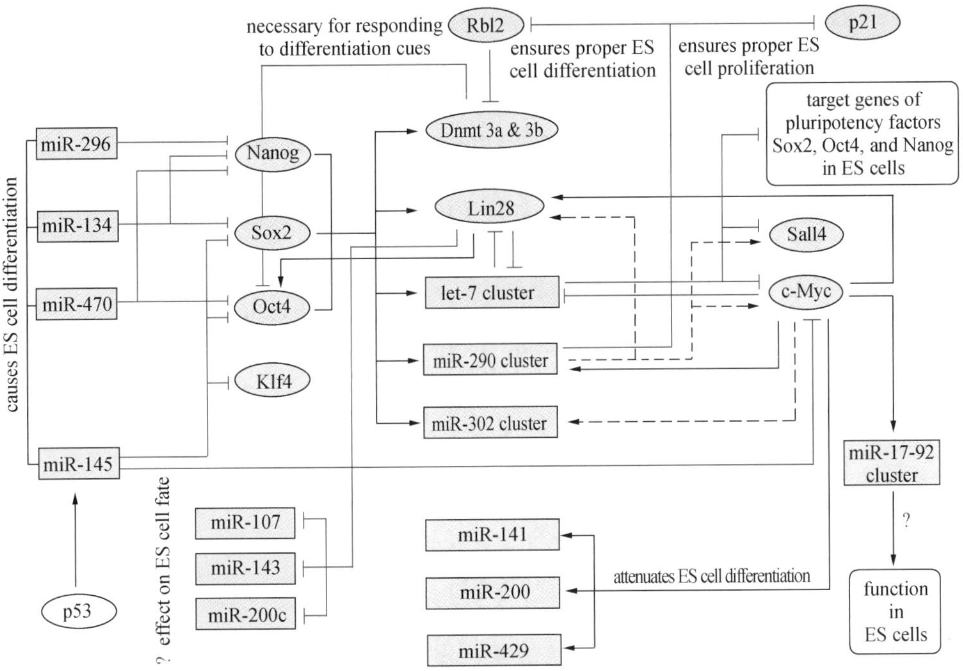

Figure 10-8 Regulatory networks of miRNAs and proteins involved in control of ES cell self-renewal and differentiation (From: Developmental Biology. 2010, 344(1):16-25)

same or similar functions in vertebrate organisms. The concerted actions of signal pathways are observed among follows: bone morphogenetic protein (BMP) pathway、transforming growth factor-β (TGF-β) pathway、fibroblast growth factor (FGF) pathway、epidermal growth factor (EGF) pathway、wingless (Wnt/Wg) pathway, and notch pathway.

The predominant signaling pathways involved in pluripotency and self renewal are TGF-β, which signals through Smad2/3/4, and FGFR, which activates the MAPK and Akt pathways. Wnt pathway also promotes pluripotency through activation of β-catenin. Signaling through these pathways results in the expression and activation of three key transcription factors: Oct-4, Sox2, and Nanog. BMP pathway promotes differentiation by both inhibiting expression of Nanog, as well as activating the expression of differentiation specific genes. Notch also plays a role in this process through the notch intracellular domain. As differentiation continues, cells further differentiate along lineage specific pathways, described in section 10.5.

Inductive pathways control the proliferation, differentiation and death of cells during organ development in both invertebrate and vertebrate organisms. The signaling pathways regulating these developmental processes are also involved in controlling the homeostasis of organs, and perturbation of the pathways often leads to disease.

10.5 The major cell differentiation systems

Adult stem cells, similar to embryonic stem cells, have the ability to differentiate into more than one cell type, but unlike embryonic stem cells they are restricted to certain lineages. The production of adult stem cells does not require the destruction of an embryo and can be isolated from a tissue sample obtained from an adult. They have mainly been studied in humans and model organisms such as mice and rats.

10.5.1 Haematopoietic stem cell

Hematopoietic stem cells (HSCs) are well characterized adult stem cells, and have been isolated from mice and humans, and have been shown to be responsible for the generation and regeneration of the blood-forming and immune (hematolymphoid) systems. Stem cells from a variety of organs might have the potential to be used for therapy in the future, but HSCs — the vital elements in bone-marrow transplantation — have already been used extensively in therapeutic settings.

(1) The properties of HSCs The haematopoietic stem cell is characterized by two cardinal properties. First, they are multipotential; that is, individual stem cells can clonally produce all of the different cell types found in the blood. Second, at least operationally, they are self-renewing; that is, during the proliferation of stem cells commitment/differentiation is balanced by the production of additional stem cells. This self-renewal ability follows logically from the lifelong production of mature cell populations.

(2) Self-renewal of hematopoietic stem cells In the hematopoietic system, stem cells are heterogeneous with respect to their ability to self-renew. Multipotent progenitors constitute 0.05% of mouse bone-marrow cells, and can be divided into three different populations: ① long-term self-renewing HSCs, short-term self-renewing HSCs, and multipotent progenitors without detectable self-renewal potential.

These populations form a lineage in which the long-term HSCs give rise to short-term HSCs, which in turn give rise to multipotent progenitors. As HSCs mature from the long-term self-renewing pool to multipotent progenitors, they progressively lose their potential to self-renew but become more mitotically active. Whereas long-term HSCs give rise to mature hematopoietic cells for the lifetime of the mouse, short-term HSCs and multipotent progenitors reconstitute lethally irradiated mice for less than eight weeks.

(3) HSCs produces mature blood cells The vertebrate haematopoietic system produces at least eight distinct lineages of mature blood cells in a continuous manner throughout adult life. These lineages include red blood cells, monocytic, granulocytic and basophilic myeloid cells, the B and T cell compartments of the antigen-specific immune system, as well as the megakaryocytes of the thrombopoietic system. Continuous replenishment of the mature cell populations is necessitated by their finite half-lives. In order to meet daily demands, the rate of mature blood cell production is very high; in the human, many billions of new cells are produced every day.

(4) CD34 as the universal marker of HSCs CD34 was discovered originally as the result of a strategy to develop antibodies that recognize small subsets of human marrow cells, but not mature blood and lymphoid cells. Cell surface expression of the CD34 antigen has rapidly become the distinguishing feature used as the basis for the enumeration, isolation, and manipulation of human stem cells, because CD34 is down-regulated as cells differentiate into more abundant mature cells.

In addition to being expressed selectively on stem cells and early progenitors during human and murine hematopoiesis, both mouse and human CD34 are expressed outside the hematopoietic system on vascular endothelial cells and some fibroblasts. This distribution suggests a function outside hematopoiesis. Transplant studies in several species, including baboons and mice, have shown that long-term marrow repopulation can be provided by CD34+ selected cells. However, several recent studies have suggested that there may be human and murine haematopoietic stem cells that do not express CD34.

(5) The differentiation capacity of HSCs HSCs have the ability to balance self-renewal against differentiation cell fate decisions; they are multipotent, a single stem cell producing

at least eight to ten distinct lineages of mature cells; they have an extensive proliferative capacity that yields a large number of mature progeny; HSCs are rare, with a frequency of 1 in 10,000 to 100,000 total blood cells; they are slowly cycling in a steady-state adult hematopoietic system(Figure 10-6).

1) **HSCs are committed to the hematopoietic lineage**: HSCs and multipotent progenitors can differentiate to the cells in haematopoietic lineage as described in figure 10-6. They differentiate to common lymphoid progenitors (CLPs; the precursors of all lymphoid cells) and common myeloid progenitors (CMPs; the precursors of all myeloid cells). Both CMPs/GMPs (granulocyte macrophage precursors) and CLPs can give rise to all known dendritic cells.

2) **HSCs are capable of differentiation to cells in non-hematopoietic lineage**: The bone marrow of adult rodents contains cells with the capacity to give rise to hepatocytes, muscle tissue, or even neurons. Some hematopoietic stem cells present in adult bone marrow co-purified with stem cells that gave rise to hepatocytes, supporting the new concept that somatic adult stem cells can change cell fate. However, these reports were based on adult rodent stem cell populations, which may differ from human stem cells in their capacity to give rise to multiple tissues.

10.5.2 Neural stem cells

Neural stem cells (NSCs) have the ability to self-renew, and are capable of differentiating into **neurons**, **astrocytes** and **oligodendrocytes**. Such cells have been isolated from the developing brain and more recently from the adult central nervous system.

After newly produced neurons and glial cells reach their final destinations, they differentiate into mature cells. For neurons, this involves outgrowth of dendrites and extension of axonal processes, formation of synapses, and production of neurotransmitter systems, including receptors and selective reuptake sites. Most axons will become insulated by myelin sheaths produced by oligodendroglial cells. Many of these events occur with a peak period from 5 months of gestation onward.

(1) Advances in NSCs research Some important considerations have emerged recently.

- The nervous tissue contains bona fide stem cells that support neuronal cell turnover throughout life.
- Despite their origin from one of the most quiescent tissues in the body, NSCs can undergo effective long-term culturing, proliferation and expansion while retaining stable functional characteristics.
- When properly challenged, the overall developmental potential appears to be broader than that observed under physiological conditions *in vivo*.
- NSCs may differentiate *in vitro*, and their possibility of transdifferentiation are confirmed.

(2) Plasticity of NSCs within the CNS Although *in vivo* NSCs appear to generate almost exclusively neuronal cells, their actual developmental potential as observed *ex vivo* is much broader than expected. In fact, cultured NSCs do give rise to neurons, astrocytes, and oligodendrocytes and are therefore normally classified as being multipotent in nature.

The concept of NSC plasticity and of their dependence on environmental cues is strengthened by transplantation and manipulation/recruitment studies *in vivo*. For example, intrahippocampus transplantation of hippocampal stem cells results in the generation of the neuronal types normally found within this region. However, when transplanted in adult brain regions in which no neurogenesis takes place, NSCs produce exclusively glial cells.

(3) Plasticity of NSCs outside the CNS NSCs (neuroectodermal in origin) can give rise to cells that normally derive from germ layers other than the neuroectoderm.

One of the most intriguing cases is the virtual pluripotency of bone marrow-derived cells; however, multiple examples of stem cells (SCs) giving rise to cells normally found in other tissues have become available. In some cases, both the original SCs and the cells to which they give rise derive from the same embryonic germ layer (intragerm layer conversion).

Nonetheless, in the presence of sustained damage as caused by lethal irradiation, adult bone marrow cells were found to integrate and differentiate within the brain tissue after systemic injection, that is "blood-into-brain" conversion. In a study, bone marrow cells from wild-type male mice were injected intraperitoneally in PU1-mutant, immunodeficient females. In these animals, 0.2% to 2.3% of the male cells found within the brain parenchyma expressed neuronal markers.

10.5.3 Mesenchymal Stem Cells

Mesenchymal stem cells (MSCs) are derived from the mesenchyme of embryonic mesoderm. It has been found that they can be isolated from many adult tissues, including adipose tissue, placenta, skin, umbilical cord blood, dental pulp, amniotic fluid and so on. Research results have proved the multi-directional differentiation potential of mesenchymal stem cells, which can differentiate into adipose tissue, bone, cartilage, tendon, nerve, liver, muscle and other tissue cells under specific induction conditions *in vivo* or *in vitro*.

The vast majority of studies on mesenchymal stem cells are based on the *in vitro* study of cultured mesenchymal stem cells, and there is no convincing method and conclusion for the identification of MSCs *in vivo*. There are some differences in biological characteristics of MSCs from different sources, which are related to tissue sources, donor species, isolation methods, culture conditions and passage times. The basic characteristics of MSCs are as follows: they can adhere and grow on plastic dishes; it has trilineage differentiation potential, namely adipocytes, chondrocytes and osteoblasts. CD105, CD73 and CD90 were usually expressed on its surface, while CD45, CD19 and HLA-DR were not expressed.

In the past two decades, the clinical application of mesenchymal stem cells has developed rapidly, and has become a hot field of stem cell research. The main applications include: hematopoietic stem cell transplantation, including enhancing hematopoietic function, promoting hematopoietic stem cell transplantation, and treating graft-versus-host disease; repair of tissue injury, including bone, cartilage, joint injury, heart injury, liver injury and nervous system disease; treatment of autoimmune diseases, including systemic lupus erythematosus, scleroderma, inflammatory enteritis; vector of gene therapy. The rapid development of MSCs in clinical treatment is not only due to the differentiation potential of stem cells, but also benefits from immunomodulatory function of MSCs through paracrine signaling.

10.5.4 Cancer stem cells

Cancer cell is a cell that divides and reproduces abnormally with uncontrolled growth. This cell can break away and travel to other parts of the body and set up another site, referred to as metastasis. Though a few cancer stem cells and the large number of cancer cells constituting the tumor are morphologically similar, they are functionally heterogeneous.

(1) Cancer cells have many of the characteristics of normal stem cells For decades, cancer researchers have wrestled with two competing visions of tumors: ①All the cells of a tumor are pretty much the same. That is, they have an equal capacity to divide and form new tumors. ②Only a few selected cells from a tumor have the capacity to initiate new, full-fledged tumors. These bad seeds are the cancer stem cells.

Cancer cells have many of the characteristics of stem cells. They share the ability for: self-renewing cell division; resistance to apoptosis (programmed cell death). several of the molecular signaling pathways associated with normal stem cell development, such as Wnt, Shh and Notch pathways, are also active in cancer development.

(2) The concept of cancer stem cells The hypothesis of the cancer stem cells is from the observation that not all the cells within a tumor can maintain tumor growth, and that large numbers of tumor cells are needed to transplant a tumor, even in an autologous context. Moreover, most cancers are not clonal, but consist of heterogeneous cell populations, similar to the hierarchical tree of stem cell lineages. Also, a hypothetical long-lived stem cell at the base of tumor outgrowth would allow progression towards malignancy through accumulation of mutations and epigenetic changes.

Researchers believe that leukemia is fueled by a small population of blood cells that divide continually, or self-renew, when they harbor certain genetic abnormalities. Continuous self-renewal prevents these cells from developing into the specialized blood cells that the body needs to function normally. According to Gilliland, previous research had established that leukemia and some other cancers were not made up of a homogeneous population of cells.

The concept of cancer stem cells could change notions of how cancers spread and how tumors should be treated. For example, some current cancer drugs may turn out to kill most cells in a tumor but to spare the stem cells, thereby setting the stage for a relapse. If we don't eliminate those cells, then they will just reform tumors.

(3) Origination of cancer stem cells There is overwhelming evidence that only a subset of cells — referred to as cancer stem cells within a tumor clone, are tumorigenic and possess the metastatic phenotype. The identification of human cancer initiating cells provided a major step forward in this field.

Recently, by applying techniques used in the stem cell field to identify self-renewing populations, has it become possible to prospectively isolate cancer stem cells. The prospective isolation and transplantation of defined cells demonstrate that only a few cells within a tumor can generate heterogeneous tumors, whereas transplantation of the rest of the cells within a tumor did not give rise to tumors upon transplantation.

1) **Cancer stem cell may come from normal stem cells:** Chronic tissue injury can result from cigarette smoke, alcohol, stomach ulcers, acid reflux, inflammatory bowel disease, and other causes. These conditions result in a continuous need to replace damaged tissues. Thus the stem cells are in a continuous state of cell division and signaling pathway activation, which resembles the conditions seen in cancers. The inability to return to a quiescent state may result in the stem cell progressing to a cancer stem cell.

2) **Abnormal activation of genes results in cancer:** Cancer cells have many of the same surface features as normal stem cells. Genes that are abnormally active in several kinds of cancers are the same ones that drive stem cells' self-renewing ability. This suggests that self-renewal, the hallmark of stem cells, may be essential to cancer. Either a normal stem cell goes awry to produce cancer, or an abnormal mature cancer cell may somehow regain stem cells special trait.

On the other hand, previous work had determined that in **chronic myeloid leukemia** (CML), hematopoietic stem cells may not be involved, but rather a progenitor cell that reacquired self-renewal capacities. Although CML is classified as a HSC disorder — the *bcr/abl* fusion gene is expressed in HSCs — direct evidence for HSC malignancy in CML has been lacking.

(4) The discovery of cancer stem cells The existence of cancer stem cells had been hypothesized for many decades, but was not until 1997 that they were isolated from patients with acute myeloid leukemia. Subsequently, cancer stem cells have been isolated from breast and brain cancers, and cancer cell lines. Cancer stem cells may derive from mutations in normal stem cells. Alternatively, differentiated tumor cells may acquire the characteristics of stem cells. Cancer stem cells are present in only very small numbers in tumors, and may not be present in all tumors.

Dysregulation of fundamental processes in developmental biology such as stem cell differentiation and proliferation may be the basis of many neoplastic diseases. Cancer stem cells have recently been identified in a number of malignancies, including leukemia, breast tumors, gliomas and cancers initiated from other tissues.

Researchers are now racing to identify tumor-forming stem cells in skin, lung, pancreatic, ovarian, prostate, and many other cancers. And researchers suspect that the dangerous cells may arise from mutations in the normal stem cells that sustain various tissues.

(5) Induced differentiation of cancer stem cells It is recognized that cancers are likely to arise from disruption of differentiation of early progenitors that fail to give birth to cell lineage restricted phenotypes. An understanding of the mechanism that controls growth and differentiation in normal cells would seem to be an essential requirement to elucidate the origin and reversibility of malignancy. But a few reports provide evidences to support the partial differentiation of cancer cell by chemicals, such as all-trans-retinoic acid (t-RA), cyclin-dependent kinase inhibitors, inhibitors of HDACs.

Drugs that affect differentiation, such as retinoic acid, have found therapeutic niches for the treatment of various forms of leukemia. As new understanding of stem cell biology and its role in cancer emerges, the development and application of drugs that target stem cell differentiation and cell fate will provide new therapeutic modalities for the management and treatment of malignant diseases. All-trans-retinoic acid administration leads to complete remission in acute promyelocytic leukemia (APL) patients by inducing growth arrest and differentiation of the leukemic clone.

Pluripotential human embryonal carcinoma (EC) cell lines undergo differentiation programs resembling those occurring in embryonal stem cells during development. Expression profiling was performed during the terminal differentiation of the EC cell line also by all-trans-retinoic acid.

10.5.5 Differentiation in plants cells

Dedifferentiation also occurs in plants cells in cell culture can lose properties they originally had, such as protein expression, or change shape. This process is also termed dedifferentiation.

(1) Plant cell types Plants have about a dozen basic cell types that are required for everyday functioning and survival. Additional cell types are required for sexual reproduction. While the basic diversity of plant cell types is low compared to animals, these cells are strikingly different.

1) **Epidermal cells**: are the most common cell type in the epidermis. These cells are often called "pavement cells" because they are flat polygonal cells that form a continuous layer, with no spaces between individual cells. Epidermal cells are transparent, because their plastids remain small and undifferentiated; hence light readily penetrates through to the photosynthetic tissues beneath the epidermis. Two more specialized cell types are also found

in the epidermis: guard cells and trichomes.

2) **Parenchyma cells**: Parenchyma cells are relatively unspecialized cells that make up the bulk of the soft internal tissues of leaves, stems, roots, and fruits. Parenchyma cells have thin, flexible cell walls and their cytoplasm typically contains a large, water-filled vacuole that fills 90 percent of the cell's volume. The vacuole may also contain compounds such as sugars.

3) **Collenchyma cells**: Collenchyma cells are long narrow cells with thick, strong, yet extensible, cell walls. Collenchyma cells are found in strands or sheets beneath the epidermis and function to provide support while a stem or leaf is still expanding.

4) **Sclerenchyma cells**: Sclerenchyma cells have extremely hard, thick walls. Two general types of sclerenchyma cells are recognized sclereids, which may be more or less isodiametric or may be branched, and fibers, which are greatly elongate cells.

5) **Vascular tissue cells**: Vascular tissue cells are highly specialized cells and aligned end to end within the xylem. The part of the cell wall between adjacent cells is degraded, so that the interior of all the vessel elements in a file becomes continuous, forming a vessel. Vessel elements are dead at maturity, leaving a hollow tube for the flow of water upward from the roots to the shoot system.

(2) Cell differentiation and development Cell differentiation is only part of the larger picture of plant development. As plant organs develop (the process of organogenesis), the precursors of the tissue systems form in response to positional signals. Then, within each tissue system precursor, cell types must be specified in the proper spatial pattern. For instance, the spacing of trichomes and stomates within the protoderm must be specified before their precursor cells begin differentiation. Exchange of signals among neighboring cells is an important aspect of the processes of spatial patterning and cell differentiation. In addition, long distance signals are required so that the strands of xylem and phloem cells within the leaf vascular bundles connect perfectly with those in the stem.

(3) Hormones control cell differentiation in plants Many of normal stem segments differentiation are controlled by hormones by regulating the proportion of auxin and cytokinin. The control of cell proliferation and differentiation during development depends, in most cases, on the concerted action of plant hormones. Among them, auxins and cytokinins are the best documented and they can impinge directly on cell cycle regulators. In addition, other hormones, e. g., abscisic acid, ethylene, jasmonic acid and brassinosteroids, whose action is much less well characterized, also have an impact on cell cycle progression and(or) arrest.

1) **Auxin**: Auxin plays an important role in the differentiation of vessel elements, both in intact and wounded plants. Auxin produced by the apical meristem and young leaves above the wound induces parenchyma cells to regenerate the damaged vascular tissue. Auxin is involved, since transdifferentiation is blocked when the sources of natural auxin (young leaves and buds) are removed or when auxin transport inhibitors are applied. If natural sources of auxin are removed, and artificial sources added, transdifferentiation of parenchyma cells will occur, regenerating the vascular bundle.

2) **Cytokinins** (CKs) : Cytokinins are a class of plant growth substances and act in concert with auxin. More cytokinin induces growth of shoot buds, while more auxin induces root formation. CKs are implicated in essential plant growth-related processes, e. g. induction of cell growth and differentiation, shoot formation and development, activation of dormant lateral buds, delayed senescence, among others. Evidence has been provided to show that in the root meristem, cytokinin acts in defined developmental domains to control cell differentiation rate, thus controlling root meristem size.

10.6 Stem cell niche

(1) stem cells require specific microenvironment The concept of stem cell niche was proposed by Schofield R in 1978. It mainly describes a kind of local micro architecture supporting stem cells. Subsequently, co-culture experiments in vitro and bone marrow transplantation experiments confirmed the hypothesis that stem cells need a special environment to perform normal physiological functions. If hematopoietic stem cells are isolated from their normal microenvironment, they will lose the ability of self-renewal. At the same time, stem cell niche can also regulate the direction of stem cell differentiation and development. In the microenvironment composed of different signaling pathways, stem cells can differentiate into different lineages of cells.

(2) stem cell niche includes multiple components Stem cell niche is a unity of structure and function. Its main components include stem cells themselves, supporting cells that interact with stem cells through surface receptors or secretory mediators, and extracellular matrix needed to maintain stem cell microenvironment. The microenvironment of different stem cells is quite different, which indicates the complexity of stem cell niche regulation. The microenvironment of stem cells is always in dynamic balance. The integration and synergy of multiple signals in the microenvironment can precisely regulate the number and biological function of stem cells.

Summary

Cell differentiation is the process by which cells acquire the specialized structures and functions. This process occurs through changes in gene expression, which controls cell differentiation. According to the potency of differentiation, cells could be classified as totipotent cells, pluripotent cells, multipotent cells and terminal differentiated cell. Stem cells have the potency of differentiation, and consist of embryonic stem cells, adult stem cells and induced pluripotent stem cells. The differentiation processes of haematopoietic stem cells, neural stem cells and mesenchymal stem cells are well described. Cancer stem cell as a special sample for cell differentiation is under investigation. Stem cells require specific microenvironment to perform normal physiological functions.

During differentiation, specialized cells take in genetic information from stem cell or progenitor cell, but they exhibit different phenotype (structure and function). It is the alternation of genes expression pattern, not the change of DNA sequence that results in the different phenotype. Cell differentiation related transcriptional factors determine the fate of cells. Epigenetic mechanisms underlying the cell differentiation include DNA methylation, histone acetylation as well as methylation, and chromatin remodeling.

Questions

1. What does the differentiation mean in cell biology?
2. What do you know the epigenetic control in cell differentiation?
3. How is cell differentiation regulated?
4. What are stem cells? What are their common features?
5. Please describe established cell differentiation systems.
6. Can cancer cells differentiate into normal cells? Why?
7. What is stem cell niche? What are the components of stem cell niche?

（吴超群）

Chapter 11

Senescence and apoptosis

Senescence (or aging) and **apoptosis** (or programmed cell death, PCD) are two common cell fates. While apoptosis leads to death and removal of the cells, senescence cells persist and survive for extended periods of time (unable to maintain their normal functions). Both senescence and apoptosis are important not only for development, but also for a cancer treatment perspective.

11.1 Senescence

The functional decline of an organism throughout its life affects multiple organs and is accompanied by the appearance of several diseases. This general decline of functional capabilities is known as aging and is fairly conserved among species. A main feature of aged organisms is the accumulation of cellular senescence, a state of permanent cell cycle arrest in response to different damaging stimuli. Hayflick and Moorhead demonstrated that primary human cells in culture have a limited capacity for replication. After undergoing a finite number of divisions, these cells entered into a permanent cell cycle arrest, subsequently termed **replicative or cellular senescence** (**Figure 11-1**).

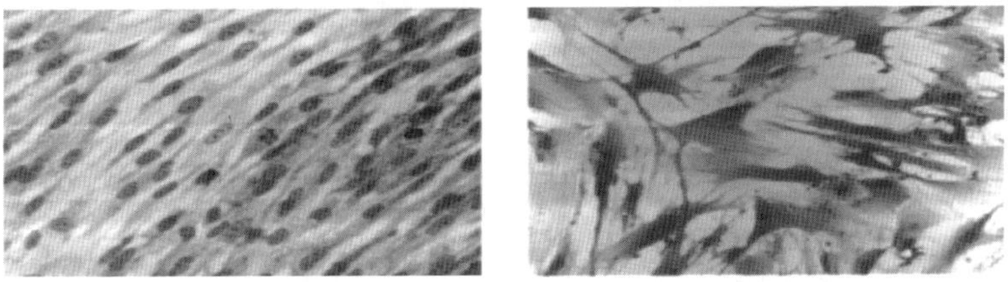

young cells after 20 population doublings(PDs)

older cells entering a senescent phase after 50 population doublings

Figure 11-1 Human fibroblast before and after replicative senescence

11.1.1 Salient Features of Senescence Cells

Senescent cells are not quiescent or terminally differentiated cells. No hallmark of senescence identified thus far is entirely specific to the senescent state. Further, not all senescent cells express all possible senescence markers. Generally, senescent cells are enlarged and have an irregular shape; Senescent cells are often detected as high senescence-associated β-galactosidase (SA-Bgal) activity, which partly reflects the increase in lysosomal mass; Cells that senesce with persistent DNA damage response (DDR) signaling harbor persistent

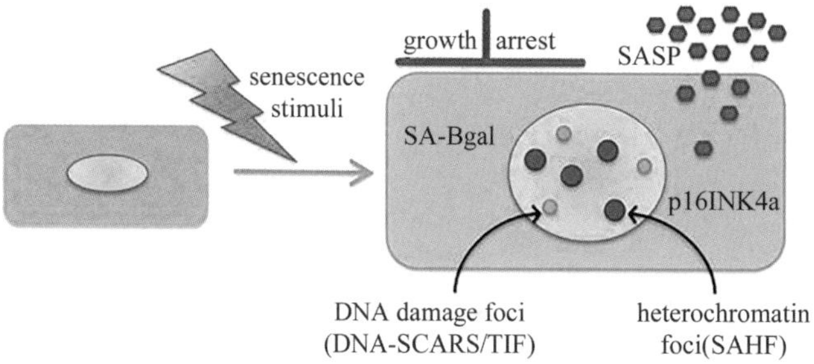

Figure 11-2 Hallmarks of senescent cells. Senescent cells differ from other nondividing (quiescent, terminally differentiated) cells in several ways, although no single feature of the senescent phenotype is exclusively specific. The pink circles in the nonsenescent cell (left) and senescent cell (right) represent the nucleus (From: Rodier F, and Campisi J. J Cell Biol, 2011).

nuclear foci, termed DNA segments with chromatin alterations reinforcing senescence (DNA-SCARS); Most senescent cells express $p16^{INK4a}$, which is not commonly expressed by quiescent or terminally differentiated cells. Activated by pRB tumor suppressor, $p16^{INK4a}$ causes formation of senescence-associated heterochromatin foci (SAHF) in some cells, which silence critical pro-proliferative genes; Senescent cells often characterized by senescence-associated secretory phenotype (SASP), which refers to secrete growth factors, proteases, cytokines, and other factors that have potent autocrine and paracrine activities (**Figure 11-2**).

11.1.2 Replicative Senescence and Telomere

Telomeres erode with each cell division due to the biochemical nature of DNA replication. Thus, telomeres have been proposed to be the "molecular clock" that determines the number of divisions a cell can undergo before reaching replicative senescence. When telomeres become critically short and lose their protective function, a DDR that upregulates inhibitors of cell cycle progression to effect and enforce the senescence growth arrest will be triggered. Besides telomere erosion, stresses that engage the DDR include exposure to oxidants (**oxidative stress**), c-irradiation, UVB light, and DNA damaging chemotherapies. The senescence phenotype is often characterized by the activation of a chronic DDR, the engagement of various cyclin-dependent kinase inhibitors (CDKi), induction of antiapoptotic genes, altered metabolic rates, and endoplasmic reticulum (ER) stress, etc.

(1) Telomere shortening The association between cellular senescence and **telomere** shortening in vitro is well established. In the laboratory, telomerase-negative differentiated somatic cells maintain a youthful state, instead of senescence, when transfected with vectors encoding telomerase. Many human cancer cells demonstrate high telomerase activity. Evidence is also accumulating that telomere shortening is associated with cellular senescence in vivo (**Figure 11-3**).

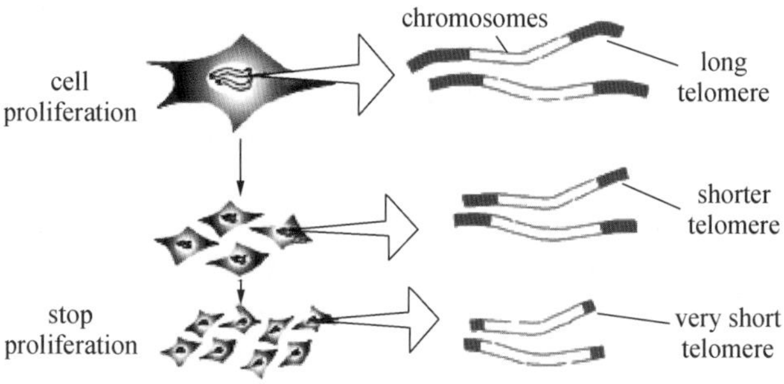

Figure 11-3 Telomeres. The linear chromosomes of all of the eukaryote cells have the telomeres at both ends. As the cells continue to proliferate, the length of the telomeres shortens and ultimately the cells stop dividing (From: PNAS. 2000, 97(23): 12407)

The telomere-based model of cell senescence suggests that the gradual loss of telomere sequences during the proliferation of cultured human somatic cells imposes a barrier on cellular replicative potential, has been strongly supported by recent genetic and biochemical studies. Evidence in support of the telomere hypothesis of senescence:

- Telomeres are shorter in most somatic tissues from older individuals compared to younger individuals.
- Telomeres are shorter in somatic cells than in germline cells.
- Children born with progeria (early senescence syndrome) have shortened telomeres compared to age-matched controls.
- Telomeres in normal cells from young individuals progressively shorten when grown in cell culture.
- Experimental elongation of telomeres extends proliferative capacity of cultured cells

(2) Telomere dynamics and cancer Telomere dynamics are a critical component of both senescence and cancer. Telomeres progressively shorten in almost all dividing cells and most human cells do not express or maintain sufficient telomerase activity to fully maintain telomeres.

There is accumulating evidence that when only a few telomeres are short, they form end-associations, leading to a DNA damage signal resulting in replicative senescence (a cellular growth arrest, also called the M1 stage). In the absence of cell-cycle checkpoint pathways (e.g. **p53** and or **p16/Rb**), cells bypass M1 senescence and telomeres continue to shorten eventually resulting in crisis (also called the M2 stage). M2 is characterized by many "uncapped" chromosome ends, end-fusions, chromosome breakage fusion-bridge cycles, mitotic catastrophe and a high fraction of apoptotic cells. In a rare M2 cell, telomerase (a cellular reverse transcriptase) can be reactivated or up-regulated, resulting in indefinite cell proliferation. This cellular immortalization is a potentially rate-limiting step in carcinogenesis that is important for the continuing evolution of most advanced cancers.

Replicative senescence occurs because, owing to the biochemistry of DNA replication, cells acquire one or more critically short telomere. Recent findings suggest that certain types of DNA damage and inappropriate mitogenic signals can also cause cells to adopt a senescent phenotype. Thus, cells respond to a number of potentially oncogenic stimuli by adopting a senescent phenotype. These findings suggest that the senescence response is a fail-safe mechanism that protects cells from tumorigenic transformation.

11.1.3 Senescence and Mitochondria ageing

Within eukaryotic cells mitochondria provide most of the ATP by oxidative phosphorylation. In addition, mitochondria are critical for other aspects of cell function, such as haem and iron sulphur centre biosynthesis and modulating calcium levels. Consequently, mitochondrial dysfunction arising from oxidative damage and mutations to mitochondrial and nuclear genes, includes the cumulative degeneration associated with ageing. Mitochondrial theory of senescence indicates the accumulation of somatic mutations of mitochondrial DNA leading to the decline of mitochondrial functionality as one of the driving forces for the process itself.

In healthy individuals, somatic mtDNA mutations accumulate with age, in the form of a wide range of different mtDNA deletions in postmitotic tissues such as skeletal muscle, myocardium, and the brain. Age associated mtDNA mutations occur more frequently and accumulate much faster in the tissues of high energy demand and low mitotic activity. When the proportion of mutant mtDNA exceeds a critical threshold concentration, a defect of mitochondrial oxidative phosphorylation appears.

Oxidative stress is a particularly important factor in mitochondrial dysfunction because the respiratory chain continually leaks the free radical superoxide. Another radical that interacts with mitochondrial metabolism is nitric oxide which can react with superoxide to form the damsenescence byproduct peroxynitrite. These reactive oxygen species lead to generalised oxidative damage to all mitochondrial components, including an increased mutation rate for mitochondrial DNA relative to nuclear DNA.

Senescent cells show an increased number of mitochondria. However, the membrane potential of these mitochondria is decreased, leading to the release of mitochondrial enzymes, such as endonuclease G (EndoG), and intensified ROS production. The main source of the extra mitochondrial content is the accumulation of old and dysfunctional mitochondria.

11.1.4 Senescence by reactive oxygen species

(1) Sources of ROS *in vivo* In a cell, approx 90% of cellular reactive oxygen species (ROS) is generated by the mitochondria. Electrons from the react with oxygen in mitochondria generate free radicals.

Superoxide that is not immediately scavenged can directly react with oxidized cytochrome c and cytochrome c oxidase. Despite the highly efficient mechanisms to mop up harmful ROS, there is still a small amount that cannot be accounted for by SOD and other peroxidases-contributing as a small factor to the ROS pool.

Proximity of mtDNA to sources of oxidants is thought to be the reason behind the observed increased sensitivity of mtDNA to oxidative damage leading to the vicious cycle by which and initial ROS-induced impairment of mitochondria leads to increased oxidant production, which in turn leads to further mitochondrial damage.

ROS generated in mitos can feed back on the organelle and damage mtDNA and other components in a vicious cycle. Similarly, oxidants can damage nuclear DNA leading to other DNA damage pathways (such as *p*53 activation). Direct oxidative modification of cellular proteins may also be an important element of ageing.

(2) ROS is increased during ageing Some researchers showed that the production of reactive oxygen species (ROS) and free radicals in mitochondria is increased during ageing as a result of the ever-increasing electron leak from the defective respiratory system, and that the ensuing oxidative damage and mtDNA mutations cause further decline in respiratory function.

During ATP production normally about 1% ~ 5% of the consumed oxygen by mitochondria are converted to superoxide anions, hydrogen peroxide, and other ROS. These ROS do have certain important physiological functions, although in excess they can be harmful to cells. Although human cells can counteract the effects of ROS, through the expression of an array of free radical scavenging enzymes, these antioxidant defence systems are by no means perfect especially in the case of excess ROS. Thus there is a known age-dependent increase in the fraction of ROS and free radicals that may escape the cellular defence mechanisms, exerting oxidative damage on cellular constituents, including DNA, RNA, proteins, and membrane lipids.

11.1.5 Cellular senescence involved in genetic errors

The role of genetics in determining life-span is complex and paradoxical. Although the heritability of life-span is relatively minor, some genetic variants significantly modify senescence of mammals and invertebrates, with both positive and negative impacts on age-related disorders and life-spans. In certain examples, the gene variants alter metabolic pathways, which could thereby mediate interactions with nutritional and other environmental factors that influence life-span. Given the relatively minor effect and variable penetrance of genetic risk factors that appear to affect survival and health at advanced ages, life-style and other environmental influences may profoundly modify outcomes of senescence.

(1) The rate of error accumulation A key prediction of the somatic mutation theory of senescence is that there is an invariant relationship between life span and the number of random mutations. Dietary restriction, which prolongs life span, results in slowed accumulation of hypoxanthine phosphoribosyltransferase (HPRT) mutants in mice. Conversely, senescence-accelerated mice, which have been bred to have a shortened life span, show accelerated accumulation of somatic mutations.

The rate of error accumulation is intrinsic and species specific. An initially low level of errors could increase exponentially and randomly through the interaction of several self-perpetuating error generating systems, cross-feeding errors into each other, resulting in the accumulation of random somatic mutations and the production of altered proteins with random amino acid substitutions at an increasing frequency with age. Environmental factors can influence the rate of error accumulation by altering DNA, RNA, and proteins.

(2) Initial alteration in DNA sequence The cellular researches about senescence have indicated the possibility that cellular senescence is conceived as the consequence of an accumulation of genetic errors. Several error-producing mechanisms have been proposed, such as DNA misduplication, gene mistranscription and mistranslation, and DNA misrepair. They have been incorporated into one model, consisting of DNA, RNA, and a few key enzymes, i. e., DNA polymerases, RNA polymerases, and aminoacyl-tRNA synthetases.

The initial alteration in DNA sequence can be introduced in a number of ways:

- Natural errors in replication. These occur at a frequency of about 1 in 1010 base pairs synthesized. For the human genome at 4×10^9 bp, about one cell division in three will contain a replication error.
- UV radiation induces dimers where there are 2 consecutive pyrimidines in the same DNA strand.
- X-radiation. Generates unstable ionic species and free radicals. High proportion of cells are killed, or carry multiple mutations. Not favoured for trying to get single mutations in cells.
- Mutagenic chemicals are both generators of CH^{3+}, potent alkylating agents, and form methyl adducts with bases that alter the H-bonding pattern of bases.
- Deaminating agents nitrous acid HONO, nitrites convert amino groups to OH groups/keto groups.

(3) Faulty repair of genetic errors Efficient repair mechanisms normally eliminate lesions in the DNA. Permanent mutations are the results of errors in repair, and this can be promoted by the use of error prone strains of organisms. Mismatch repair replaces mismatched bases resulting from errors in replication. Cells containing defective components of the mismatch repair process have high spontaneous rates of mutation. Faulty repair equals a decline in DNA repair which causes genetic errors which causes faulty information which, in turn, causes senescence.

11.1.6 Epigenetic regulation of senescence

Epigenetic entails the study of the switching on and off of genes during development, cell proliferation, senescence and also by environmental insults. Epigenetic changes can result in the stable inheritance of a given spectrum of gene activities in specific cells. Genome modifications resulting from epigenetic changes appear to play a critical role in the cellular senescence.

(1) Histone deacetylases and chromatin remodeling Chromatin remodeling is thought to mediate senescence in human melanocytes. According to one hypothesis, the histone acetyltransferase p300 is removed from promoters of certain cell-cycle regulatory genes, and histone deacetylases (HDACs) are added. The resulting repression of those genes leads to a halt in mitosis.

As cells senesce, there is increased association of HDAC1 with the retinoblastoma protein RB, and a parallel decrease in the levels of the histone acetyltransferases p300 and CBP proteins. Elevated HDAC activity appears causally related to cellular senescence, as overexpression of a p300 mutant protein, or treatment with a specific chemical inhibitor of p300, results in irreversible growth arrest and senescence of normal human cells.

(2) Euchromatin formation Heterochromatin represses gene activity, but many genes in senescent cells actually display higher expression levels, therefore euchromatin, which facilitates gene transcription, must also be involved. Data from a cDNA microarray study provide tentative support for this contention. The study identified transcriptional fingerprints unique to senescence, not those merely correlated with cell-cycle arrest. It also found that upregulated senescence-specific genes were physically clustered, an arrangement consistent with euchromatin formation.

11.1.7 Senescence pathway

p53 and Rb, paradigmatic tumor suppressor proteins, have been shown to play critical roles in the induction of senescence. Both p53 and Rb are activated upon the entry into senescence (Figure 11-4). The p53 protein is stabilized and proceeds to activate its transcriptional targets, such as p21CIP1/WAF1. Rb is found at senescence in its active, hypophosphorylated form, in which it binds to the E2F protein family members to repress their transcriptional targets. These targets constitute the majority of effectors required for cell-cycle progression.

Senescence requires activation of the Rb and/or p53 proteins, and expression of their regulators such as $p16^{INK4a}$, $p21^{CIP}$, and $p14^{ARF}$. P53 can activate senescence by activating Rb through p21 and other unknown proteins, and also, in human cells, can activate senescence independently of Rb. P53 activation is achieved by phosphorylation, performed by the ATM/ATR and Chk1/Chk2 proteins, and by the p19ARF product of the INK4a locus, which sequester Mdm2 in the nucleolus. The transcriptional control of the INK4a products is not fully elucidated, indicated are some of these regulators (Figure 11-5).

(1) The $p19^{ARF}/p53/p21^{Cip1}$ pathway One form of senescence is mediated by p53, which induces a variety of anti-proliferative activities including transcriptional activation of $p21^{CIP}$, a non-INK4 CDKI. The activity of p53 is regulated predominantly at the post-translational level and is induced by inhibiting its MDM2-mediated degradation in response to diverse stimuli including DNA damage signals and oncogene activation. The stabilization of p53 by oncogenes is in part mediated by ARF.

The INK4a/ARF locus encodes two tumor suppressors, $p16^{INK4a}$ and $p19^{ARF}$, that are transcriptionally activated by the accumulation of cell doublings. The induction of cell-cycle arrest by $p19^{ARF}$ requires the functional activity of p53, being $p21^{Cip1}$ one of the most important targets of p53. Two oncogenes are known to interfere with the $p19^{ARF}/p53$

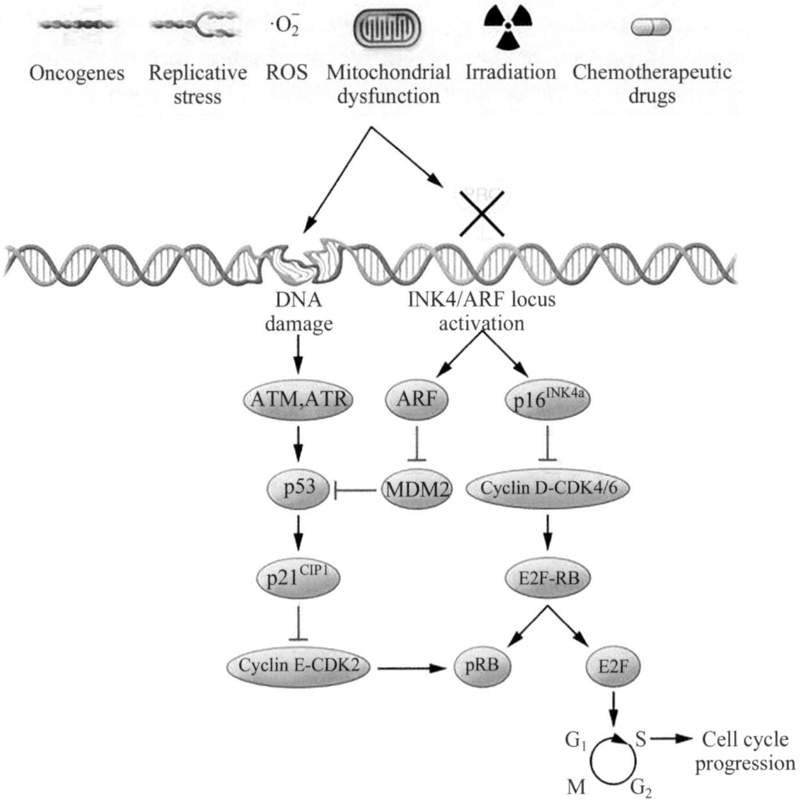

Figure 11-4 Molecular pathways controlling growth arrest during senescence. A variety of stressors induce senescence-associated growth arrest. Cell cycle exit is regulated by induction of the $p16^{INK4a}$/Rb and $p53/p21^{CIP1}$ pathways (From: Nicolás Herranz and Jesús Gil, J Clin Invest. 2018, 128(4):1238-1246)

pathway: overexpression of Bmi-1 constitutively represses the $p19^{ARF}$ gene, and overexpression of MDM2 destabilizes p53.

(2) The $p16^{INK4a}$/Rb pathway $p16^{INK4a}$ also increases markedly in senescent cells, and correlates with increasing Rb hypophosphorylation during this process. While the regulation of $p16^{INK4a}$ in senescence is not as well understood as that of p53, it appears to be induced by several stimuli including MAP kinase signaling, oncogene activation, and growth in culture. Thus, the p16/pRB pathway appears to be particularly important for ensuring that the senescence growth arrest is essentially irreversible and re-fractory to subsequent inactivation of p53, pRB, or both.

The $p16^{INK4a}$ gene seems to be a critical effectors of senescence. The effective establishment of cell-cycle arrest by $p16^{INK4a}$ requires the functional activity of Rb. Several oncogenes may interfere with this pathway, delaying or disabling senescence. Specifically, Bmi-1 overexpression constitutively represses the $p16^{INK4a}$ gene, whereas overexpression of CDK4 or cyclin D1 bypasses $p16^{INK4a}$ by inactivating Rb.

11.2 Apoptosis

2002 year's Nobel Prize in medicine was for discoveries about how genes regulate organ growth and a process of programmed cell suicide. Britons Sydney Brenner, John E. Sulston and H. Robert Horvitz shared the prize.

Apoptosis is often referred to as programmed cell death. Cells can die due to external factors (i. e., injury, infection) or by a natural death which is intrinsic in the programming of the cell itself. The study of the apoptotic process has led to a greater understanding of the pathogenesis of disease.

A coordination and balance between cell proliferation and apoptosis is crucial for normal development and tissue-size homeostasis in the adult. Cancer results when clones of mutated cells survive and proliferate inappropriately, disrupting this balance.

11.2.1 Apoptosis

Apoptosis begins when a cell activates it's own destruction by initiating a series of complex cascading events. Apoptotic cells have a decrease in cell volume. It is the ordered disassembly of the cell from within. Disassembly creates changes in the phospholipid content of the plasma membrane outer leaflet. **Phosphatidy-lserine** (PS) is exposed on the outer leaflet and phagocytic cells recognizing this change, may engulf the apoptotic cell or cell-derived, membrane-limited apoptotic bodies. The apoptotic cells are removed by tissue phagocyte by phagcytosis, and no inflammation is initiated.

11.2.2 Caspases

Apoptosis relies on a cascade of enzymes. Most of these are caspases (cytosolic aspartate-specific proteases), enzymes that are expressed as inactive precursors (procaspases) and are activated by being cleaved at specific peptide bonds. Likewise, activated caspases in turn activate procaspases at the next point in the enzyme chain by cleaving them at specific bonds.

(1) Caspase family Caspases are synthesized as pro-caspases with an N-terminal pro-domain. The active caspase is a heterotetramer of two large and two small subunits. Caspases involved in apoptosis. "Initiator" caspases are responsible for kick-starting the process of cell death in response to some external signal, and for activating the "effector" caspases. Effector caspases actually execute the process — they are responsible for breaking down the cell. It is important to note that caspases may not be the only enzymes involved in cell death, so even if all the caspases are blocked, cell death may still occur.

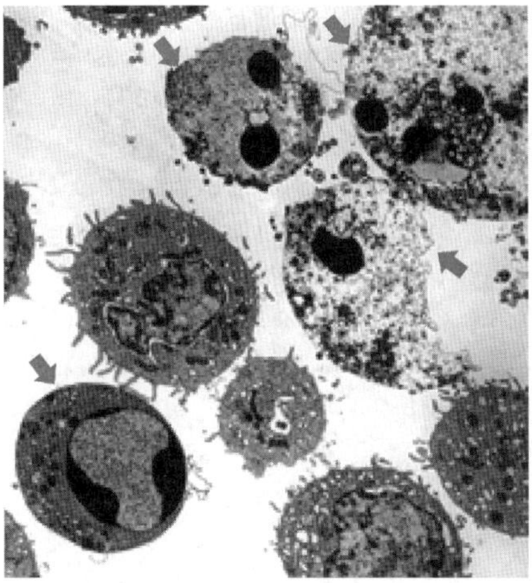

Figure 11-5 Dexamethasone induced apoptosis of Burkitt cells (arrows indicate the apoptotic cells)

Caspases have been found in organisms ranging from C elegans to humans. The 14 identified mammalian caspases (named caspase-1 to caspase-14) have distinct roles in apoptosis and inflammation. All caspases are produced in cells as catalytically inactive zymogens and must undergo proteolytic processing and activation during apoptosis.

There are three major groups of caspases: ①group 1-inflammatory caspases: caspase-1 (human and mouse), caspase-4 (human), caspase-5 (human) and caspase-11 (mouse); ② group 2-initiator caspases: caspase 2, 8, 9, 10; ③group 3-

effector caspases: caspase 3, 6, 7.

The initiator caspase activation takes place in large protein complexes bringing together several caspase zymogens. The effector caspases are activated by initiator caspases. Active effector caspases cleave substrates escalating the death signal and executing apoptosis. Caspase-3 is the main effector caspase that cleaves the majority of the cellular substrates in apoptotic cells, while caspase-7 is highly similar to caspase-3. Inflammatory caspases are part of the innate immune system, and have also been suggested to bridge the innate immune responses to the adaptive immune system.

(2) Activation of caspase Caspases are responsible for the deliberate disassembly of a cell into apoptotic bodies. *In vivo*, caspases are present as inactive pro-enzymes, most of which are activated by proteolytic cleavage. Caspase-8, caspase-9, and caspase-3 are situated at pivotal junctions in apoptotic pathways. Caspase-8 initiates disassembly in response to extracellular apoptosis-inducing ligands and is activated in a complex associated with the receptors cytoplasmic death domains. Caspase-9 activates disassembly in response to agents or insults that trigger release of cytochrome C from the mitochondria and is activated when complexed with dATP, APAF-1, and extramitochondrial cytochrome C. Caspase-3 appears to amplify caspase-8 and caspase-9 signals into full-fledged commitment to disassembly. Both caspase-8 and caspase-9 can activate caspase-3 by proteolytic cleavage and caspase-3 may then cleave vital cellular proteins or activate additional caspases by proteolytic cleavage. Many other caspases have been described.

The active enzyme is composed of a large (~20 kD) and small (~10 kD) subunit, each of which contributes amino acids to the active site. In the two structures that are available, two heterodimers associate to form a tetramer. In common with other proteases, caspases are synthesized as precursors that undergo proteolytic maturation. The NH_2-terminal domain, which is highly variable in length (23 to 216 amino acids) and sequence, is involved in regulation of these enzymes (Figure 11-6).

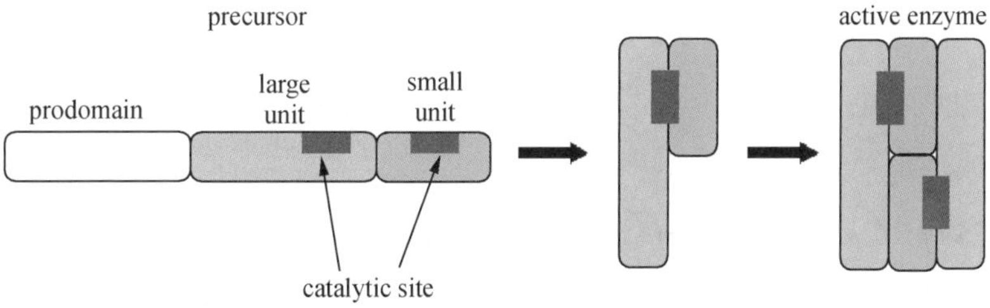

Figure 11-6 A model for caspase activation

11.2.3 The extrinsic pathway

In deed, the extrinsic death pathways trigger receptor-mediated apoptosis. Much is known about the biochemical action of many apoptotic players, and key components are being assembled into pathways.

The major components of the extrinsic apoptotic signal transduction pathway are: A proapoptotic ligand; A receptor, such as those membrane bound receptors with homology to the TNF receptor that recognize and bind the apoptotic ligand; The adapter protein tumor necrosis factor receptor (TNFR) 1-associated death domain (TRADD) plays an essential role in recruiting signaling molecules to the TNFR1 receptor complex at the cell membrane; Adaptor proteins that bind at the cytoplasmic face of the receptor and recruit additional molecules; Initiator caspases to carry out the apoptotic process. The intracellular domain of these receptors contains an 80 amino acid stretch termed the death domain. A death-inducing signaling complex (DISC) formed by the death domain, the adaptor protein Fas-associating protein with death domain (FADD), and procaspase 8 produces active caspase 8. Active caspase 8, in turn, activates downstream or executioner caspases.

The extrinsic death pathway is death receptor-dependent, which activated by the binding of secreted ligands such as Fas ligand (FasL) to "death receptors" of the tumor-necrosis-factor-receptor family, such as Fas. Receptor aggregation recruits other proteins, including the adaptor protein FADD and pro-caspase-8, into the "death-inducing signaling complex" (DISC) (Figure 11-7).

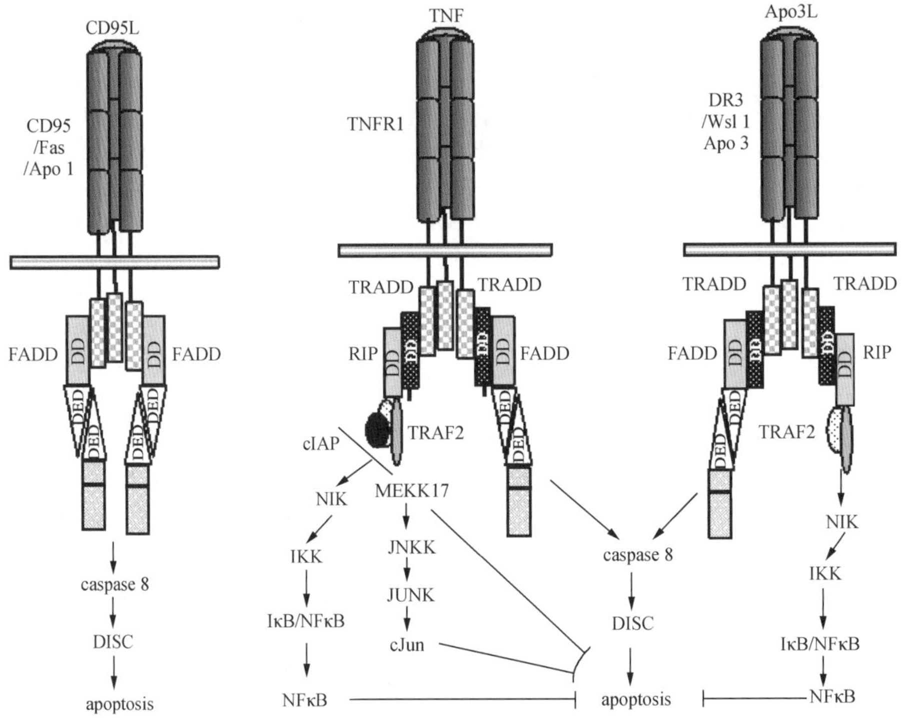

Figure 11-7 Simplified model of extrinsic apoptosis pathways (→indicates activation; ⊣ means inhibition)

(1) Death receptors Death receptors are cell surface receptors that transmit apoptosis signals initiated by specific ligands. They play an important role in apoptosis and can activate a caspase cascade within seconds of ligand binding. Induction of apoptosis via this mechanism is therefore very rapid. Death receptors belong to the tumour necrosis factor (TNF) gene superfamily and generally can have several functions other than initiating apoptosis. The best characterized of the death receptors are CD95 (or Fas), TNFR1 (TNF receptor-1) and the TRAIL (TNF-related apoptosis inducing ligand) receptors DR4 and DR5.

Death receptor members are defined by similar, cysteine-rich extracellular domains. The death receptors contain in addition a homologous cytoplasmic sequence termed the "death domain". Death domains typically enable death receptors to engage the cell's apoptotic machinery, but in some instances they mediate functions that are distinct from or even counteract apoptosis. Some molecules that transmit signals from death receptors contain death domains themselves.

(2) **Signaling by CD95** CD95 and CD95 ligand (also termed as Fas and Fas ligand) to "death receptors" of the tumor-necrosis-factor-receptor family, such as Fas. Receptor aggregation recruits other proteins, including the adaptor protein Fas-associated death-domain protein (FADD) and pro-caspase-8, into the death-inducing signaling complex (DISC). Cleavage and activation of caspase-8 in the complex triggers the effecter caspase cascade, which leads to cell death. Caspases are inactivated by inhibitor of apoptosis (IAP) proteins, which can bind to and block the active site on the caspase.

(3) **Signaling by TNFR1** Upon binding, TNF-α trimerizes its ligand TNFR1 and results in the subsequent recruitment of the signal transducing molecules TRADD through conserved protein interaction regions known as "death domains". TRADD recruits RIP and TNF receptor-associated factor (TRAF-2), leading to activation of nuclear factor κB (NF-κB), which suppresses TNF-α-induced apoptosis. The recruitment of FADD by TRADD results in apoptosis through activation of a cell death protease, caspase-8. Activated caspase-8 initiates a protease cascade that cleaves cellular targets and results in apoptotic cell death. Hence, disruption of FADD can prevent activation of caspase-8, thereby producing defects in receptor-mediated cell death.

(4) **Signaling by TRAIL receptors** In a number of ways TRAIL is similar in action to CD95. Binding of TRAIL to its receptors DR4 or DR5 triggers rapid apoptosis in many cells, however unlike CD95, its expression has been shown to be constitutive in many tissues. The DR4 and DR5 receptors contain death domains in their intracellular domain. Work in FADD-deficient mice has indicated that FADD is not essential for triggering apoptosis via these receptors.

(5) **The death-inducing signaling complex (DISC)** The DISC refers to a complex of proteins that is assembled around the cytosolic domains of certain TNF-family death receptors that contain the death-domain structure. Invariant proteins of the DISC include the adaptor protein FADD and caspase-8. Other proteins can be found in some circumstances, depending on the TNF-family receptor and the cell type. Binding of the proenzyme pro-caspase-8 to the DISC results in its activation and the release of active caspase-8 into the cytoplasm where it can cleave a number of death substrates including caspase-3, BID and proteins of the cytoskeleton such as plectin.

Cleavage and activation of caspase-8 in the complex triggers the effector caspase cascade, which leads to cell death. Caspases are inactivated by inhibitor of apoptosis (IAP) proteins, which can bind to and block the active site on the caspase. IAPs are targeted for degradation by the activity of pro-apoptotic proteins like second mitochondria-derived activator of caspase(SMAC)/Diablo and its functional homologues in flies, which include Grim, Reaper and Hid.

11.2.4 The intrinsic pathway

The intrinsic death pathway is the mitochondrial pathway, which is activated by the release of cytochrome C from mitochondria in response to various stresses and developmental-death cues.

The major components of the intrinsic apoptotic signal transduction pathway are(Figure 11-7): ①A extra or/and intracellular proapoptotic signal. ②Cytochrome C release from the mitochondria into the cytosol. ③Cytochrome C binds with apoptotic protease activating factor-1(Apaf-1) proteins. ④A apoptosome where procaspase-9 inactivated, and hydrolysis of adenosine triphosphate/deoxyadenosine triphosphate (ATP/dATP) and binding of cytochrome C to Apaf-1 drives oligomerization of Apaf-1 and therefore complex formation. ⑤Executioners (caspases-3, 6, 7) to carry out the apoptotic process.

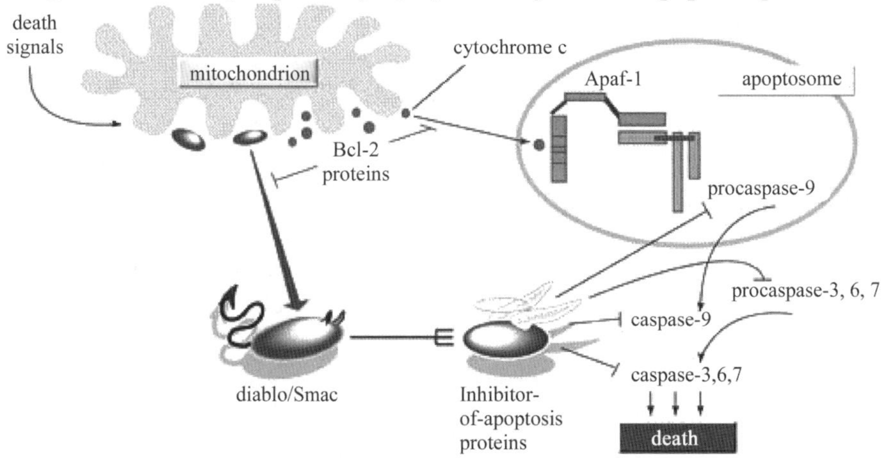

Figure 11-8 Simplified model of intrinsic apoptosis pathways (From: David R. Hipfner and Stephen M. Cohen, *Nature Reviews Molecular Cell Biology*, 2004, **5**:805-815)

This pathway is activated by a variety of extra and intracellular stresses, including oxidative stress and treatment with cytotoxic drugs. The apoptotic signal leads to the release of cytochrome C from the mitochondrial intermembrane space into the cytosol, where it binds to the Apaf-1, a mammalian CED-4 homologue. The resulting complex, termed the apoptosome, recruits and activates pro-caspase-9, which in turn activates downstream caspases. Bcl-2, the mammalian homolog of Ced-9, prevents apoptosis by inhibiting the release of cytochrome C from mitochondria. Inhibitors of apoptosis proteins(IAPs), second mitochondrial activator of caspases(smac), apoptosis-inducing factor(AIF), and Omi/Htr A2 also have important roles in the apoptotic process (Figure 11-8).

Apoptosome is activated by binding of ATP. Hydrolysis of adenosine triphosphate/deoxyadenosine triphosphate (ATP/dATP) and binding of cytochrome c to Apaf-1 drives oligomerization of Apaf-1 and therefore complex formation (**Figure 11-9**). The size of the inactive apoptosome complex is very large (on the order of approximately 1.4 MDa) and common active complex forms have a size of approximately 700 kDa. The assembly of the apoptosome complex is central to the activation of cell death by apoptosis as it is involved in the oligomerization of larger multimers that are capable of activating caspase-9.

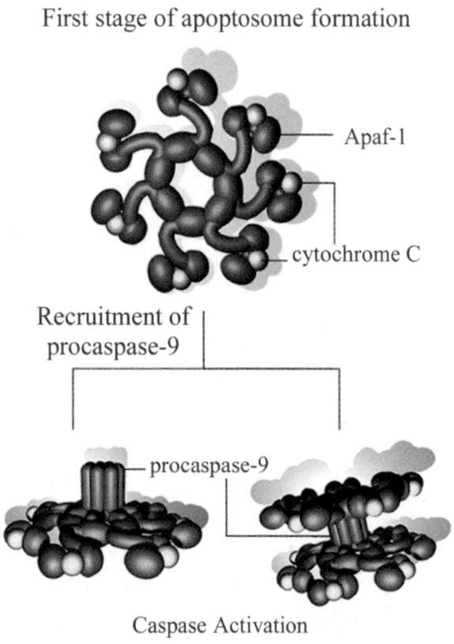

Figure 11-9 Apoptosome formation (From: The School of medicine, St. George's, University of London)

The exact mechanism of caspase activation is still uncertain although two possibilities have been proposed. In one case the Apaf-1, cytochrome c and procaspase-9 complex can act as a stage to activate cytosolic procaspase-9 as it is recruited to the apoptosome. In the other scenario two apoptosome have been proposed to interact with each other and to activate the caspase-9 located on the other apoptosome.

11.2.5 DNA damage-induced apoptosis

DNA damage-induced apoptosis is common form of apoptosis, which belongs to the intrinsic death pathway. DNA damage can also induce apoptosis through a central player, p53. DNA strand breaks are sensed by kinases such as the DNA dependent protein kinase (DNA-PK) or the ataxia-telangiectasia mutated gene product (ATM), leading to phosphorylation and activation of p53. p53 transmits the apoptotic signal by a complicated mechanism that involves, at least in part, its ability to transactivate target genes such as *bax* and a series of p53-inducible genes. The next step may involve changes in mitochondrial membrane potential and release of cytochrome c. In any case, once in the cytosol, cytochrome c can interact with Apaf1 and pro-caspase-9, leading to caspase-9 activation and the initiation of a protease cascade similar to that described above (Figure 11-10).

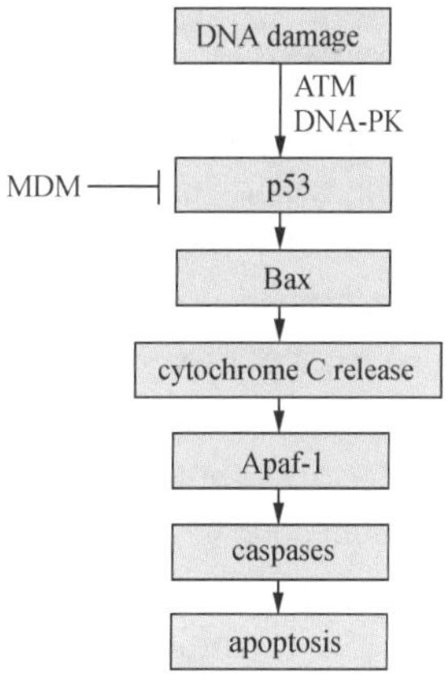

Figure 11-10 DNA damage initiates pathway involved in apoptosis (→ indicates activation; ⊣ indicates suppression)

11.2.6 Important role of mitochondria in apoptosis

Mitochondria have the ability to promote apoptosis through release of cytochrome C. On the other hand, anti-apoptotic members of the Bcl-2 family of proteins, such as Bcl-2 and Bcl-XL, are located in the outer mitochondrial membrane and act to promote cell survival. Many of the pro-apoptotic members of the Bcl-2 family, such as Bad and Bax also mediate their effects though the mitochondria, either by interacting with Bcl-2 and Bcl-XL, or through direct interactions with the mitochondrial membrane (Figure 11-11).

(1) Bcl-2 family members modulate the mitochondrial function though the permeability transition pore (PTP)

Growth factors, cytokines and DNA damage appear to signal cell death through the mitochondria, and this pathway is the target of many oncogenic mutations. These diverse signals affect the function of Bcl-2 family members which, in turn, can modulate the mitochondrial function though the permeability transition pore (PTP), a proposed channel evolved in mitochondria following necrotic or apoptotic signals.

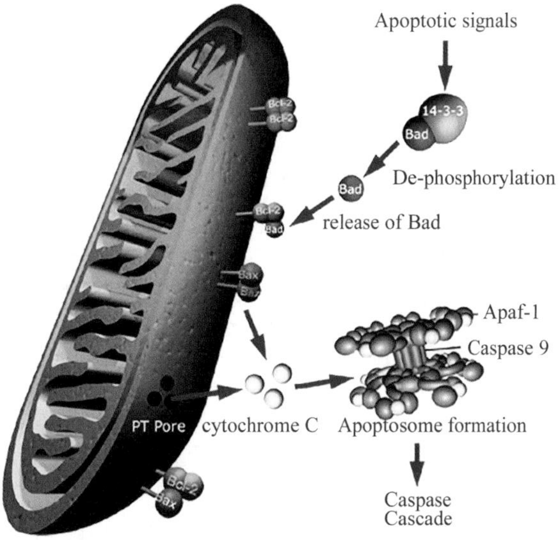

Figure 11-11 The roles of the mitochondria in intrinsic apoptosis pathways (From: The School of medicine, St. George's, University of London)

The PTP is thought to be composed of clustered components of the mitochondrial membranes, including the voltage-dependent anion channel and adenine nucleotide translocator; and the opening of PTP results in the release of cytochrome C from mitochondria. Consistent with this idea, the crystal structure of Bcl-XL is reminiscent of pore-forming proteins of some bacterial toxins. Enforced expression of the pro-apoptotic molecules Bax or Bak can result in increased mitochondria membrane potential and release of cytochrome C, which can be blocked by overexpression of Bcl-2. Cytosolic cytochrome C can interact with Apaf-1 and pro-caspase-9 to initiate a protease cascade similar to that described above.

Since Bax, and other Bcl-2 proteins, show structural similarities with pore-forming proteins. It has therefore been suggested that Bax can form a transmembrane pore across the outer mitochondrial membrane, leading to loss of membrane potential and efflux of cytochrome C and AIF (apoptosis inducing factor). It is thought that Bcl-2 and Bcl-XL act to prevent this pore formation. Heterodimerisation of Bax or Bad with Bcl-2 or Bcl-XL is thought to inhibit their protective effects. It is also thought that proteins such as Bax and Bad can promote the formation of the large diameter PT pore, with subsequent loss of cytochrome C and initiation of apoptosis.

(2) **Cytochrome C** Suppression of the anti-apoptotic members or activation of the pro-apoptotic members of the Bcl-2 family leads to altered mitochondrial membrane permeability resulting in release of cytochrome C into the cytosol. In the cytosol, or on the surface of the mitochondria, cytochrome C is bound by the protein Apaf-1 (apoptotic protease activating factor), which also binds caspase-9 and dATP. Release of cytochrome C from mitochondria has been established by determining the distribution of cytochrome C in subcellular fractions of cells treated or untreated to induce apoptosis. Cytochrome C was primarily in the mitochondria-containing fractions obtained from healthy, non-apoptotic cells and in the cytosolic non-mitochondria-containing fractions obtained from apoptotic cells. Using mitochondria-enriched fractions from mouse liver, rat liver, or cultured cells it has been shown that release of cytochrome C from mitochondria is greatly accelerated by addition of Bax, fragments of BID, and by cell extracts.

(3) **Bcl-2 family** Cellular commitment to apoptosis is partly regulated by the Bcl-2 family proteins, which includes the death antagonists Bcl-2 and Bcl-XL, and death agonists Bax and Bad. They were divided into three subfamilies: ①Bcl-2 subfamily (anti-apoptotic): Bcl-2, Bcl-XL, Bcl-w, Mcl-1 and A1. ②Bax subfamily (pro-apoptotic): Bax, Bak and Bok. ③ BH3 subfamily (pro-apoptotic): Bad, Bid, Bik, Blk, Hrk, BNIP3 and BimL.

1) **Bcl-2 and Bcl-XL bind to Apaf-1**: A central checkpoint of apoptosis is the activation of caspase-9 by mitochondria. The BH4 domain of Bcl-2 and Bcl-XL can bind to the C terminal part of Apaf-1 (to the CED-4 like part and the WD-40 domain), thus inhibiting the association of caspase-9 with Apaf-1. This process seems to be conserved from nematodes to humans since in *C. elegans* CED-9 binds to CED-4, preventing it from binding and activating CED-3.

2) **Bcl-2 family members regulate cytochrome C release**: The pro-survival proteins also seem to maintain organelle integrity since Bcl-2 directly or indirectly prevents the release of cytochrome C from mitochondria. On the other hand, the pro-apoptotic BH3 subfamily member Bid was reported to mediate the release of cytochrome C (without evoking mitochondrial swelling and permeability transition). Interestingly, Bid is able to bind to pro-apoptotic members of the Bcl-2 family (e. g. Bax) as well as to pro-survival members Bcl-2 and Bcl-XL.

3) **Bax maybe involved in caspase-independent death**: The BH1 and BH2 domains of Bcl-2 family members (Bcl-2, Bcl-XL and Bax) resemble membrane insertion domains of

bacterial toxins: hypothetically they can form pores in organelles such as mitochondria, what at least was demonstrated in lipid bilayers *in vitro*. In yeast which lack Bcl-2 like proteins, CED-4, caspases -Bax and Bak were shown to induce cell death, while Bcl-2 can protect, apparently by preventing mitochondrial disruption. Bax and Bax-like proteins might mediate caspase-independent death via channel-forming activity, which would promote the mitochondrial permeability transition.

11.2.7 Suppressors of apoptosis

Apoptosis, or programmed cell death, is an active process of self-destruction, described a long time ago. However, the understanding of the molecular pathways which regulate programmed cell death is more recent. Apoptosis occurs during embryonic and foetal development, and tissue remodeling, and its purpose is to assure homeostasis of cells and tissues. Several genes affecting various steps in programmed cell death must be expressed to trigger apoptosis, but the genes suppressing apoptosis were also identified.

(1) Inhibitors of apoptosis (IAPs) The induction of apoptosis or progression through the process of apoptosis is inhibited by a group of proteins called IAPs. These proteins contain a BIR (baculovirus IAP repeat) domain near the amino-terminus. The BIR domain can bind some caspases. Many members of the IAP family of proteins block proteolytic activation of caspase-3 and -7. XIAP, cIAP-1 and cIAP-2 appear to block cytochrome C-induced activation of caspase-9, thereby preventing initiation of the caspase cascade. Since cIAP-1 and cIAP-2 were first identified as components in the cytosolic death domain-induced complex associated with the TNF family of receptors, they may inhibit apoptosis by additional mechanisms.

(2) Survivin Survivin (also termed *Birc5*) belongs to the family of genes known as inhibitors of apoptosis, and it has been implicated in both prevention of cell death and control of mitosis.

Survivin contains a baculovirus inhibitor of apoptosis repeat (BIR) protein domain that classifies it as a member of the IAP family. Survivin inhibits apoptosis, via its BIR domain, by either directly or indirectly interfering with the function of caspases. Survivin is also a chromosomal passenger protein that is required for cell division. Survivin is expressed in embryonic tissues as well as in the majority of human cancers.

(3) Bcl-2 members with anti-apoptotic potentials The Bcl-2 family of proteins regulates apoptosis, some antagonizing cell death and others facilitating it. Bcl-2 subfamily member Bcl-2, Bcl-XL, Bcl-w, Mcl-1 and A1 are anti-apoptotic proteins.

Human Bcl-2 gene and related Bcl-2 genes were described as anti-apoptotic genes. It is clear that Bcl-2 protects many cell lines from induced apoptosis. Other proteins, like Bcl-XL, A1 or Mcl-1 have the same anti-apoptotic function. It is known that Bcl-2 can interact with other proteins. For example, Bax, which can exist as a homodimer, is also able to form a heterodimer with Bcl-2. A surexpression of Bax in several cell lines allows to counteract the effect of Bcl-2.

R-ras p23 is another example, among others, of a protein interacting with Bcl-2, and this results in an interruption of the apoptotic signal transduction pathway when Bcl-2 is overexpressed. Many interesting results suggest that Bcl-2 is a death repressor molecule functioning in an anti-oxydant pathway, but other data seem to claim the contrary.

(4) Flip/Flame FLIP/FLAME is highly homologous to caspase-8. It does not, however, contain the active site required for proteolytic activity. FLIP appears to compete with caspase-8 for binding to the cytosolic receptor complex, thereby preventing the activation of the caspase cascade in response to members of the TNF family of ligands. The exact *in vivo* influence of the IAP family of proteins on apoptosis is still under investigation.

11.3 Senescence or apoptosis

As described above, senescence and apoptosis share numerous characteristics at molecular and cellular levels, and one of the characteristics of the senescent cell is apoptotic resistance, indicating complicate linkage between these two phenotypes.

11.3.1 Dysregulation of the apoptosis involves in senescence

Recent evidence shows that dysregulation of the apoptotic process may be involved in some senescence processes. Apoptosis and susceptibility to apoptosis in several types of intact cells are enhanced by senescence.

While senescence is characterized as a relatively undefined process, apoptosis on the other hand is well defined and conserved from *C.*

11.3.2 Senescence and apoptosis result from antagonistic pleiotropy

The cellular and molecular changes that occur with senescence or apoptosis, specifically changes that occur as a result of cellular responses to environmental insults, are evolved. Because the force of natural selection declines with age, it is likely that these processes were never optimized during their evolution to benefit old organisms. That is, some age-related changes may be the result of gene activities that were selected for their beneficial effects in young organisms, but the same gene activities may have unselected, deleterious effects in old organisms, a phenomenon termed antagonistic pleiotropy. These two cellular processes, apoptosis and cellular senescence, may be examples of antagonistic pleiotropy. Both processes are essential for the viability and fitness of young organisms, but may contribute to aging phenotypes, including certain age-related phenotypes.

11.3.3 Mitochondria mediates both senescence and apoptosis

Apoptosis and senescence share common mechanisms in oxidative stress and mitochondrial involvement. The mechanism involved in the aging process is free radical attacks. Mitochondria are the primary source of free radicals due to inevitable leakage of electrons from the electron transport chain. Since free radicals are highly reactive, they may interact with DNA and proteins to alter cellular functions. A very destructive target is the voltage-dependent anion channel (VDAC) which plays the central role in mitochondria-mediated apoptosis.

Several diseases associated with senescence appear to be the direct result of cells containing dysfunctional mitochondria. The typical human cell has several hundred mitochondria. Mitochondria are the sites of three important processes which are involved in the progression of senescence: energy conversion (ATP)、production of reactive oxygen species (ROS)、initiation of cell death.

Mitochondria function to convert the chemical energy from food into ATP, which cellular enzymes use to do work. As a by product of energy conversion, ROS are produced.

11.3.4 Oxidative stress causes both senescence and apoptosis

Oxidative stress is a major form of endogenous and exogenous damage to cells. Interestingly, the same genes (e. g., p53 and Rb) are involved in cell senescence and apoptosis. A better understanding of the molecular mechanism of signal transduction leading to cell senescence or cell death through divergent pathways is required to understand cellular responses to environmental stresses, particularly oxidative damage.

11.3.5 Opting for senescence or apoptosis

Cell could choose senescence or apoptosis under certain stresses. For example, one obvious mechanism that could regulate the choice between senescence and apoptosis is if the apoptotic process was inhibited and senescence occurred instead as a default mechanism. The mitochondria anti-apoptotic protein Bcl-2 has been shown to induce senescence, judged by SA-βGAL staining, yet this may more resemble quiescence as p27 was overexpressed.

(1) p21 overexpression refers to senescence When cell shifted from apoptosis to senescence upon Bcl-2 overexpression, p21 was also found to be upregulated. This should be the reason for the shift from apoptosis to senescence as p21 expression after DNA damage leads to senescence while absence of p21 induction after DNA damage leads to apoptosis. Similarly, apoptosis was associated with low p21 levels whereas senescence was associated with high p21 levels in a cancer cell treated with interferon-γ.

p21 is a molecule downstream of p53 induction, then an important question is why p53 sometimes induces p21 expression and sometimes not. Some of the regulation could be a result of the convergence of several pathways that directly regulate p21. Such as cisplatin, a chemotherapy that is given as treatment for certain types of cancer, can induce growth arrest at low concentration and this growth arrest is associated with translational activation of p21 by an unknown mechanism.

(2) Higher levels of p53 facilitate apoptosis It is also possible that the decision, of whether or not to induce p21, occurs at the activation level of p53. Interestingly the phosphorylation pattern of p53 is different during induction of senescence and DNA damage-induced apoptosis. It appears that a large DNA damage response leads to apoptosis while a low but persistent activation of p53 induces senescence. For example, upon hydrogen peroxide treatment, both senescence and apoptosis are possible outcomes; apoptosis was associated with higher levels of p53 and lower levels of p21, while senescence was associated with lower levels of p53 and higher levels of p21. Similarly, a TRF2 (telomeric-repeat binding factor 2) mutant that causes telomere dysfunction induced apoptosis or senescence, depending upon the expression level and thereby the extent of telomere damage; substantial overexpression of p53 induced apoptosis while lower overexpression induced senescence. Therefore, it seems like that the nature of p53 response is the major determinant whether p53 response will induce apoptosis or senescence.

(3) BMI1 regulates senescence and apoptosis The polycomb group protein BMI1 has been linked to proliferation, senescence and apoptosis. BMI1 regulates both $p19^{ARF}$-MDM2-

P53 and p16^{INK4A}-pRB pathways in mouse cells, while it primarily regulates the p16IN K4A-pRB pathway in human cells.

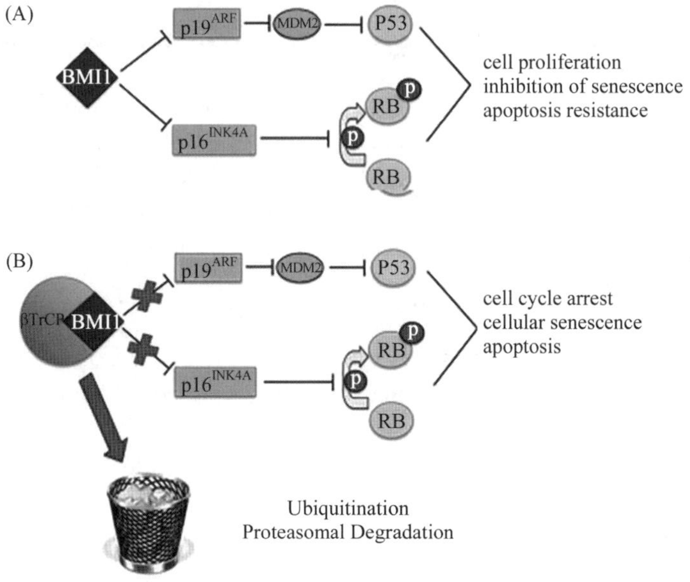

Figure 11-12 Regulation of cell cycle, apoptosis, and senescence by BMI1. Modulation of BMI1 expression by bTrCP, regulates cell proliferation, apoptosis, senescence, and oncogenesis. (A) BMI1 inhibits signaling through the IN K4A-ARF tumor suppressor locus through repression of p16 and p19, resulting in loss of p53 and RB cell cycle regulation, inhibition of senescence and reduced apoptosis. (B) Association of BMI1 with bTrCP results in ubiquitination and proteasomal degradation of BMI1, senescence, inhibition of the cell cycle, and increased apoptosis.

Bmi1 is downregulated when the cells undergo replicative senescence, but not when they are quiescent. Additionally, *Bmi1* extends replicative lifespan but does not induce immortalization when overexpressed. βTRCP (beta-transducin repeat containing protein) belongs to E3 ubiquitin ligases, which regulates BMI1 protein turnover via ubiquitination and degradation. In the absence of Bmi1, both the $p16^{Ink4a}$ and the $p19^{Arf}$ genes from the Ink4a locus are expressed. Lifespan extension by *Bmi1* is mediated in part by suppression of the $p16^{Ink4a}$-dependent senescence pathway and requires an intact pRB pathway (**Figure 11-12**).

Summary

Senescence and apoptosis are two common cell fates. While apoptosis leads to death and removal of the cells, senescence cells persist and survive for extended periods of time. Both senescence and apoptosis are important not only for development, also for a cancer treatment perspective.

Senescence is defined as progressive deterioration in physiological functions and metabolic processes of a cell or organism. It is a program activated by normal cells in response to various types of stress. Several hypotheses explain senescence, such as mitochondria ageing, reactive oxygen species stress, replicative ageing and telomere shortening. Apoptosis is often referred to as programmed cell death, which mainly results from the extrinsic apoptotic signal transduction, the intrinsic death signal transduction and DNA damage. The mitochondria plays important role in cellular apoptosis. Cell senescence and apoptosis share numerous characteristics at molecular and cellular levels, and there is complicate linkage between these two phenotypes.

Questions

1. What does the senescence mean in cell biology?
2. What do you know the mechanisms in cell apoptosis?
3. What are the roles of mitochondria in senescence and apoptosis?
4. How does cell choose senescence or apoptosis?
5. How can cancer cells undergo senescence or apoptosis?

（吴超群,黄　燕修改）

Chapter 12

Cells in immune response

Loving organisms provide ideal habitats in which other organisms can grow. It is not surprising that animals are subject to infection by viruses, bacteria, fungi, and parasites. Vertebrates have evolved several mechanisms that able to recognize and destroy these infectious agents. As a result, vertebrates are able to develop immunity against invading pathogens. Immunity results from the combined activities of immune response with many different cells, lymphoid organs. It is often divided into the two sections of innate and adaptive immunity, the former encompassing unchanging mechanisms that are continuously in force to ward off noxious influences, and the latter responding to new influences by mounting an immune response. In this chapter, we will focus on the basic events in the body's response to the presence of an intruding microbe.

12.1 The immune system

The body is protected from infectious agents, their toxins, and the damage they cause by a variety of effector cells and molecules that together make up the **immune system**. Both innate and adaptive immune responses depend upon the activities of many cell types including the **macrophages**, **dendritic cells**, **nature killer cells** (**NK cells**), **neutrophils**, **basophils**, **eosinophils**, **mast cells**, **B cells**, **T cells**, and so on. Most of these cells arise from **bone marrow**, where many of them develop and mature, and communicate with interdependent cells by cytokines, which control proliferation, differentiation and function of cells in the immune response. They are also involved in processes of inflammation and in the neuronal, hematopoietic and embryonal development of an organism.

The immune system is composed of related organs which harbor immune cells, and where the cells undergo development, differentiation and maturation. Lymphoid organs can be divided broadly into the central or primary lymphoid organs, where lymphocytes are generated, and the peripheral or secondary lymphoid organs, where mature naïve lymphocytes are maintained and adaptive immune responses are initiated (**Table 12-1**).

Table 12-1 Central and peripheral lymphoid organs

Central lymphoid organs	Peripheral lymphoid organs
Bone marrow	Spleen
Thymus	Lymph nodes
	Mucosal associated lymphoid tissues

12.1.1 Central lymphoid organs

Central lymphoid organs are the sites of place for lymphocyte development. In humans, B lymphocytes develop in bone marrow, whereas T lymphocytes develop within the thymus from bone marrow-derived progenitors.

(1) Bone marrow New blood cells are produced from bone marrow of large bones. All the immune cells are initially derived from the bone marrow by hematopoiesis. During hematopoiesis, bone marrow contains two types of pluripotent stem cells. A common lymphoid stem cells gives rise to the lymphoid lineage — the T and B lymphocytes and the innate lymphoid cells (ILCs) and natural killer (NK) cells. A common myeloid stem cells gives rise to the myeloid lineage — the leukocytes, the erythrocytes and the megakaryocytes that produce platelets for blood clotting (**Figure 12-1**). The most of them mature in bone marrow except T lymphocytes that mature in thymus.

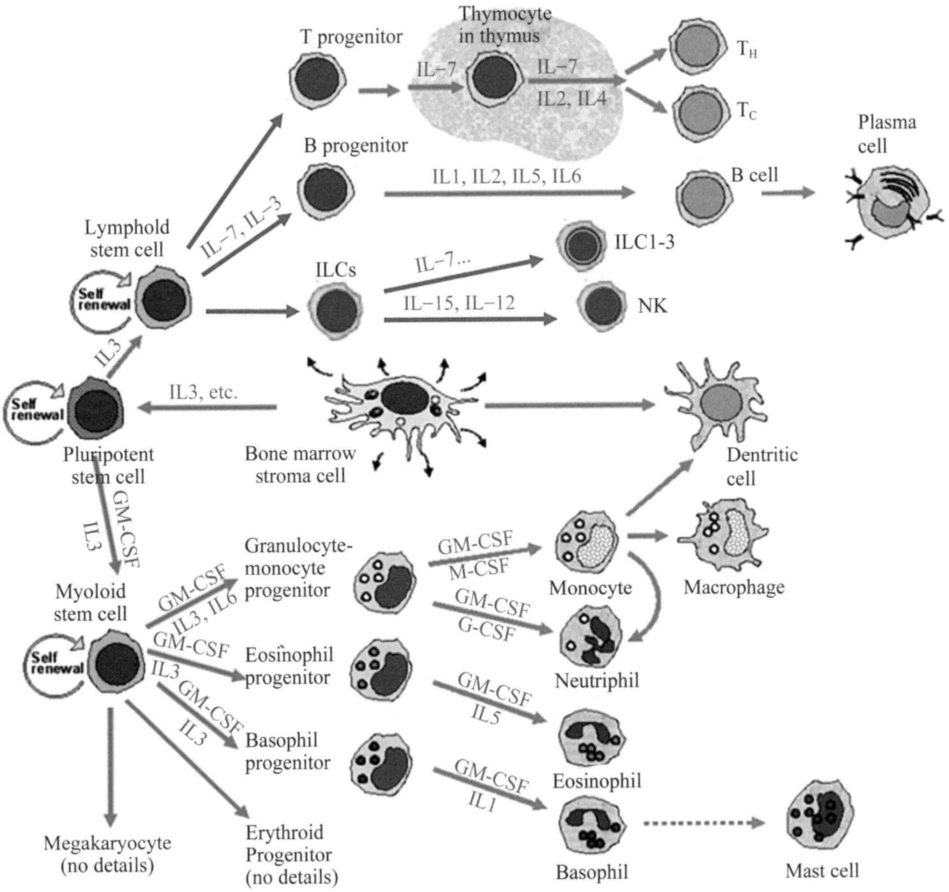

Figure 12-1 The immune cells are differentiated from pluripotent stem cell. The stem cell can give rise to lymphoid stem cell and myeloid stem cells, and differentiate into most of the various blood cells and the various types of lymphocytes.

(2) Thymus In human anatomy, the **thymus** is a ductless gland located in the upper chest. It is most active during puberty, after which it shrinks in size and activity in most individuals and is replaced with fat. The thymus plays an important role in the development of the immune system in early life, especially for T cell differentiation to produce mature T

cells. In general, the most immature — T cell leaves the bone marrow and enter the thymic cortex through the blood vessels. Through a remarkable maturation process sometimes referred to as thymic education, T cells that are beneficial to the immune system are spared, while those T cells that might evoke a detrimental autoimmune response are eliminated. In the thymus lobules lymphocytes mature into naïve T cells that behave in different ways according to their type. Maturation begins in the cortex, and as thymocytes mature, they migrate toward the medulla, so that the medulla contains mostly mature T cells. Only naïve $CD4^+$T or naïve $CD8^+$ T cells exit the thymus and enter the blood and peripheral lymphoid tissues.

12.1.2 Peripheral lymphoid organs

The peripheral lymphoid organs are the sites of place for mature B and T lymphocyte activation by antigen, and lymphocytes recirculate between the blood and these organs until they encounter their specific antigen.

(1) Lymph nodes and the lymphatic system Lymph nodes are small nodular organs and components of the lymphatic system, and are encapsulated nodular aggregates of lymphoid tissues located along lymphatic channels throughout the body. Clusters of lymph nodes are found in the underarms, groin, neck, chest, and abdomen. Lymph nodes act as filters for antigens, with an internal honeycomb of connective tissue filled with T cells, B cells, dendritic cells and macrophages, that collect and destroy bacteria and viruses. When the body is fighting an infection, these lymphocytes multiply rapidly and produce a characteristic swelling of the lymph nodes.

The lymphatic vessels (or lymphatics) are a network of thin tubes that branch, like blood vessels, into tissues throughout the body (Figure 12-2). Lymphatic vessels carry lymph, a colorless, watery fluid originating from interstitial fluid (fluid in the tissues). The lymphatic system transports infection-fighting cells-lymphocytes, is involved in the removal of foreign matter and cell debris by phagocytes and is part of the body's immune system. It also transports fats from the small intestine to the blood.

Unlike the circulatory system, the lymphatic circulatory system is not closed and has no central pump; the lymph moves slowly and under low pressure. Like veins, lymph vessels have one-way valves and depend mainly on the movement of skeletal muscles to squeeze fluid through them. Rhythmic contraction of the vessel walls may also help draw fluid into the lymphatic capillaries.

(2) Spleen Spleen is an organ derived from mesenchyme and lying in the mesentery. It is the major site of immune responses to blood-borne antigens. The organ consists of masses of lymphoid tissue of granular appearance located around fine terminal branches of veins and arteries. The spleen is an immunologic filter of the blood. In addition to capturing foreign materials(**antigens**) from the blood that passes through the spleen, migratory macrophages and dendritic cells bring antigens to the spleen via the bloodstream. An immune response is initiated when the macrophage or **dendritic cells** present the antigen to the appropriate B or T cells. This organ can be thought of as an immunological conference center. In the spleen, B cells become activated and produce large amounts of antibody. Also, old red blood cells are destroyed in the spleen.

(3) Other important lymphoid tissues The cutaneous immune system and mucosal immune system are specialized collections of lymphoid tissues and **antigen-presenting cells** (**APCs**) located in and under the epithelia of the skin and the gastrointestinal and respiratory tracts, respectively. Although most of the immune cells in these tissues are diffusely scattered beneath the epithelial barriers, there are discrete collections of

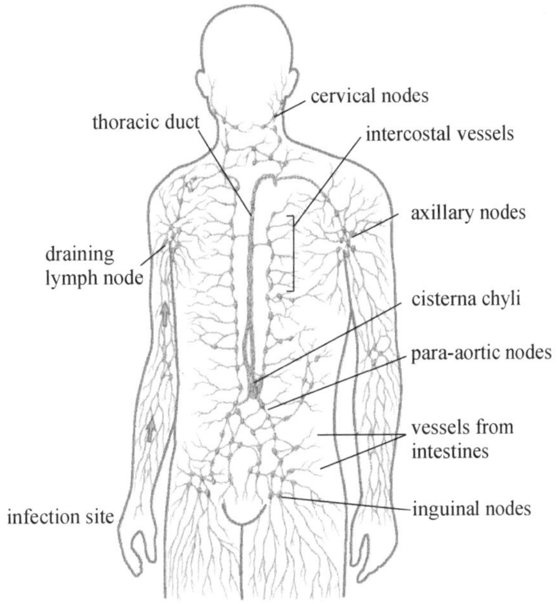

Figure 12-2 The lymphatic system. The major lymphatic vessels and collections of lymph nodes are illustrated. One-way valves in the lymphatic vessels prevent a reverse flow, and the movements of one part of the body in relation to another are important in driving the lymph along.

lymphocytes and APCs organized in a similar way as in lymph nodes. The principal cell populations within the epidermis are keratinocytes, melanocytes, epidermal Langerhans cells, and intraepithelial T cells. Many foreign antigens gain an entry into the body through the skin, **Langerhans cells** is an important immature dendritic cell for cutaneous immune system.

The bulk of the body's lymphoid tissue (>50%) is found associated with the mucosal system. Mucosal epithelia are barriers between the internal and external environments and are an important site of entry of microbes. The mucosal immunity is essential for life and requires continual and effective protection against invasion. Mucosal associated lymphoid tissue (MALT) play crucial role in the immune system. The organized lymphoid tissues in the gut are called the gut-associated lymphoid tissues (GALT), which including the Peyer's patches and solitary lymphoid follicles of the intestine, the appendix, the tonsils and adenoids in throat, and the mesenteric lymph nodes. The organized lymphoid tissues in the wall of the upper respiratory tract are called the bronchus-associated lymphoid tissues (BALT), and in the lining of the nose are called the nasal-associated lymphoid tissues (NALT). These lymphoid tissues contain large numbers of effector lymphocytes even in the absence of disease. They also include the exocrine glands associated with these organs, such as the pancreas, the conjunctivae and lachrymal glands of the eye, the salivary glands, and the lactating breast.

12.2 Cells in the innate immune system

12.2.1 Innate immune cells

Innate immune system is the first line of defense against infections. It comprises epithelial barriers, macrophages, neutrophils, dendritic cells, natural killer cells, circulating proteins and cytokines, etc. Innate immunity serves protection against infections that relies on mechanisms that exist before infection. It is capable of rapid responses to microbes, and react in essentially the same way to repeat infections.

(1) Epithelial barriers Epithelial barriers are major component of innate immune system to prevent the interaction between the host and pathogen. Pathogens most commonly enter the body by across the mucosa lining the respiratory, digestive, and urogenital tracts, or through damaged skin. The skin and the extensive epithelial lining of the mucosal tissues act

as nonspecific barriers impeding, if not preventing, infectious agents from entering the body. Epithelial barriers encounter constitutively expressed antimicrobial molecules such as lysozyme that restrict the infection.

(2) Classical innate immune cells

1) **Phagocytes**: The two types of circulating phagocytes, neutrophils and monocytes, are blood cells that are recruited to sites of infection, where they recognize and ingest microbes for intracellular killing.

Phagocytes can recognize danger signals by using **pattern recognition receptors** (**PRRs**) (such as Toll- like receptor) on their cell membranes during their journey through the tissues and their encounters with danger signal of **pathogen-associated molecular patterns** (**PAMPs**) from pathogens or damaged cells. Once they are stimulated through their receptors, phagocytes kill and clear pathogens through phagocytosis, respiratory burst, and the release of proteolytic enzymes and various cytokines. More details will be discussed later.

2) **Monocytes and macrophages**: Monocytes are abundant in the blood and ingest microbes in the blood and in tissues. During inflammatory reactions, monocytes enter extravascular tissues and differentiate into cells called macrophages, which, unlike neutrophils, survive in these sites for long periods. Thus, blood monocytes and tissue macrophages are two stages of the same cell lineage, which often is called the mononuclear phagocyte system. Macrophages are cells found in tissues that are responsible for phagocytosis of pathogens, dead cells and cellular debris. Once through the endothelium, the monocytes are distributed to different parts of tissue and further sub-divided depending on their location (**Table 12-2**).

Table 12-2 Cells of the mononuclear phagocytic system and their respective locations

Cells	Locations
Monocytes	Blood stream
Alveolar macrophages	Lungs
Sinus macrophages	Lymph nodes and spleen
Kupffer cells	Liver
Osteoclast	Bone
Microglia	Nerve system

Macrophages are large cells with a horseshoe-shaped nucleus. Its manly functions are phagocytosis and antigen presentation to T cells, while functional as APCs (see later). Macrophages typically respond to microbes nearly as rapidly as neutrophils do, but macrophages survive much longer at sites of inflammation. Unlike neutrophils, macrophages are not terminally differentiated and can undergo cell division forming classically activated macrophages (also called M1) or alternatively activated macrophage (also called M2) at an inflammatory site. M1 are involved in destroying microbes and in triggering inflammation. M2 occurs in the absence of strong TLR signals and is induced by the cytokines IL-4 and IL-13, which appear to be more important for tissue repair and to terminate inflammation. Therefore, macrophages are the dominant effector cells of the later stages of the innate immune response, 1 or 2 days after infection.

3) **Dendritic cells**: Dendritic cells (DC) are immune cells and form part of the innate immune system. They are present at a low frequency in those tissues which are in contact with the environment, such as in the skin (called **Langerhans cells**) and the lining of nose,

lungs, stomach and intestines. Especially in immature state, they can also be found in blood. They have long spiky arms, so called dendritic cells. As one of an important APCs, immature dendritic cells can travel through the blood stream to the spleen or through the lymphatic system to a lymph node and characterized by high endocytic activity and low T cell activation potential. Once they have come into contact with a pathogen, they become activated into mature dendritic cells activating naïve $CD4^+$ T cell and $CD8^+$ T cells by presenting antigen-peptide derived from the pathogen. Every T cell only recognizes one particular antigen-peptide. Only dendritic cells are able to activate a naïve T cell which has never encountered its antigen before. Dendritic cells form an important bridge between innate and adaptive immunity as the cells present the antigenic peptide to the $CD4^+$ T cell to form T helper cell or to $CD8^+$ T cell to form cytotoxic T lymphocytes (CTL). Dendritic cells are the most potent of all APCs and served a critical function in adaptive immune responses by capturing and displaying microbial antigens to T lymphocytes. So, dendritic cells constitute an important bridge between innate and adaptive immunity.

4) **Neutrophils**: Neutrophils, also called polymorphonuclear leukocytes (PMNs), are the most abundant leukocytes in the blood, which are much more numerous than the longer-lived monocyte/macrophages. In response to certain bacterial and fungal infections, the production of neutrophils from the bone marrow increases rapidly, and their numbers in the blood may rise up to 10 times the normal. They have a pivotal role to play in the development of acute inflammation as a part of the immune response. Neutrophils activate as phagocytes and capable of only one phagocytic event, expending all of their glucose reserves in an extremely vigorous respiratory burst. The respiratory burst involves the activation of an NADPH oxidase enzyme, which produces large quantities of superoxide, a reactive oxygen species. Being highly motile, neutrophils quickly congregate at a focus of infection, attracted by cytokines expressed by activated endothelium, mast cells and macrophages. In addition, neutrophils release their granules including acidic and alkaline phosphatases, defensins and peroxidase — all of which represent the requisite molecules required for successful elimination of the unwanted microbes. However, neutrophil antimicrobial products can also damage host tissues at site of inflammation.

5) **Eosinophils**: Eosinophils are granulocytes named because of their intense staining with "eosin-loving" granules. Under the microscope, eosinophils typically have a bilobed nucleus and contain many basic crystal granules in their cytoplasm. The granules are eosinophil mediators that are toxic to many organisms and also to tissues as in asthma. Eosinophils are motile and phagocytic, and are particularly active in parasitic infection, such as schistosome infection, hay fever etc., by releasing cytoplasmic granules to provide molecules that potent weapons against parasitic worms.

6) **Basophils**: Basophils are a type of bone marrow-derived circulating granulocytes with structural and functional similarities to mast cells. Their granules containing many of the same inflammatory mediators as mast cells and expresses a high-affinity Fc receptor for IgE, which are involved in Type I hypersensitivity responses. Crosslinking of the IgE causes the basophils to release pharmacologically active mediators such as heparin and histamine. Basophils are recruited into tissue sites where antigen is present may contribute to immediate hypersensitivity reactions. Basophils, therefore act very much like mast cells except that they are in the blood instead of the tissues.

7) **Mast cells**: Mast cells are bone marrow-derived cells with abundant cytoplasmic granules that are present throughout the skin and mucosal barriers. Mast cells can be activated by microbial products binding to TLRs and by components of the complement

system as part of innate immunity or by an antibody-dependent mechanism in adaptive immunity. Mast cells contain large distinctive cytoplasmic granules in their cytoplasm that contain a mixture of chemical mediators, including histamine, chymase, tryptase, and serine esterases, etc., that cause vasodilation and increased capillary permeability, as well as proteolytic enzymes that can kill bacteria or inactivate microbial toxins. High affinity Fc receptors specific for IgE and IgG are expressed on the surface of mast cells and can be activated to release their granules that act rapidly to make local blood vessels permeable, swelling and mucus secretion. Mast cells also synthesize and secrete lipid mediators (e.g., prostaglandins and leukotrienes) and cytokines (e.g., tumor necrosis factor, TNF), which stimulate both the acute and the chronic inflammatory response. Mast cell products provide defense against helminths and other pathogens, as well as protection against snake and insect venoms, and they are responsible for symptoms of allergic diseases.

(3) Innate lymphoid cells Innate lymphoid cells (ILCs) are a newly classified subgroup cells of common lymphoid stem cells, which includes natural killer (NK) cells, group 1 ILCs (ILC1), group 2 ILCs (ILC2) and group 3 ILCs (ILC3) (**Table 12-3**). They migrate from the bone marrow and populate lymphoid tissues and peripheral organs, notably the dermis, liver, small intestine, and lung. They secrete various cytokines that induce inflammation, against various infections.

Table 12-3 Subgroup of innate lymphoid cells and its functions

Subgroup	Inducing cytokine	Effector molecules produced	Functions
NK cells	IL-12	IFN γ, perforin, granzyme	Against viruses, intracellular pathogens
ILC1	IL-12	IFN γ	Against viruses, intracellular pathogens
ILC2	IL-25, IL-33, TSLP	IL-5, IL-13	Expulsion of extracellular parasites
ILC3, ITi cells	IL-23	IL22, IL-17	Immunity to extracellular bacteria and fungi

1) **Natural killer cells**: Nature killer (NK) are developmentally related to group 1 ILCs (ILC1) and rise from the lymphoid lineage as T and B lymphocytes, but their size is larger than T and B cells. By surface phenotype and lineage, NK cells are neither T nor B lymphocytes, and they do not express somatically rearranged, clonally distributed antigen receptors like immunoglobulin or T cell receptors. Many receptors expressed on the surface of NK cell are responsible for NK cell activation, which is regulated by a balance between signals that are generated from activating receptors and inhibitory receptors.

Natural killer cells play several important roles in defense against intracellular microbes. They kill virally infected cells during the first few days after viral infection. They recognize infected and/or stressed cells and respond by directly killing these cells and by secreting inflammatory cytokines. When NK cells are activated, granule exocytosis releases perforin and granzyme proteins adjacent to the target cells and eliminate reservoirs of infection. The IFN-γ secreted by NK cells activates macrophages to destroy phagocytosed microbes. This IFN-γ-dependent NK cell-macrophage reaction can control an infection with intracellular bacteria such as *Listeria monocytogenes* for several days or weeks.

2) **Other cells in Innate lymphoid cells**: Other innate lymphoid cells (ILCs) are tissue-resident cells that produce cytokines similar to those secreted by helper T lymphocytes but do not express T cell receptors (TCRs). These ILCs have been divided into three major groups based on their secreted cytokines. These groups correspond to the Th1, Th2, and Th17 subsets of $CD4^+$ T cells. The responses of ILCs are often stimulated by cytokines produced by damaged epithelial and other cells at sites of infection. ILCs likely provide

early defense against infections in tissues, but their essential roles in host defense or immunological diseases, especially in humans, are not clear.

(4) Innate-like lymphocytes Innate-like lymphocytes (ILLs) group of cells include natural killer T cell (NK-T), γδ T cells and B-1 cells, which all respond to infections in ways that are characteristic of adaptive immunity (e. g., cytokine secretion or antibody production) but have features of innate immunity (rapid responses, limited diversity of antigen recognition).

1) **Natural killer T cells**: There smaller subpopulation of T cells, comprising less than 5% of all T cells, express αβ TCRs and surface molecules found on natural killer cells, and are therefore called natural killer T cells (NKT cells). NKT cells express αβ TCRs with limited diversity, and they recognize lipid antigens displayed by nonpolymorphic class I MHC-like molecules called CD1. The physiologic functions of NKT cells are not well defined.

2) **γδT cells**: About 5% to 10% of T cells in the body express receptors composed of γδ TCRs. These receptors are structurally similar to the αβ TCR but have very different specificities. The γδTCR may recognize a variety of protein and nonprotein antigens by class I MHC-like molecules and others apparently requiring no specific processing or display, usually not displayed by classical MHC molecules. T cells expressing γδTCRs are abundant in epithelia. This observation suggests that γδT cells recognize microbes usually encountered at epithelial surfaces, but neither the specificity nor the function of these T cells is well established. Intraepithelial T lymphocytes may be effector cells of innate immunity.

3) **B-1 cells**: B-1 cells, a distinct population found at mucosal sites and the peritoneal cavity, develop earlier from fetal liver-derived hematopoietic stem cells. Many B-1 cells express the CD5 molecule and play greater roles in T-independent responses. B-1 cells express antigen receptors of limited diversity and make predominantly T-independent IgM responses. IgM antibodies may be produced spontaneously by B-1 cells, without overt immunization. These antibodies, called natural antibodies, may help to clear some cells that die by apoptosis during normal cell turnover and may also provide protection against some bacterial pathogens. B-1 cells also appear to be important sources of IgA antibody in mucosal tissues, especially against nonprotein antigens.

12.2.2 Adaptive immune cells

The principal cells of the adaptive immune system include B and T lymphocytes, APCs, and effector cells. Lymphocytes are the cells that specifically recognize and respond to foreign antigens and are therefore the mediators of humoral and cellular immunity. There are distinct subpopulations of lymphocytes that differ in how they recognize antigens and in their functions. As discussed earlier in this chapter, the cells of the lymphoid lineage B and T cells arise from common lymphoid precursor cells in the bone marrow (Figure 12-1) but they mature in different lymphoid organs. Adaptive immunity is based on clonal selection from a repertoire of lymphocytes bearing highly diverse antigen-specific receptors that enable the immune system to recognize any foreign antigen. It is characterized by exquisite specificity for distinct macromolecules, and by "memory", which is the ability to respond more vigorously to repeated exposures to the same microbe.

(1) B cells The development of a B lineage cell proceeds through several stages marked by the rearrangement and expression of the immunoglobulin genes, and B cells receptors (BCRs). The stages for B cell development include stem cells, pro-B cells, pre-B cells,

follicular (FO) B cells, marginal zone B-2 cells and B-1 cells. B-1 cells are discussed earlier. B-2 cells (as called B cells) undergo a selection process (referred to as negative selection) in an attempt to ensure that their antigen receptor does not display self-reactivity. Those B cells surviving negative selection disperse to the peripheral lymphoid organs as naïve B cells.

Naïve B cells must migrate to secondary lymphoid organs to meet antigen for their functions. The molecules on the surface of naïve B cells are responsible for B cell activation, including BCR, CD79α/β, and other co-stimulators. BCRs are unique in the early cellular events that induced by antigen initiate B cell activation. Once encountering the foreign antigen, B cells are activated and, usually with Th2 cell help, proliferate and differentiate into antibody-producing cells (plasma cells) and memory cells.

The diversity in the B cell population is generated by immunoglobulin gene re-arrangements that occur at the pro-B cell stage. Each B cell clone secretes antibody with a single antigen specificity. The affinity of the antibody for the cognate antigen can be modified by further gene re-arrangements that take place during the maturation of the humoral response. The isotype (IgM, IgG, IgE, IgA or IgD) is influenced by direct and indirect (i. e., mediated via cytokines) interactions with Th1 and Th2 cells.

Not all proliferating B cells develop into plasma cells. Indeed, a significant proportion remains as memory B cells through a process known as clonal selection. This process is vital in eliminating the antigen and the body should become re-exposed to the same antigen in the future.

(2) T cells T lymphocytes are derived from common lymphoid stem cells in bone marrow and generated in the thymus. The stages for T cell development include stem cells, pro-T cells, mature T cells, effector T cells, and memory T cells. Pro-T cells may commit to either the αβT or γδT cell lineages. The αβT cell lineages further developed into mature $CD4^+$ T and $CD8^+$ T cells. The γδT cell acts as a part of the first line of defense and involved in innate immune response (see above). T cell development in the thymus is a multistep process with several built-in check points to ensure that the appropriate differentiation has taken place. As a result of selection (including positive and negative selection), only a small proportion of pro-T cells, actually exit the thymus to the periphery as different subsets of mature T cells.

The mature T cells that leave the thymus have not yet encountered the antigen that they have specificity for. These cells are called naïve T cells and mainly circulate from the bloodstream to the peripheral lymphoid organ for years before they die or encounter antigen. Important molecules on the surface of T cells are TCR, CD4/CD8, and other co-stimulators. The TCRs are unique in that it is only able to identify antigen when it is associated with a MHC molecule on the surface of the cell. T cells described by cluster of differentiation (CD) can be broadly divided into helper T (Th, or $CD4^+$ T cells) and cytotoxic T cells (CTL, or $CD8^+$ T cells).

$CD4^+$ T cells only recognize antigen displayed on MHC class Ⅱ molecules of APCs and differentiate into a subsets of T cells (Th1, Th2, Th17, Tfh or Treg), depending on the type of immune response taking place and in particular the cytokines present in the environment of the T cells. These subsets of T cells perform different functions in immune response. $CD8^+$ T cells recognize antigen displayed on MHC class Ⅰ molecules. Because MHC class Ⅰ molecules are found on essentially all nucleated cells of the body, $CD8^+$ T cells can monitor all cells for signs of infection. Hence, activated $CD8^+$ T cells (also called CTL), unlike Th cells, are primarily involved in the destruction of infected cells, notably viruses.

When naïve T cells recognize microbial antigens displayed by MHC and also receive additional signals induced by microbes, the clonally selected antigen-specific T cells proliferate and then differentiate into effector T cells and memory T cells.

12.3 Immune responses

Host defenses are grouped under innate response and adaptive response. Innate response is phylogenetically older and response early, and the more specialized and powerful adaptive immune response evolved later. Innate response provides immediate protection against microbial invasion by epithelial barriers, innate immune cells and several plasma proteins. The innate immune responses are required to initiate adaptive immune responses against the infectious agents. Adaptive response is essential for defense against infectious microbes that may have evolved to resist innate response. Adaptive immune responses develop more slowly and provide more specialized defense against infections.

12.3.1 Innate immune response

The innate immune system eliminates microbes mainly by inducing **the acute inflammatory response** and by antiviral defense mechanisms. Different microbes may elicit different types of innate immune reactions, each type of response being particularly effective in eliminating a particular kind of microbe. Innate immune response refers to various physical and cellular attributes that collectively represent the first lines of defense against an intruding pathogen and infectious disease. The response evolved is therefore rapid, and is unable to "memorize". The body can be exposed to the same pathogen again without any memory in the future.

The response to an initial infection occurs in three phases. These are the innate phase, the early induced innate response, and the initiating adaptive immune response (**Table 12-4**). The first two phases rely on the recognition of pathogens by germline-encoded receptors of the innate immune system, whereas adaptive immunity uses variable antigen-specific receptors that are produced as a result of gene segment rearrangements. Adaptive immunity occurs late, because the rare B cells and T cells specific for the invading pathogen must first undergo clonal expansion before they differentiate into effector cells that migrate to the site of infection and clear the infection. The effector mechanisms that remove the infectious agent are similar or identical in each phase.

Table 12-4 Innate immune response to infection occurs in three phases

Phases	Immediate innate response	Early innate response	Initiating adaptive response
Time	<4 hours	4~96 hours	>96 hours
Reaction	Recognition by non-specific and broadly specific effectors	Recognition of PAMPs. Activation of effector cells and inflammation	Recognition by naïve B and T cells. Clonal expansion and differentiation to specific effector cells
Result	Clear the infection agent. If fail, recruitment of innate cells	Clear the infection agent. If fail, transport of antigen to lymphoid organs	Clear the infection agent

The immediate innate defenses include several classes of preformed soluble molecules that are present in extracellular fluid, blood, and epithelial secretions and that can either kill the pathogen or weaken its effect. Infection agents are lysis by antimicrobial enzymes, antimicrobial peptides, or phagocytic cells. If these fail, innate immune cells become activated by PRRs that detect PAMPs molecules from typical of microbes. The activated innate cells can engage various effector mechanisms to eliminate the infection. Only if an infectious organism breaches these first two lines of defense will mechanisms be engaged to induce an adaptive immune response. This leads to the expansion of antigen-specific lymphocytes that target the pathogen specifically and to the formation of memory cells that provide long-lasting specific immunity.

Chemical barriers are the initial defenses against infection. A variety of antimicrobial proteins that act as natural antibiotics to prevent microbes from entering the body. **Complement** not only acts in conjunction with antibodies, but also can target foreign organisms in the absence of a specific antibody; thus it contributes to both innate and adaptive responses.

Innate immunity serves two important functions: ①The initial response to microbes that prevents, controls, or eliminates infection of the host; ②Microbes stimulates adaptive immune responses and can influence the nature of the adaptive responses to make them optimally effective against different types of microbes. Thus, innate immunity not only serves defensive functions early after infection but also provides the "warning" that an infection is present against which a subsequent adaptive immune response has to be mounted. Moreover, different components of the innate immune response often react in distinct ways to different microbes (e. g. , bacteria versus viruses) and thereby influence the type of adaptive immune response development.

12.3.2 Role of innate response in stimulating adaptive immune responses

Innate immune responses generate molecules that provide signals, in addition to antigens, that are required to activate naïve B and T cell if an infectious organism breaches these first two lines of defense. B and T cells express different receptors that recognize antigens. Membrane-bound antibodies (so-called BCRs) on B cell and T cell receptors (TCRs) on T cells are important for their functions in adaptive immunity. Although these receptors have many similarities in terms of structure and mechanisms of signaling, there are fundamental differences related to the types of antigenic structures that B cells and T cells recognized.

Membrane-bound antibodies, which serve as the antigen receptors of B lymphocytes, can recognize many types of chemical structures. B lymphocyte antigen receptors and the antibodies that B cells secrete can directly recognize the shapes, or conformations, of macromolecules, including proteins, lipids, carbohydrates, and nucleic acids, as well as simpler, smaller chemical moieties. This broad specificity of B cells for structurally different types of molecules in their native form enables the humoral immune system to recognize, respond to, and eliminate diverse microbes and toxins.

The initiation and development of T cell immune responses require that antigens be captured and displayed by MHCs molecules on APCs since T cell cannot bound to antigen directly. It is necessary for the cells of APCs "present" antigen in a form that could be recognized by T cell. All nucleated cells are express MHC class Ⅰ and can be functional as APCs. Some specialized APCs, express not only MHC class Ⅰ but also MHC class Ⅱ, are called "professional APCs" that include dendritic cells, macrophages and B cells. These cells are an important for communication with T cells and subsequent T cell activation.

Dendritic cells express both antigen and the co-stimulatory molecules, which mainly activate a naïve T cell proliferation and differentiation. The primary function of macrophages is to capture and present protein antigens to effector T cells. Macrophages are also capable of capturing and presenting protein antigens to naïve T cells although they are not as important in this function. Stimulated macrophages exhibit increased levels of phagocytosis and secretion. B cells capture and present protein antigens to effector $CD4^+$ T cell. This interaction between B-T cells eventually triggers the $CD4^+$ T cell to produce and secrete various cytokines that enable that B cell to proliferate and differentiate into antibody-secreting plasma cells.

TCRs only recognize peptide-MHC complexes and bind these with relatively low affinity, which may be why the binding of T cells to APCs has to be strengthened by additional cell surface adhesion molecules. So, antigen capture and antigen presentation are important steps for linking innate and adaptive immune response together for T cell activation.

The innate immune response provides signals that function in concert with antigen to stimulate the proliferation and differentiation of antigen-specific T and B lymphocytes. As the innate immune response is providing the initial defense, it also sets in motion the adaptive immune response. The activation of lymphocytes requires two distinct signals, antigen recognition by lymphocytes provides 1st signal for the activation of the lymphocytes, and molecules induced on host cells during innate immune responses to microbes provide 2nd signal. This idea is called the two-signal hypothesis for lymphocyte activation. In infected tissues, activated dendritic cells and macrophages also secrete cytokines such as IL-12, IL-1, and IL-6, which stimulate the differentiation of naïve T cells into different type of effector cells for cell-mediated adaptive immunity.

12.3.3 Adaptive immune response

Adaptive immune response is exquisite specificity for distinct molecules and an ability to "remember" and respond more vigorously to repeated exposures to the same microbe. It is able to recognize and react to a large number of microbial and nonmicrobial substances. The mechanisms of innate immunity provide an effective initial defense against infections. However, many pathogenic microbes have evolved to resist innate immunity, and their elimination requires the more powerful mechanisms of adaptive immunity. The innate immune response to microbes also stimulates adaptive immune responses and influences the nature of the adaptive responses. The differences between innate and adaptive immunity are summarized in the **Table 12-5**.

Table 12-5 Differences between innate and adaptive immunity

Type	Innate Immunity	Adaptive Immunity
Antigen	Independent	Dependent
Time	No lag	A lag period
Antigen specific	No	Yes
Memory	No	Yes
Cells involved	Phagocytes, NK cells, mast cells, dendritic cells	T and B cells
Molecules involved	Cytokines, acute phase proteins, complement	Cytokines, Specific antibodies

As mentioned previously, there is a great deal of synergy between the adaptive immune response and its innate counterpart. The adaptive immune system frequently incorporates cells and molecules of the innate immune system in its fight against harmful pathogens. For example, complement (molecules of the innate immune system) may be activated by antibodies (molecules of the adaptive immune system) thus providing a useful addition to the adaptive system's armamentaria. Although the cells and molecules of the adaptive immune system possess slower temporal dynamics, they possess a high degree of specificity and evoke a more potent response on secondary exposure to the pathogen.

The adaptive immune system may take days or weeks after an initial infection to have an effect. However, most organisms are under constant assault from pathogens, which must be kept in check by the faster-acting innate immune system. There are two types of adaptive immune responses, called humoral immune response and cell-mediated immune response, that are mediated by different components of the adaptive immune response and function to eliminate different types of microbes.

(1) Humoral immune response

1) B cell activation and clonal expansion: Humoral immunity is mediated by secreted antibodies, produced by B cell, thus it is also named as antibody-mediated immunity. The humoral immune response combats microbes in many ways. Many polysaccharide and lipid antigens have multiple identical antigenic determinants that are able to engage many antigen receptor molecules on each B cell and initiate the process of B cell activation. The activation of naïve B cells results in their proliferation, leading to expansion of antigen-specific clones, and their differentiation into plasma cells, which secrete antibodies. Antibodies bind to the antigens of extracellular microbes and function to neutralize and eliminate these microbes.

Naïve B cell functional as APC will uptake and ingest protein antigens, degrade them, and display peptides bound to MHC class Ⅱ molecules for recognition by $CD4^+$T cells, which then activate the T cells. If BCR on the naïve B cell bound protein antigens as 1st signal for B cell activation. The response of B cells to protein antigens requires 2nd activating signals ("help") from Th2 cells. The activated B cells undergo proliferation and secrete different classes of antibodies with distinct functions. Some of the progeny of the expanded B cell clones differentiate into antibody-secreting plasma cells. Each plasma cell secretes antibodies that have the same antigen binding site as the cell surface antibodies (B cell receptors) that first recognized the antigen.

2) Types of antibodies: Antibodies are the secreted form of the BCRs. The basic structure antibody consists of a glycosylated protein consisting of two heavy and two light chains with Y-shaped (**Figure 12-3**). The region which binds to the antigen is known as the variable region for neutralization, while the constant region, not only determines the isotype but

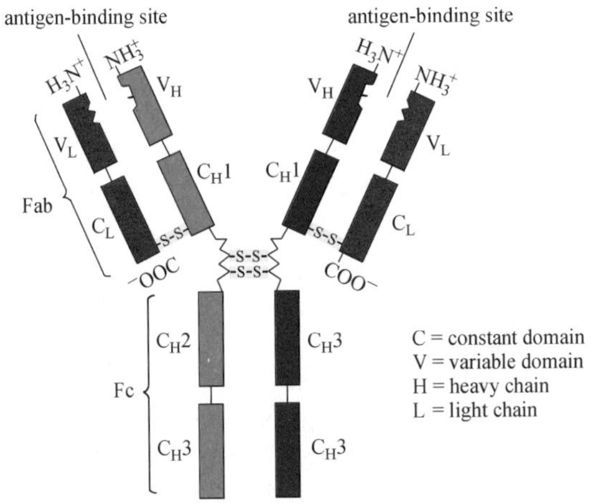

Figure 12-3 Schematic structure of antibody molecule. An antibody is composed of two identical heavy chains (blue) and two identical light chains (red). Each chain has a variable part and a constant part. The variable region is created by a light chain and a heavy chain of variable parts, which contains the antigen-binding site. The constant parts of the two heavy chains form constant region, which is involved in the elimination of the bound antigen.

is the region responsible for evoking antibody functions, such as binding with Fc receptors present on the surface of phagocytic cells to facilitate the phagocytosis. Antibodies can protect the body in a variety of ways. For instance, by binding tightly to a toxin or virus, an antibody can prevent it from recognizing its receptor on a host cell. Antibody also can activate complements and induce cell-mediated cytotoxicity (ADCC).

There are five different classes of antibodies in the human immune system — namely IgG, IgM, IgA, IgE and IgD. Each antibody can occur as transmembrane antigen receptors or secreted antibodies. The different classes of antibodies are adapted to function in different compartments of the body. Their functional activities are listed in the **Table 12-6**.

Table 12-6 Antibody isotypes and corresponding functions

Antibody	Characteristics
IgG	Predominant antibody in blood and tissue fluid Four subclasses (IgG1~4), With a high affinity Activate complement Cross placenta thus providing newborn with useful humoral immunity
IgM	Large pentameric structure in circulation Present in monometric form on B cell surface Secreted form is predominant antibody in early immune response against antigen Reach 75% of adult levels at 12 months of age
IgA	Exist in both a monometric and dimeric form Two subclasses (IgA1~2), Secretory IgA (dimeric form) represents 1st line of defence against microbes invading the mucosal surface (e.g., tears)
IgE	Low level in circulation Fc region has high affinity for mast cell and basophil thus involved in allergy Increased levels in worm infections
IgD	Antigen receptor on B cells Absent from memory cells

3) **Functions of antibody**: Antibodies bind to microbes and prevent them from infecting cells, thus "neutralizing" the microbes. In this way, antibodies are able to prevent infections. In fact, antibodies are the only mechanisms of adaptive immunity that block an infection before it is established; this is why eliciting the production of potent antibodies is a key goal of vaccination. IgG antibodies coat microbes and target them for phagocytosis, since phagocytes (neutrophils and macrophages) express receptors for the tails of IgG. IgG and IgM activate the complement system, by the classical pathway, and complement products promote phagocytosis and destruction of microbes. Some antibodies serve special roles at particular anatomic sites. IgA is secreted and translocated to mucosal epithelia and neutralizes microbes in the lumens of the respiratory and gastrointestinal tracts (and other mucosal tissues). IgG is actively transported across the placenta, and protects the newborn until the immune system becomes mature. Most antibodies have half-lives of about 3 weeks. However, some antibody-secreting plasma cells migrate to the bone marrow and live for years, continuing to produce low levels of antibodies. The antibodies that are secreted by these long-lived plasma cells provide immediate protection if the microbe returns to infect the individual. More effective protection is provided by memory cells that are activated by the microbe.

4) **Phases and type of humoral immune responses**: Antibody responses generated during the first exposure to an antigen, called primary responses, differ quantitatively and qualitatively from responses to subsequent exposures, called secondary responses. In a primary immune response, naïve B cells are stimulated by antigen, become activated, and

differentiate into antibody-secreting cells that produce antibodies specific for the eliciting antigen. Some of the antibody-secreting plasma cells survive in the bone marrow and continue to produce antibodies for long periods. Long-lived B cells, so called memory B cells, are also generated during the primary response. A secondary immune response is elicited when the same antigen stimulates these memory B cells, leading to more rapid proliferation and differentiation and production of greater quantities of specific antibody than are produced in the primary response.

The amounts of antibody produced in the primary immune response are smaller than the amounts produced in secondary responses. In secondary responses to protein antigens, there is increased heavy-chain isotype switching and affinity maturation, because repeated stimulation by a protein antigen leads to an increase in the number and activity of helper T lymphocytes. The principal characteristics of primary and secondary antibody responses are summarized in the **Table 12-7**.

Table 12-7 The comparison of primary and secondary humoral immune responses

Type of response	primary response	secondary response
Time lag after immunization	usually 5~10 days	usually 1~3 days
Peak response	smaller	larger
Antibody isotype	usually IgM >IgG	IgG>IgM
Antibody affinity	lower average affinity more variable	higer average affinity
Induced by	all immunogens	only protein antigens
Required immunization	high doses of antigen optimally with adjuvants	low doses of antigen Adjuvants not necessary

(2) Cell-mediated immune response T cells mainly involved in cell mediated immunity. These immune responses are directed principally at intracellular pathogens. Unlike B cells, T cells involve the destruction of infected cells (also called target cells) or intracellular pathogens in macrophages.

1) T cells activation and clonal expansion: Naïve T cells leave thymus and home to secondary lymphoid organs, where they may encounter antigens presented by mature dendritic cells with MHC class Ⅰ or Ⅱ molecules. TCR on the surface of T cells is an important membrane receptor for binding MHC on the surface of mature dendritic cells. $CD4^+$ T cells only bind with MHC Ⅱ-antigen peptide while $CD8^+$ T cells bind with MHC Ⅰ-antigen peptide. Antigen-stimulated T cells that have received both 1st signal through the antigen receptor and 2nd signal via costimulatory molecules may be induced to secrete cytokines and to express cytokine receptors. The cytokine interleukin-2 (IL-2) provides autocrine signals to activated T cells, leading to expansion of antigen-specific clones. Only the T-cell clones that activated by the specific antigen are divided and proliferation (**Figure 12-4**).

2) T cells proliferation and differentiation: After the antigen-specific T clonal expansion, the T cells are further proliferation and differentiation into many subsets of effector cells. The activated $CD4^+$ T cells differentiate into many subset of T cells, such as Th1, Th2, Th17. Some $CD4^+$ T cells that migrate into B cell-rich follicles are called T follicular helper (Tfh) cells. Some of $CD4^+$ T cells tend to function as down-regulating immune response. These cells are called regulatory T cells (Treg). The activated $CD8^+$ T cells differentiate into $CD8^+$ cytotoxic T lymphocytes (CTLs). These subsets of T cells all involved in cell-mediated immune response.

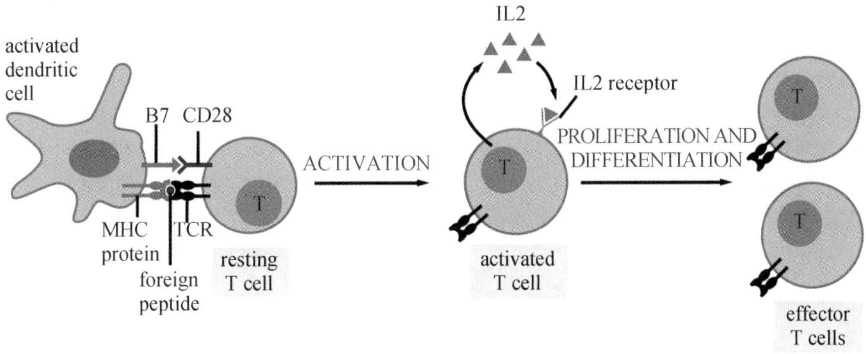

Figure 12-4 Activation and clonal expansion of T cells. The cytokine IL-2 is an important for T-cell clone expansion.

3) **Functions of subsets T cells:** T cells mainly involve the destruction of infected cells and tumor cells. Because subsets of effector T cells use distinct mechanisms to combat infections, the functions of these effector cells are classify individually. **Th1** cells produce the cytokines interferon-γ (IFN-γ), which activates macrophages to kill phagocytosed microbes. **Th2** cells contribute to humoral immunity by stimulating the production of antibodies from B cells. **Th17** cells help to recruit neutrophils to sites of infection early in adaptive immune response, which is also a response aimed mainly at extracellular pathogens. **Tfh** cells encounter the antigens to B cell in germinal center, which then provide critical survival signals for antibody-producing B cells. **Treg** cells tend to suppress the adaptive immune response and are important in preventing immune response from becoming uncontrolled and in preventing autoimmunity. **CTL** cells recognize class Ⅰ MHC-peptide complexes on the surface of infected cells and kill these cells, thus eliminating the reservoir of infection or destruction of tumor cells. More functions of effector T cells are summarized in the **Table 12-8**.

Table 12-8 The role of effector T cells in adaptive immunity

Effector T cells	CTL	Th1	Th2	Th17	Tfh	Treg
Markers on the surface	CD8	CD4	CD4	CD4	CD4	CD4
Recognized molecules	Class Ⅰ	Class Ⅱ	Class Ⅱ	Class Ⅱ	CXCR5, ICOS	Class Ⅱ
Effector molecules	Perrorin Granzymes Granulysin Fas/FasL	IFN-γ TNF-α CD40 ligand	IL-4 IL-5 CD40 ligand	IL-17 IL-6	IL-4, IFN-γ IL-17	IL-10 TGF-β
Main functions	Kill virus -infected cells; tumor cells	Activate infected macrophages, help B cells produce antibody	Help B cells produce antibody	Enhance neutrophil response	Help B cell survival in germinal center	Suppress T-cell responses
Pathogens targeted	viruses; some intracellular bacteria	microbes that persist in macrophage vesicles, extracellular bacteria	helminth parasites	extracellular bacteria		

(3) **immunological memory** A contraction phase is followed after the effective immune response eliminates the microbes that initiated the response. During this process the expanded lymphocyte clones die and homeostasis is restored. Some long-lived B cells and memory T cells are remained for many years. These memory cells are more effective in combating microbes than are naïve lymphocytes and could be rapidly recruited in recall, or memory, responses. Thus, generating memory response is an important goal of vaccination. Immunologic memory is one mechanism by which vaccines confer long-lasting protection against infections.

Summary

The immune system is the organ system, which is composed of many interdependent cell types that collectively protect the body from invaders and the growth of tumor cells. Many of these cell types have specialized functions. It is often divided into the two sections of innate and adaptive immunity, the former, representing a non-specific (no memory) response to antigen, and the latter, responding to new influences by mounting an immune response, which displays a high degree of memory and specificity. Among the innate defensive cells the most notable are macrophages, nature killer cells, and dendritic cells. The adaptive immune defense consists of B and T cells. There is a great deal of synergy between the innate and adaptive immunity. Antigen presentation is required to initiate both forms of immune response. Humoral immunity is mediated by secreted antibodies, produced in the B cells, whilst the cell-mediated immunity protects against foreign organisms, and also protects the body from itself by controlling cancerous cells through different subsets of T cells. Memory lymphocytes also are important cells for protecting body.

Questions

1. What is immune system? What are the components of immune system?
2. What are the different lines of defense against infection?
3. What is innate immune response? Which types of cells are involved in?
4. What is antigen-presenting cells? What is its functions?
5. What is humoral immune response? What type of cells produce antibody?
6. Which lymphocytes are involved in the cell mediated immunity?
7. How does adaptive immunity differ from innate immunity?
8. What is immunological memory?

(戴亚蕾)

Chapter 13

Cancer cells

Cancer is a very common disease which is one of the leading causes of death in the world. Since being one of the highest occurrence frequency and mortality rate among all the disease, cancer has been the focus of massive research effort for decades. and these studies have led to a remarkable breakthrough in our understanding of the cellular and molecular basis of cancer.

13.1 Basic knowledge about cancer cell

All cancer cells share one important characteristic: they are abnormal cells in which the processes regulating normal cell division are disrupted.

How a specific cancer cell behaves depends on which processes are not functioning properly. Some tumor cells simply divide and produce more tumor cells, the tumor mass remains localized, and the disease can usually be cured by surgical removal of the tumor. Malignant tumors tend to metastasize, that is, to break away from the parent mass, enter the lymphatic or vascular systems, and metastasize to a remote site in the body where they establish lethal secondary tumors that are on longer amenable to surgical removal.

13.1.1 Cancer types

Cancer comprises more than 100 different diseases. Cancer cells within a tumor are usually the descendants of a single cell, even after it has metastasized. Hence, a cancer can be categorized based on the type of cell in which it originates and by the location of the cell. Generally, there are classified into four major types of tumors.

(1) Carcinoma **Carcinomas** are cancers derived from epithelial (lining) cells, which cover the surface of our skin and internal organs. This is the most common cancer type and represents about 80% ~ 90% of all cancer cases reported. **Adenocarcinoma** refers to carcinoma derived from cells of glandular origin. One can, for instance, have an adenocarcinoma of the pancreas, or an adenocarcinoma of the lung.

(2) Sarcoma **Sarcomas** are cancers of the connective tissue, cartilage, bone, fat, muscle, and so on.

(3) Blood cancers **Lymphoma** is a group of blood malignancies that develop from lymphocytes. **Leukemias** are cancers derived from bone marrow and affects the lymphatic system. In leukaemia, lymphocytes cells do not mature properly and become too numerous in the blood and bone marrow. Leukemias may be acute or chronic. The most common type is acute lymphoblastic leukemia. **Myelomas** is a subtype of lymphoma, which refers to

specialized white blood cells responsible for the production of antibodies.

13.1.2 The environmental factors of carcinogenesis

Cancer cells behave as independent cells, growing without control to form tumors. Tumors are invariably found to have arisen from a single cell. They grow in a series of steps. The process of forming cancer cells from normal cells to carcinomas is called **oncogenesis** (onco = cancer), or **tumorigenesis**, or **carcinogenesis**.

Carcinogenesis is interplay between genetics and the environment. Most cancers arise after genes are altered by environmental factors or by errors in DNA replication process.

(1) The concept of carcinogens Several environmental factors affect one's probability of acquiring cancer. These factors are considered as **carcinogens** when there is a consistent correlation between exposure to an agent and the occurrence of a specific type of cancer.

Carcinogens can induce inherited mutations and finally result in carcinogenesis, so carcinogen is also called mutagen. A known human carcinogen means that there is sufficient evidence of a cause and effect relationship between exposure to the material and cancer in humans. Such determination requires evidence from epidemiologic (demographic and statistical), clinical, and(or) tissue(cell) studies involving humans who were exposed to the substance in question.

Carcinogen-induced cancer generally develops in many years after exposure to a toxic agent. A latency period of as much as thirty years has been observed between exposure to asbestos and incidence of lung cancer. Some cancers associated with environmental factors are preventable. Simply understanding the danger of carcinogens and avoiding them can usually minimize an individual's exposure to these agents.

(2) Carcinogens affect the genes The effect of environmental factors is not independent of cancer genes. Sunlight alters **tumor suppressor genes** in skin cells; causes changes in lung cells, making them more sensitive to carcinogenic compounds in smoke. The environmental factors probably act directly or indirectly on the genes that are already known to be involved in cancer. Individual genetic differences also affect the susceptibility of an individual to the carcinogenic affects of environmental agents. About ten percent of the population has an alteration in a gene, causing them to produce excessive amounts of an enzyme that breaks down hydrocarbons presented in smoke and various air pollutants. The excess enzyme reacts with these chemicals, turning them into carcinogens. These individuals are about twenty-five times more likely to develop cancer from hydrocarbons in the air than others are.

(3) The classification of carcinogens Carcinogens are classified into three main types: Radiation carcinogens, chemical carcinogens and virus. Besides carcinogens from environment, physical condition is also contributed to carcinogenesis, such as chronic inflammation and oxygen radicals.

1) **Radiation carcinogens**: Radiation is one of the first known mutagens, it's a potent inducer of mutations. Different types of radiation cause different types of genetic changes. Ultra-violet (UV) radiation causes point mutations. X-rays can cause breaks in the DNA double-helix and lead to translocations, inversions and other types of chromosome damage. Exposure to the UV rays in sunlight has been linked to skin cancer.

2) **Chemical carcinogens**: There are hundreds of distinct types of chemical carcinogens capable of inducing cancer after prolonged or excessive exposure. Many chemical carcinogens usually exert their effect by binding to DNA or the building blocks of DNA and interfering with the replication or transcription processes. Many industrial chemicals are

carcinogenic. The fumes of the metals cadmium, nickel, and chromium are known to cause lung cancer. Vinyl chloride causes liver sarcomas. Exposure to arsenic increases the risk of skin and lung cancer. Aflatoxin, a mutagen most often found on improperly stored agricultural products, induces hepatocellular carcinoma.

3) **Viruses**: In addition to chemicals and radiation, the third source of mutation is viruses. Many viruses infect humans but only a few viruses play a significant role in the development of particular cancers in many different animals, including humans. Viruses associated with cancer include human papillomavirus (genital carcinomas), hepatitis B (liver carcinoma), Epstein-Barr virus (Burkitt's lymphoma and nasopharyngeal carcinoma), human T-cell leukemia virus (T-cell lymphoma); and, probably, a herpes virus HHV-8 (Kaposi's sarcoma and some B cell lymphomas), SV40 (mesothelioma). Viruses can disrupt cell behavior in several different ways. These include both DNA viruses and retroviruses.

Oncogenic DNA viruses contribute to cancer by inserting their genomes into the DNA of the host cell. Insertion of the virus DNA directly into a **proto-oncogene** may mutate the gene into an **oncogene**, resulting in a tumor cell. Insertion of the virus DNA near a gene in the chromosome that regulates cell growth and division can increase transcription of that gene, also resulting in a tumor cell.

The viruses may contain their own oncogenes (v-onc comparing to c-onc) that disrupt the regulation of the infected cell. This process may be beneficial to the virus if it allows for rapid production of progeny, but can be seriously detrimental to the host. For example, human papillomavirus makes proteins that bind to two tumor suppressors, p53 protein and RB protein, transforming these cells into tumor cells. Remember that these viruses contribute to cancer, but they do not by themselves cause it.

The ability of retroviruses to promote cancer is associated with the presence of oncogenes in these viruses. Retroviruses have acquired the proto-oncogene from infected animal cells, and altered the versions of genes by some ways. These altered genes no longer function properly, and when they are inserted into a new host cell, they cause disregulation and can lead to cancerous growth. An example of this is the normal cellular c-SIS proto-oncogene, which makes a cell growth factor. The viral form of this gene is an oncogene called v-SIS. Cells infected with the virus that has v-SIS overproduce the growth factor, leading to high levels of cell growth and possibly to tumor cells.

4) **Chronic inflammation**: Chronic inflammation can lead to DNA damage due to the production of mutagenic chemicals by the cells of the immune system. For example, the long-term inflammation caused by infection with the hepatitis virus will cause hepatocellular carcinoma.

5) **Oxygen Radicals**: During the capture of energy from food, which occurs in our mitochondria, oxygen radicals may be generated, which are very reactive and are capable of damaging cell membranes and DNA itself. These **reactive oxygen intermediates** (ROI) may also be generated by exposure of cells to radiation.

The mutagenic activity of ROI is associated with the development of cancer as well as the activities of several anticancer treatments, including radiation and chemotherapy.

13.1.3 Stages of tumor progression

The cells, as the basic components of the tissues and organs, experience several stages from the normal to the cancerous. This process described below is applicable mainly for a solid tumor such as a carcinoma or a sarcoma (Figure 13-1). Blood cell tumors go through a similar process but since the cells are free-floating, they are not limited to one location in the body.

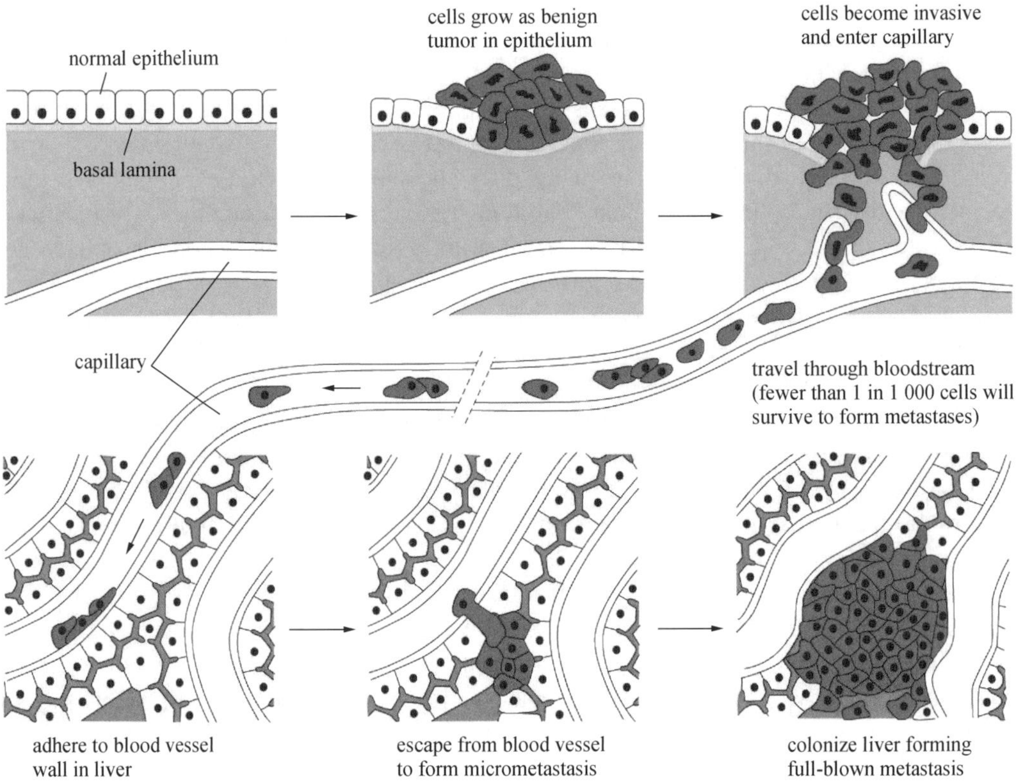

Figure 13-1 Stages of tumor progression (From: Alberts B, et al. Molecular Biology of the Cell. 2008)

(1) Hyperplasia This is the first step. These cells appear normal, but genetic changes have occurred that result in some loss of control of growth. leading to an excess of cells in that region of the tissue.

(2) Dysplasia In this stage, the additional genetic changes in the hyperplastic cells lead to the even more abnormal growth. The cells and the tissue no longer look normal and become disorganized.

(3) Carcinoma *in situ* (benign tumor) Further genetic changes are required in the stage of benign tumor, which result in cells that are even more abnormal and can now spread over a larger area of tissue. These cells often "regress" or become more primitive in their capabilities. They begin to lose their original function, for example a liver cell that no longer makes liver-specific proteins. Such cells are called de-differentiated or anaplastic. At this stage, because the benign tumor is still sited within its original location (called *in situ*), and the tumor cells have not yet crossed the basal lamina to invade other tissues. These growths are often considered to have the potential to become invasive and are treated as malignant growths.

Benign tumors are not considered to be cancerous because they do not progress to the point where they invade distant tissues. Though they are less often lethal than malignant tumors, they can still cause serious health problems in two ways. First, large benign tumors can put pressure on organs and cause other problems. In the case of brain tumors, the limited space within the skull means that a large growth in the brain cavity can be fatal. Secondly, they secrete excess amounts of biologically active substances like hormone which will bring serious healthy problem.

(4) Cancer (malignant tumors) The last step occurs when the cells in the tumor metastasize, which means that they can invade surrounding tissue, and spread to other locations. These tumors arc the most dangerous and account for a large percentage of cancer deaths.

13.1.4 Basic properties of cancer cells

Tumor cells display a characteristic set of features that distinguish them from normal cells. These traits allow the individual cells to form a tumor mass and eventually to metastasize to other parts of the body.

The behavior of cancer cells is more easily studied when the cells are growing in culture. Cancer cells can be obtained by removing a malignant tumor, dissociating the tissue into separate cells, and culturing the cells *in vitro*. Alternatively, normal cells can be transformed *in vitro* to cancer cells by treatment with carcinogens which will be discussed in the next section. These transformed cells can generally cause tumors when they are introduced into host animal. A wide range of changes occur during the **transformation** of a normal cell to a cancerous cell.

There are many differences in properties from one type of cancer cell to another. Meanwhile, there are several common basic properties that are shared by cancer cells, regardless of their tissue of origin.

(1) Loss of growth control The most important characteristic of a cancer cell is its unlimited number of cell divisions and loss of growth control. Normal culture cells depend on growth factors, such as epidermal growth factor and insulin. Cancer cells can proliferate even in the absence of these external factors and are therefore immune to many of the normal regulations on cell division. The cancer cells no longer function as a part of a larger organism, but behave more like independent entities living without regard for the organism as a whole.

The division of normal cells is restricted. Cells will stop dividing when they are in contact with neighboring cells, a phenomenon known as **contact inhibition**. When the normal cells grown in tissue culture under suitable conditions proliferate to the point where they occupy the entire substratum, their growth rate decreases dramatically, and they tend to grow into a layer one cell thick (monolayer). This is because the normal cells respond to inhibitory influences from their environment.

Decrease of proteins in cell surface and loss of gap junction allow the reduction of cell: cell and cell: extracellular matrix adhesion, and let cancer cells do not exhibit contact inhibition and are able to continue to grow even when surrounded by other cells.

The cell: cell contacts send signals into the dividing cells that cause them to stop dividing. In contrast, tumor cells show density and anchorage independent growth. When tumor cells are cultured under the same conditions, they continue to grow, piling on top of one another to form clumps. Cancer cells failure to respond to "stop" signals.

(2) Dedifferentiation of the cancer cells An alternative mechanism to stop cell division is differentiation. This is the process by which a precursor cell acquires its final functional capabilities. The maturation process involves the coordinated regulation of gene expression that results in differential morphological and biological properties for the cells. Often, the differentiation process results in cells that have very limited potential for cell division. The cancer cell acquires dedifferentiation in some extension and keep dividing.

(3) Avoidance of cell death In normal tissue, there is a balance between the generation

of new cells via cell division and the loss of cells through cell death. Old cells become damaged over time and are eliminated by apoptosis. This process is a normal and necessary means of maintaining ourselves. There are safeguards built into the cell cycle that allow for the identification and elimination of cells that are dividing in an aberrant manner. Cancer cells that get past the safeguards have acquired the ability to avoid the cell death signals triggered by their abnormal behavior. Avoidance of cell death, coupled with continued cell division leads to the growth of the tumor.

(4) The morphological and chemical changes of cancer cells Cancer cells can often be distinguished from normal cells by microscopic examination. They are usually less well differentiated than normal cells. There are other detectable changes in the physical and chemical properties of the cells. These main changes include the following.

1) **Cytoskeletal changes**: The distribution and activity of the microfilaments and microtubules may change. These alterations change the ways in which the cell interacts with neighboring cells and alter the appearance of the cells. Changes in the cytoskeleton also affect cell adhesion and movement (motility).

2) **Nuclear changes**: The shape and organization of the nuclei of cancer cells may be markedly different from that of the nuclei of normal cells of the same origin, such as a high nucleus-to-cytoplasm ratio, prominent nucleoli, and relatively little specialized structure. This change in appearance may be useful in the diagnosis and staging of tumors.

(5) Replicative Immortality Telomerase is generally produced by Germ-line cells and rapid dividing somatic cells, it will gradually lose its activity during differentiation of somatic cells. However, tumor cells keep strong telomerase activity, and telomerase expression is essential for a tumor cell to be immortal.

13.1.5 Angiogenesis and metastasis

All of the above properties of cancer cells can be studied *in vitro*, as well as *in vivo* cancer cells. Besides these, there are two important properties of cancer *in vivo*: **angiogenesis** and **metastasis**.

(1) Angiogenesis

1) **Tumor cells also need requiring nutrients and oxygen in order to grow to a large mass**: As tumors enlarge, the cells in the center no longer receive nutrients from the normal blood vessels. To provide nutrients and oxygen for all the cells in the tumor, it must form new blood vessels. In a process called angiogenesis, tumor cells make growth factors (or cause nearby cells to produce), which activate the blood vessel cells to divide and stimulate formation of new capillary blood vessels. One well-described angiogenic factor vascular endothelial growth factor (VEGF)-a common inductive signal. In some tumors, dominant oncogenes operating within tumor cells, such as Ras and Myc, can upregulate expression of angiogenic factors, whereas in others, such inductive signals are produced indirectly by immune inflammatory cells. Angiogenesis is a prerequisite for the growth and metastasis of solid tumors.

2) **Angiogenesis is an essential step in the growth of a tumor**: Dormant tumors are those that do not have blood vessels, they can only grow into a mass of about 10^6 cells, roughly a sphere1 ~ 2 mm in diameter. Several autopsy studies in which trauma victims were examined for such very small tumors revealed that thirty-nine percent of women aged forty to fifty have very small breast tumors, while forty-six percent of men aged sixty to seventy have very small prostate tumors. Amazingly, ninety-eight percent of people aged fifty to seventy have very small thyroid tumors. However, for those age groups in the general population,

the incidence of these particular cancers is only one-tenth of a percent (thyroid) or one percent (breast or prostate cancer). The conclusion is that the incidence of dormant tumors is very high compared to the incidence of cancer. Therefore, angiogenesis is critical for the progression of dormant tumors into cancer.

(2) Tumor metastasis In most cases tumors have little risk to their host because they are localized and small size. Tumors that grow only in their original location are said to be benign. Benign tumors can cause health problems, and they will become serious medical problem only when they become malignant tumors which metastasize to other locations outside its site of origin in the body.

Metastasis is the process by which cancer cells migrate from the initial or primary tumor to a distant location through the blood or lymphatic system or via direct contact to another location, which divide and form a secondary tumor at the new site. The distant growths are termed metastases. Metastatic tumors often interfere with the functions of the organs involved and lead to the morbidity and mortality seen in cancer.

The tumor cells retain the characteristics they had when they were in their original location. As an example, breast cancer that metastasizes to the liver is still breast cancer, NOT liver cancer.

1) **MMP in metastasis**: In order for cells to move through the body, they must first climb over/around neighboring cells. They do this by rearranging their cytoskeleton and attaching to the other cells or extracellular matrix via proteins on the outside of their plasma membrane. By extending part of the cell forward and letting to go at the back end, the cells can migrate forward. The alterations in cell adhesion also impact on the ability of the cells to move. The cells can crawl until they hit a blockage which cannot be bypassed. Often this block is a thick layer of proteins and glycoproteins surrounding the tissues, called the basal lamina or basement membrane. In order to cross this layer, cancer cells secrete a mixture of enzymes that degrade the proteins in the basal lamina and allow them to crawl through.

The enzymes secreted by cancer cells contain a group of enzymes called matrix metalloproteases(MMP). These enzymes digest proteins that inhibit the movement of the migrating cancer cells. Once the cells have traversed the basal lamina, they can spread through the body in several ways.

2) **Blood stream and lymphatic metastasis**: Metastatic cancer cells can enter the bloodstream by squeezing between the cells that make up the blood vessels. Once in the blood stream, the cells float through the circulatory system until they find a suitable location to settle and re-enter the tissues. The cells can then begin to grow in this new location, forming a new tumor.

An alternative but similar route would be entry of the cancer cells in to the vessels of the lymphatic system. The fluid in this extensive network flows throughout the body, much like the blood supply. It is the movement of cancer cells into the lymphatic system, specifically the lymph nodes, that is used in the detection of metastatic disease. Some cancers may spread through direct contact with other organs, as in the gut cavity.

A developmental regulatory program, referred to as the "**epithelial-mesenchymal transition**" (**EMT**), has become prominently implicated as a means by which transformed epithelial cells can acquire the abilities to invade, to resist apoptosis, and to disseminate. By co-opting a process involved in various steps of embryonic morphogenesis and wound healing, carcinoma cells can concomitantly acquire multiple attributes that enable invasion and metastasis. A **mesenchymal-epithelial transition** (**MET**) is the reverse process of EMT that involves the transition from motile, multipolar or spindle-shaped mesenchymal cells to planar arrays of polarized epithelial cells, which is believed to participate in the establishment and stabilization of distant metastases by allowing cancerous cells to regain epithelial properties and integrate into distant organs.

(3) Angiogenesis and metastasis have close relationship Angiogenesis contribute largely to metastases. Without a blood supply, tumor cells cannot spread to new tissues. Tumor cells can cross through the walls of the capillary blood vessel at a rate of about one million cells per day. However, not all cells in a tumor are angiogenic. Both angiogenic and non-angiogenic cells in a tumor cross into blood vessels and spread; however, non-angiogenic cells give rise to dormant tumors when they grow in other locations. In contrast, the angiogenic cells quickly establish themselves in new locations by growing and producing new blood vessels, resulting in rapid growth of the tumor.

13.1.6 Enabling characteristics of tumor cells

(1) Genome instability and mutation The extraordinary ability of genome maintenance systems to detect and resolve defects in the DNA ensures that rates of spontaneous mutation are usually very low during each cell generation. However, cancer cells often have increased rates of mutation. A large number of genome maintenance and repair defects that have already been documented in human tumors, together with abundant evidence of widespread destabilization of gene copy number and nucleotide sequence, suggesting that instability of the genome is inherent to the great majority of human cancer cells. This mutability is achieved through increased sensitivity to mutagenic agents, through a breakdown in one or several components of the genomic maintenance machinery, or both. The loss of telomeric DNA in many tumors generates karyotypic instability and associated amplification and deletion of chromosomal segments.

Genomic instability and thus mutability endow cancer cells with genetic alterations that drive tumor progression.

(2) Reprogramming Energy Metabolism The chronic and often uncontrolled cell proliferation that represents the essence of neoplastic disease involves not only deregulated control of cell proliferation but also corresponding adjustments of energy metabolism in order to fuel cell growth and division. Under aerobic conditions, normal cells process glucose, first to pyruvate via glycolysis in the cytosol and thereafter to carbon dioxide in the mitochondria; under anaerobic conditions, glycolysis is favored and relatively little pyruvate is dispatched to the oxygen-consuming mitochondria. Otto Warburg first observed an anomalous characteristic of cancer cell energy metabolism: even in the presence of oxygen, cancer cells can reprogram their glucose metabolism, and thus their energy production, by limiting their energy metabolism largely to glycolysis, leading to a state that has been termed "aerobic glycolysis." The reprogramming of cellular energy metabolism in order to support continuous cell growth and proliferation, replacing the metabolic program that operates in most normal tissues and fuels the physiological operations of the associated cells.

(3) Cancer-related inflammation It is accepted that chronic inflammation is a critical hallmark of cancer, with at least 25% of cancers associated with it, and possible underlying causes include microbial infections, autoimmunity, and immune deregulation. For example, human papilloma viruses (HPVs) induce inflammation and are responsible for 90%~100% of all cervical cancers. Similarly, chronic infection with Helicobacter pylori elevates the risk for gastric cancer. In addition, the immune deregulation seen in inflammatory bowel disease increases colorectal cancer incidence. Diet may also play a causal role in induction of cancer-associated inflammation, such as intake of nonhuman form of sialic acid — N-glycolylneuroaminic acid (Neu5Gc) — in red meat. Inflammatory cells can release chemicals, notably reactive oxygen species, that are actively mutagenic for nearby cancer cells, accelerating their genetic evolution toward states of heightened malignancy.

While chronic inflammation has an important role in cancer, less is known about the impact of acute inflammation on tumor progression. For example, inducing acute inflammation locally in the bladder with a vaccine containing an attenuated Mycobacterium bovis strain successfully treats squamous cancer of the bladder. Hence, with the infiltration of leukocytes and subsequent inflammation, the impact from inflammatory mediators can both initiate and, in certain cases, eliminate tumor cells and prevent tumor development.

(4) Evading Immune Destruction Cancers have complex interactions with the immune system. Pathologists have long recognized that some tumors are densely infiltrated by cells of both the innate and adaptive arms of the immune system. Historically, such immune responses were largely thought to reflect an attempt by the immune system to eradicate tumors. During the early stages of tumor development, cytotoxic immune cells such as natural killer (NK) and $CD8^+$ T cells recognize and eliminate the more immunogenic cancer cells. This first phase of elimination selects the proliferation of cancer cell variants that are less immunogenic and therefore invisible to immune detection. As the neoplastic tissue evolves to a clinically detectable tumor, different subsets of inflammatory cells impact tumor fate. High levels of tumor-infiltrated T cells might correlate with good prognosis in many solid cancers; on the other hand, high levels of macrophage infiltration might correlate with a worse prognosis. Thus, various components of immune system may sometimes help as well as hinder tumor progression.

13.2 The genes involved in cancer

There are a lot of genes participating in the initiation, development and metastasis of the cancers. These malfunctioning genes which play key roles in cancer induction can be broadly categorized into two broad classes, depending on their normal functions in the cell.

The first group, called proto-oncogenes in normal cells, produces protein products that normally enhance cell division or inhibit normal cell death. The mutated forms of these genes are called oncogenes. The second group, called tumor suppressors, makes proteins that normally prevent cell division or lead to cell death. Uncontrolled cell growth occurs either when active oncogenes are expressed (dominant) or when tumor suppressor genes (recessive) are lost. In fact, for a cell to become malignant, numerous mutations are necessary. In some cases, both types of mutations may occur.

13.2.1 Proto-oncogenes and oncogenes

An oncogene is any gene that encodes a protein being able to transform cells in culture or to induce cancer in animals. Almost all the oncogenes are derived from normal cellular genes which is called proto-oncogenes.

The proto-oncogenes that have been identified so far have many different functions in the cell, they usually code for the proteins that participate in signal transduction or pathway and promote cell proliferation.

(1) The proto-oncogenes includes (Figure 13-2) the extracellular signaling molecules、the membrane receptors for the signal molecules (e.g., HER-2/neu (erbB-2)、the growth factor receptor、The signal-transduction proteins(ras and src)、the transcription factors in the nucleus(myc)、The enzymes that function in DNA replication (hTERT)、the genes that resist apoptosis and promote survival (Bcl-2), a membrane associated protein that functions to prevent apoptosis.

Despite the differences in their normal roles, most proto-oncogenes are responsible for providing the positive signals that lead to cell division. Some proto-oncogenes work to regulate cell death. The defective versions of these genes, known as oncogenes, can cause a cell to divide in an unregulated manner. This growth can occur in the absence of normal pro-growth signals such as those provided by growth factors.

(2) Activation of a proto-oncogene into an oncogene The oncogene derived from proto-oncogene is the gene that encodes the protein being able to transform cells in vivo or vitro to induce cancer in animals. Conversion or activation of a proto-oncogene into an oncogene generally involves a *gain-of-function* mutation, because the cells with the mutant form of the proteins have gained a new function which does not present in cells with the normal gene. Most oncogenes are dominant mutations; and a single altered copy is sufficient to cause gain of function.

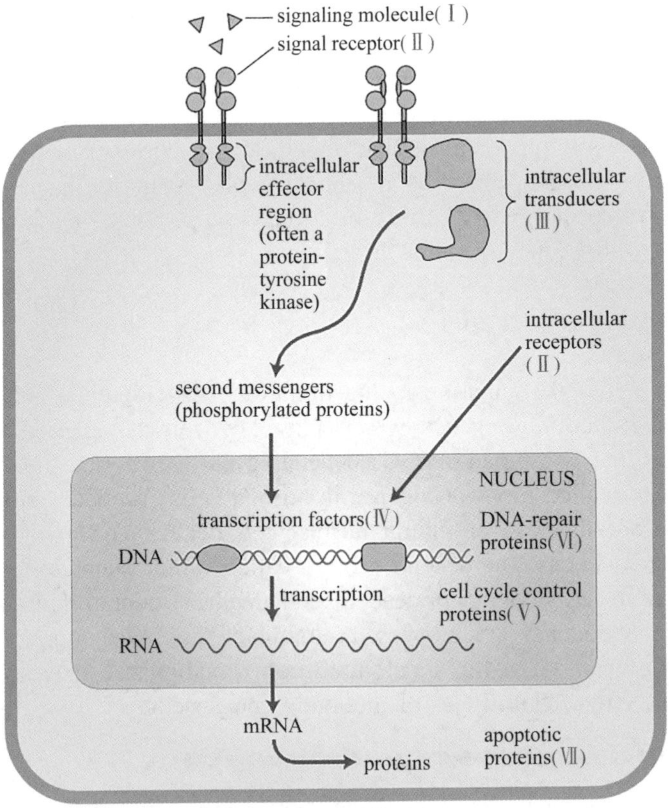

Figure 13-2 Seven types of proteins that participate in carcinogenesis. Mutations of Ⅰ-Ⅳ part of V are in oncogene category. Ⅵ, Ⅶ and part of Ⅴ are belong to tumor-suppressor genes (From: Lodish H, et al. Molecular Cell Biology. 2004)

The myc protein acts as a transcription factor and it controls the expression of several genes. Mutations in the myc gene have been found in many different cancers, including Burkitt's lymphoma, B cell leukemia, and lung cancer. RAS is another oncogene that normally functions as an "on-off" switch in the signal cascade. Mutations in RAS cause the signaling pathway to remain "on", leading to uncontrolled cell growth. Mutant RAS has been identified in cancers of many different origins, including: pancreas (90%), colon (50%), lung (30%), thyroid (50%), bladder (6%), ovarian (15%), breast, skin, liver, kidney, and some leukemias.

(3) Some oncogenes induce cells to produce angiogenic factors Oncogenes in some tumor cells allow the cells to produce angiogenic factors. An oncogene called BCL-2 has been shown to greatly increase the production of a potent stimulator of angiogenesis. It appears, then, that oncogenes in tumor cells may cause an increased expression of genes that make angiogenic factors. There are at least fifteen angiogenic factors and their production is greatly increased by a variety of oncogenes.

(4) Telomerase The gene that codes for the active component of the **telomerase** enzyme, hTERT, is considered a proto-oncogene because abnormal expression contributes to unregulated cell growth. Cancer cells have the ability to replicate without reaching a state of senescence. In many cancers, the ability to divide without limitation is achieved by the production of telomerase. Telomerase maintains the ends of chromosomes so that they do not shorten. Telomerase is a normal protein that is present in cells during fetal development. In most cells of an adult human, telomerase does not present as the gene for the enzyme is not being expressed (transcribed and translated).

(5) Oncogenes in the germ line cells The presence of an oncogene in a **germ line cell** (egg or sperm) results in an inherited predisposition for tumors in the offspring. However, a single oncogene is not usually sufficient to cause cancer, so inheritance of an oncogene does not necessarily result in cancer.

13.2.2 Tumor suppressor genes

Tumor suppressor genes generally encode proteins that inhibit cell proliferation and tumor formation in certain ways.

(1) Groups of tumor suppressor genes Tumor suppressor genes function in many key cellular processes, mainly grouped in the following 5 classes: ①Intracellular proteins that regulate or restrain cell cycle (p16 and Rb); ②Receptors or signal transducers for secreted hormones or developmental signals that inhibit cell proliferation (TGFβ); ③Checkpoint-control proteins that arrest the cell cycle if DNA is damaged or chromosomes are abnormal (P53); ④Proteins that promote apoptosis; ⑤Enzymes that participate in DNA repair (BRCA).

(2) Mutations of tumor suppressor genes Mutations in these genes result in loss-of-function. The cell loses the ability to prevent division only when both alleles of tumor-suppressor gene in the cell are lost.

How is it that both genes can become mutated? The majority of cancers are sporadic with no indication of a hereditary component, but oncogenic loss-of-function mutations in tumor-suppressor genes are genetic recessive.

In some cases, heredity is an important factor in cancer. When a parent provides a germ line mutation in one copy of the gene, since the mutation is recessive, the trait is not expressed. Later a mutation occurs in the second copy of the gene in a somatic cell. In that cell both copies of the gene are mutated and the cell develops uncontrolled growth. This may lead to a higher frequency of loss of both genes in the individual who inherits the mutated copy than in the general population. However, mutations in both copies of a tumor suppressor gene can occur in a somatic cell, so these cancers are not always hereditary.

(3) pRB Hereditary retinoblastoma, a serious cancer of the retina that occurs in early

childhood. When one parent carries a mutation in one copy of the RB tumor suppressor gene, it is transmitted to offspring with a fifty percent probability. Because retinoblasts are rapidly dividing cells and there are thousands of them, about ninety percent of the offsprings who receive the one mutated RB gene from a parent, and who would also develop a mutation in the second allele, producing no functional RB protein. These individuals then develop retinoblastoma in both eyes usually very early in life. Thus the tendency to develop retinoblastoma is inherited as a dominant trait. Not all cases of retinoblastoma are hereditary: sporadic retinoblastoma can also occur by mutation of both copies of RB in the somatic cell of the individual. Because losing two copies of RB is far less likely than losing one, sporadic retinoblastoma is rarely, developed late in life, and usually affects only one eye.

(4) *p53* The *p*53 (or T*p*53) is one of the most important tumor suppressor genes to date. The gene produces a protein that functions as a transcription factor to prevent unregulated cell growth. Genes controlled by *p*53 are involved in cell division and viability. The *p*53 gene is found to be defective in about half of all tumors, regardless of their type or origin.

The *p*53 protein interacts directly with DNA. It also interacts with other proteins that direct cellular action. The *p*53 protein is at the center of a large network of proteins that "sense" the health of a cell and cellular DNA. It is the conductor of a well orchestrated system of cellular damage detection and control. When DNA damage or other cellular insults are detected, the activity of the *p*53 protein aids in the decision between repair and the induction of apoptosis. The role of T*p*53 is central here, leading to its being called the "guardian of the genome".

13.3 The genetic and epigenetic changes of cancer

Cancer is a highly variable disease with multiple heterogeneous genetic and epigenetic changes. For almost all types of cancer studied up to date, it seems as if the transition from a normal, healthy cell to a cancer cell is step-wise progression that requires an accumulation of genetic changes in several different oncogenes and tumor suppressors. The cells become progressively more abnormal as more genes become damaged.

13.3.1 The genetic changes in cancers

It has been concluded that several independent genetic changes are required to create cells that become cancerous. The genes involved may differ from each other, and the order in which the genes become abnormal may also vary. As an example, colon cancer tumors from two different individuals may involve very different sets of tumor suppressors and oncogenes, even though the outcome is the same (Figure 13-3).

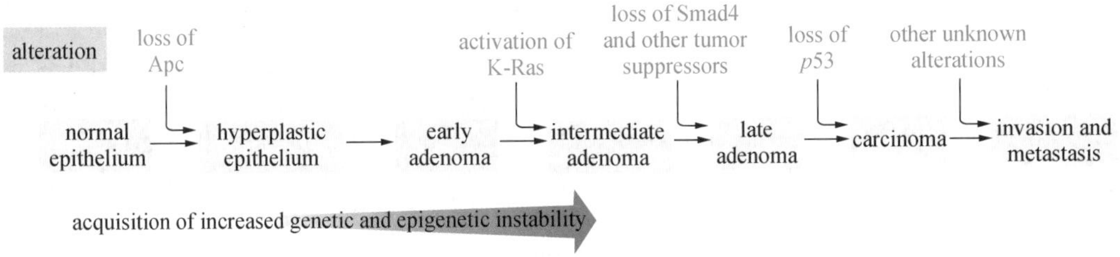

Figure 13-3 The development and metastasis of human colorectal cancer and its genetic alterations (From: Alberts B, et al. Molecular Biology of the Cell. 2008)

The genetic changes that lead to unregulated cell growth may be acquired in two different ways. The mutation in somatic cells can occur gradually over a number of years, leading to the development of a "sporadic" case of cancer. Alternatively, inherit dysfunctional genes lead to the development of a familial form of a particular cancer type. Hereditary cancers constitute about 10 percent of human cancers, including breast cancer, colon cancer and retinoblastoma.

(1) The ways to cause genetic alterations There are several ways to cause genetic alterations. It can be placed into two large categories. The first category is comprised of changes that alter only one or a few nucleotides along a DNA strand. A second category involves alterations of larger amounts of DNA, often at the chromosome level, such as translocation, amplification and deletion.

1) **DNA point mutations**: Changes that alter only one or a few nucleotides along a DNA strand are termed point mutations. The altered gene may lead to the production of a protein that no longer functions properly, or alter gene expression levels. This includes both *gain-of-function* mutation of proto-oncogene (Figure 13-4) and *loss-of-function* mutation of tumor-suppressor gene.

The mechanisms by which the changes are induced is varied. It includes spontaneous mutations and induced mutations.

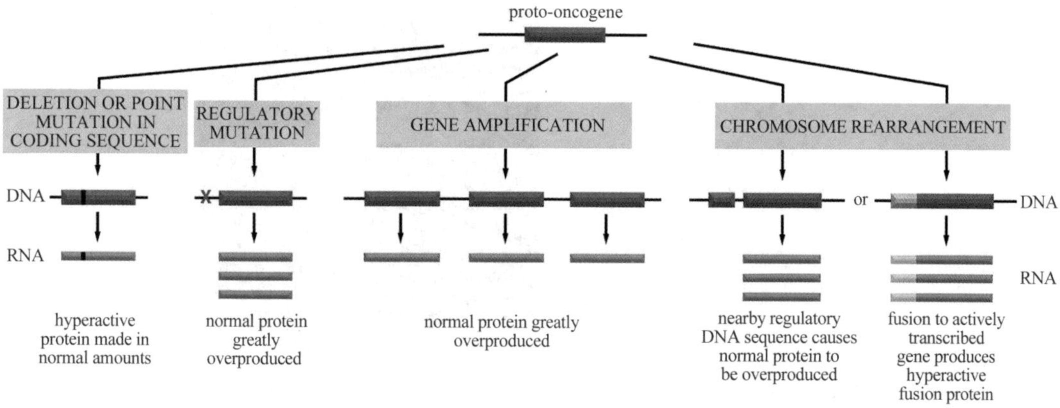

Figure 13-4 The ways for *gain-of-function* mutation of proto-oncogene (From: Bruce Alberts, et al. Molecular Biology of The Cell (6th edition), Fig 20-33)

2) **Alterations of larger amounts of DNA**: The larger amounts of DNA occur on the basis of chromatin, or fragments of chromatin. It contains several different ways.

Translocations: Translocations involve the breakage and movement of chromosome fragments between two chromosomes. The process can lead to change cell growth in different ways: ①The genes may not be transcribed and translated appropriately in their new location. The movement of a gene can lead to an increase or a decrease in its level of transcription. ②The breakage and rejoining may also occur within a gene, leading to form a new protein with changed structure and activation.

Translocations are common in leukemias and lymphomas and have been less commonly identified in cancers of solid tissues. For example, an exchange between chromosomes 9 and 22 was observed in over 90% of patients with chronic myelogenous leukemia (CML). The exchange leads to the formation of a shortened form of chromosome 22 called the Philadelphia chromosome (after the location of its discovery). This translocation leads to the formation of an oncogene from the *abl* proto-oncogene.

Inversions: In these alterations, DNA fragment is released from a chromosome and then re-inserted in the opposite orientation. As in the previous examples, these rearrangements can lead to abnormal gene expression, either by activating an oncogene or de-activating a tumor suppressor gene.

Duplications/deletions: Through replication errors, a gene or group of genes may be copied more than one time within a chromosome. This is different from gene amplification in that the genes are not replicated outside the chromosome and they are only copied one extra time, not hundreds or thousands of times. Genes may also be lost due to failure of the replication process or other genetic damage.

Aneuploidy: Cancer cells often have highly aberrant chromosome complements. Aneuploidy is genetic changes that involve the loss or gain of entire or part of chromosomes. Aneuploidy is a common feature of cancer cells. Cancer cells often have sometimes more than 100 chromosomes. The presence of the extra chromosomes makes the cells unstable and severely disrupts the controls on cell division (Figure 13-5). There is controversy as to whether aneuploidy occurs at an early stage in tumor formation, and is a cause of genetic instability that characterizes cancer cells, or is a late event and is simply a consequence of abnormal cancer growth.

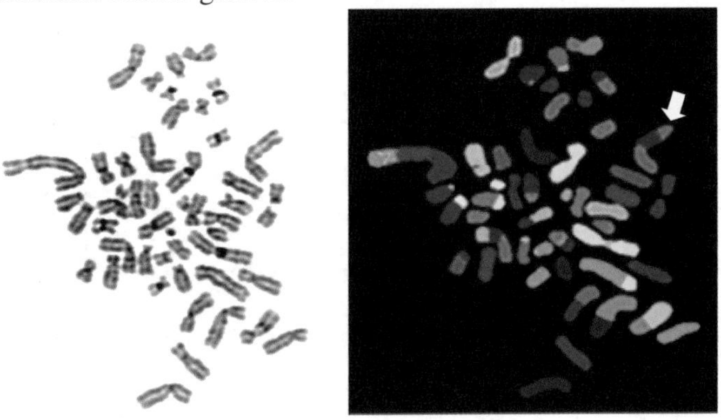

(A) (B)

Figure 13-5 Chromosomes from a breast tumor displaying abnormalities in structure and number. Chromosomes were prepared from a breast tumor cell in metaphase, spread on a glass slide, and stained with (A) a general DNA stain or (B) a combination of fluorescently labeled DNA molecules that color each normal human chromosome differently. The staining (displayed in false color) shows multiple translocations, including a doubly translocated chromosome (white arrow) that is made up of two pieces of chromosome 8 (green-brown) and a piece of chromosome 17 (purple). The karyotype also contains 48 chromosomes, instead of the normal 46. (From: Bruce Alberts, et al. Molecular Biology of The Cell (6th edition), Fig 20-13)

(2) Gene amplification When the normal DNA replication process is seriously flawed, it leads to DNA amplification from a single copy of a region of a chromosome to many copies. it occurs quite often in cancer cells.

The genes on each of the copies can be transcribed and translated, leading to an overproduction of the mRNA and encoded protein. If an oncogene is included in the amplified region, then the resulting *gain-of-function* mutation of this proto-oncogene. Examples of this include the amplification of the *myc* oncogene in a wide range of tumors and the amplification of the *ErbB-2* or *HER-2/neu* oncogene in breast and ovarian cancers.

Gene amplification also contributes to one of the biggest problems in chemotherapy. A gene commonly involved in multiple drug resistance (MDR). The protein product of this gene acts as a pump located in the membrane of cells. It is capable of selectively ejecting molecules from the cell, including chemotherapy drugs. In the presence of chemotherapy drugs, and such multiple resistances do occur. The amplification of MDR let the cancer cells make large amounts of this protein, allowing them to keep multiple chemotherapy drugs with similar structure outside the cell.

(3) Loss of heterozygosity of tumor-suppressor genes In hereditary cancers, the progeny receives one mutated copies of tumor-suppressor genes from his parents and it means that he is heterozygous for the mutation. Subsequent loss or inactivation of the normal copy in somatic cells, referred to as **loss of heterozygosity** (LOH), is a prerequisite for cancer development. There are three ways for LOH, mis-segregation during mitosis; mitotic recombination between two homologous chromatin and subsequent segregation, mutation or deletion of normal copy.

(4) Genetic changes contribute to cancer Most cancers are thought to arise from a single mutant precursor cell. As that cell divides, the resulting "daughter" cells may acquire different mutations and different behaviors over a period of time. Those cells that gain an advantage in division or resistance to cell death will tend to take over the population. In this way, the tumor cells are able to gain a wide range of capabilities that are not normally seen in the healthy version of the cell type represented. This is also the reason why cancer is much more prevalent in older individuals. The cells in a 70 year old body have had more time to accumulate the changes needed to form cancer cells but those in a child are much less likely to have acquired the requisite genetic changes.

13.3.2 The epigenetic changes in cancer

In addition to actual alterations in DNA sequence, gene expression can also be altered by modification of the DNA and chromatin that do not change the sequence. Since these changes do not alter the sequence of the DNA in the genes, they are termed **epigenetic changes** or **epigenetic DNA modification**.

13.3.3 Aberrant gene expression pattern of cancer cells

Genetic and epigenetic changes in cancer cell result in changes in **expression pattern** of many genes which finally cause aberrant cell behavior.

Because the heterogeneity of cancer, expression profiles are diverse between normal cells and corresponding cancer cells and among different types of tumors, even in different individual of same kind of cancer. Some of these differences can be correlated with biological differences between the tumors or diverse of tumorigenesis. Even though most of the differences can not be explained, they can provide cancer researchers with a list of genes look at more closely as potential targets for therapeutic drugs and **diagnosis biomarkers**.

13.4 Treatment of cancer

Cancer is such a heterogeneous disease, even those of the same kind of cancer, means that diagnosis and treatment are complicated. Current advances in the molecular classification of tumors should allow the rational design of treatment protocols based on the actual genes involved in any given case. New diagnostic tests may involve the screening of hundreds or thousands of genes to create a personalized profile of the tumor in an individual. This technical revolution is accelerating the process of identifying the genes of diagnostic, prognostic or therapeutic significance in cancer.

13.4.1 Traditional treatments

Because cancer comprises many diseases, doctors use many different treatments, mainly including **surgery**, **chemotherapy**, and **radiation.** The course of treatment depends on the type of cancer, its location, and its state of development.

Surgery, often the first treatment, is used to remove solid tumors. It may be the only treatment necessary for early stage cancers and benign tumors. Radiation-based cancer treatments intended to kill all the rapid dividing cancer cells based on the DNA damaging properties of radiation, therefore radiation treatment can cause some of the side effects by non-specifically killing the normal cells. Surgery and radiation treatment are often used together.

Chemotherapy drugs are toxic compounds that target rapidly growing cells by interfering their cell cycle or by promoting apoptosis. The most common class of drugs are designed to interfere S phase with the synthesis of precursor molecules needed for DNA replication, or interfere with the ability of the cell to complete the S phase of the cell cycle, or cause extensive DNA damage, which stops replication. The second class of drugs called spindle inhibitors stops cell replication early in mitosis. During mitosis, chromosome separation requires spindle fibers made of microtubules; spindle inhibitors stop the synthesis of microtubules. Because most adult cells don't divide often, they are less sensitive to these drugs than are cancer cells. The third kind of drugs works by forcing the cancer cells to undergo apoptosis. They kill certain adult cells that divide more rapidly, such as those that line the gastrointestinal tract, bone marrow cells, and hair follicles. This also causes some of the side effects of chemotherapy, including gastrointestinal distress, white blood cell count down, and hair loss.

13.4.2 New strategies for cancer treatment

Different kind of new anticancer strategies and new anticancer drugs appear along with the new discoveries of cancer and development of new technologies, most of them are still in the clinic trials. The main new strategies are separated into 4 groups: ①**Immunotherapy**; ②Gene therapy; ③Inhibiting the activity of oncogenes; ④Preventing angiogenesis.

(1) Immunotherapy We have all heard or read cancer patients who was expected to survive several months by the doctors remain alive and cancer-free years later, yet they defy the prognosis. This is because of the capability to destroy a tumor with his own immune system. But in most cases, one contributing factor in cancer is the failure of the immune system to destroy cancer cells. Several types of immunotherapy are used to treat cancer. These include:

1) **Immune checkpoint inhibitors**: These inhibitors block immune checkpoints. The checkpoints are a normal part of the immune system and keep immune responses from being too strong to destroy healthy cells in the body. By blocking them, these drugs allow immune cells to respond more strongly to cancer. Immune checkpoints engage when proteins on the surface of immune cells called T cells recognize and bind to partner proteins on other cells, such as some tumor cells. These proteins are called immune checkpoint proteins. When the checkpoint and partner proteins bind together, they send an "off" signal to the T cells. This can prevent the immune system from destroying the cancer. Immune checkpoint inhibitors work by blocking checkpoint proteins from binding with their partner proteins. This prevents the "off" signal from being sent, allowing the T cells to kill cancer cells. One such drug acts against a checkpoint protein called CTLA-4. Other immune checkpoint inhibitors act against a checkpoint protein called PD-1 or its partner protein PD-

L1.

2) **T-cell transfer therapy**: It is a treatment that boosts the natural ability of your T cells to fight cancer. In this treatment, immune cells are taken from your tumor. In recent years, cancer treatment involving adoptive cell therapy with chimeric antigen receptor (CAR)-modified patient's immune cells has attracted growing interest. Using gene transfer techniques, the patient's T cells are modified ex vivo with a CAR which redirects the T cells toward the cancer cells through an antibody-derived binding domain. The T cells are activated by the CAR primary signaling and costimulatory domains. Through targeting of the CD19 antigen on B cells durable remissions have been achieved in patients with B cell non-Hodgkin lymphoma and acute lymphoblastic lymphoma.

3) **Immune system modulators**: Immune-modulating agents are a type of immunotherapy that enhance the body's immune response against cancer, which includes cytokines, interleukin 2 and alpha interferon et al.

4) **Chemo-immunotherapy**: A technique that chemotherapy drugs are attached to antibodies that are specific for cancer cells. The antibody then delivers the drug directly to cancer cells without harming normal cells, reducing the toxic side effects of chemotherapy. These molecules contain two parts: the cancer-cell-specific antibody and a drug that is toxic once it is taken into the cancer cell.

5) **Radio-immunotherapy**: A similar strategy to chemo-immunotherapy, which couples specific antibodies to radioactive atoms, thereby targeting the deadly radiation specifically to cancer cells.

(2) Gene therapy Gene therapy was first conducted in inherited disease with single gene mutation. In recent years, the focus of gene therapy development has shifted toward the cancer treatment. Since *loss-of-function* change of tumor-suppressor genes in cancer cells, using gene therapy, a wild-type tumor suppressor gene can be introduced to tumor cells through infection with engineered vector such as virus, so that it either kills tumor cells or converts them to nonmalignant state.

(3) Inhibiting the activity of oncogenes Cancer cells contain *gain-of-function* mutation of oncogenes, if these proteins can be selectively blocked, it should be possible to stop the uncontrolled growth and invasive properties of malignant cells. Oncogenes are usually the valid targets for drug screen, small-molecular-weight compounds that specifically bind the cancer-promoting proteins and inhibit its activity are screened for cancer drug.

Hormone therapy is another strategy for cancer treatment. Many of the factors that affect normal cell growth are hormones. Although cancer cells have lost some of the normal responses to growth factors, some cancer cells still require hormones for growth. Hormone therapy for cancer attempts to starve the cancer cells of these hormones. This is usually done with drugs that block the activity of the hormone, although some drugs can block synthesis of the hormone. For example, some breast cancer cells require estrogen for growth. Drugs that block the binding site for estrogen can slow the growth of these cancers. These drugs are called selective estrogen receptor modulators (SERMs) or anti-estrogens. Tamoxifen and Raloxifene are examples of this type of drug.

(4) Preventing angiogenesis Another promising target for cancer therapy is angiogenesis. Cancer cells promote angiogenesis by secreting growth factors, such as VEGF. The first naturally occuring inhibitor for angiogenesis discovered was thrombospondin, identified in 1989 by Dr. Noel Bouck. More agiogenic inhibitors have been developed using different approaches, including antibodies and synthetic compounds directed against the targets.

The anti-angiogenesis treatments share two important advantages: ①Because they are natural products of the body, they should be much less toxic than conventional chemotherapy drugs. ② Because they act on normal (blood vessel) cells instead of attacking the tumors directly, there is less chance than the cancer cells will develop resistance to the drug.

All treatments are far from representing an ultimate cure. It's hoped, over the next few years, the advances in genomics and proteomics will lead to the development of new strategies for cancer diagnosis and therapy.

Summary

In this chapter, we introduce the basic knowledge about cancer cells. All cancer cells share one important characteristic: they are abnormal cells in which the processes regulating normal cell division are disrupted. That is, cancer develops from violations of the basic rules of social cell behavior. Unlike normal cells, cancer cells ignore signals to stop proliferation, to differentiation, or to apoptosis and be shed. Some tumor cells simply divide and produce more tumor cells, the tumor mass remains localized, and the disease can usually be cured by surgical removal of the tumor. Other malignant tumors tend to metastasize.

Cancer is a highly variable disease with multiple heterogeneous genetic and epigenetic changes. Carcinogenesis is interplay between genetics and the environment. Most cancers arise after genes are altered by environmental factors or by errors in DNA replication process. For almost all types of cancer studied up to date, it seems as if the transition from a normal, healthy cell to a cancer cell is step-wise progression that requires an accumulation of genetic changes in several different oncogenes and tumor suppressors. There are several ways to cause genetic alterations, which is placed into two large categories, one is comprised of changes that alter only one or a few nucleotides along a DNA strand; another involves alterations of larger amounts of DNA, often at the chromosome level, such as translocation, amplification and deletion.

Cancer is such a heterogeneous disease, even those of the same kind of cancer, means that diagnosis and treatment are complicated. Current advances in the molecular classification of tumors should allow the rational design of treatment protocols based on the actual genes involved in any given case. Different kind of new anticancer strategies and new anticancer drugs appear along with the new discoveries of cancer and development of new technologies, most of them are still in the clinic trials.

Questions

1. Give short definitions of the following terms: cancer, tumor, transformation, angiogenesis, metastasis, carcinogen, proto-oncogene, oncogene, tumor suppressor gene, gain-of-function mutation, Loss of function mutation.
2. What characteristics distinguish benign from malignant tumor?
3. What's angiogenesis? Explain the mechanism of angiogenesis, and its effect on carcinogenesis.
4. Comparing proto-oncogenes and tumor suppressor genes.
5. *Loss-of-function* occurs in the majority of human tumors. Name two ways to cause loss of *p*53 function.
6. Is cancer hereditary?
7. Example two new strategies to anti-cancer treatments.

（李　瑶，黄　燕修订）

References

[1] Alberts B, Bray D. and Hopkin K, *et al.* Essential Cell Biology. Second edition. New York and London: Garland Science, 2004

[2] Alberts B, Lewis J, Johnson A, *et al.* Molecular Biology of the Cell. 5th ed. New York: Gardand Science, 2008

[3] Alizadeh AA, Eisen MB, Davis RE, *et al.* Distinct types of diffuse large B-cell lymphoma identified by gene expression profiling. Nature, 2000, 403 (6769): 503-511

[4] Pombo A. Cellular genomics: which genes are transcribed, when and where? Trends Biochem Sci, 2003, 28 (1):6-9

[5] Steimer A, Schöb H, Grossniklaus M. Epigenetic control of plant development: new layers of complexity. Curr Opin Plant Biol, 2004, 7:11-19

[6] Antoine HFMP, Dirk S. Methylation of histones: playing memory with DNA. Curr Opin Cell Biol, 2005, 17: 230-238

[7] Avi As, Vishva MD, Death R. Signaling and modulation. Science, 1998, 281 (5381):1305

[8] B Alberts B, Lewis J, Johnson A, *et al.* Molecular Biology of the Cell. Fourth Edition. New York: Garland Science, 2002

[9] Beachy P, et al. Tissue repair and stem cell renewal in carcinogenesis. Nature, 2004, 18(432): 324

[10] Bonfanti ALV. Neural stem cells. Circ Res. 2003, 92: 598

[11] Brown TA. Genomes. 2nd ed. Oxford UK: BIOS Scientific Publishers Ltd, 2002

[12] Brush up on your 'omics. Chemical & Engineering News, 2003, 81 (49): 20. http://pubs. acs. org/cen/coverstory/8149/8149genomics1. html

[13] Burt R, Neklason DW. Genetic testing for inherited colon cancer. Gastroenterology, 2005, 128(6):1696-1716

[14] Classon M, Harlow E. The retinoblastoma tumor suppressor in development and cancer. Nature Rev Cancer, 2002, 2:910-917

[15] Cooper GM. The Cell – A Molecular Approach. 2nd ed. Sunderland (MA): Sinauer Associates Inc, 2000

[16] Cozzolino R, Palladino P, Rossi F, *et al.* Antineoplastic cyclic astin analogues kill tumour cells via caspase-mediated induction of apoptosis. Carcinogenesis, 2005, 26 (4): 733

[17] Craig LP. Chromatin remodeling enzymes: taming the machines. EMBO Reports, 2002, 3 (4):319-322

[18] David RH, Stephen MC. Connecting proliferation and apoptosis in development and disease. Nat Rev Mol Cell Biol, 2004, 5 : 805

[19] David S. Cancer stem cell refined. Nat Immunol, 2004, 5(7):701

[20] Dedrick J, Emmanuel ML. Signaling networks: the origins review of cellular multitasking. Cell, 2000, 103: 193

[21] Fulda S, Debatin KM. Sensitization for tumor necrosis factor-related apoptosis-inducing ligand-induced apoptosis by the chemopreventive agent resveratrol. Cancer Res, 2004, 64(1): 337

[22] Gautier J, et al. Cyclin is a component of maturation-promoting factor from *Xenopus*. Cell, 1990, 60:487-494

[23] Gordon K. Embryonic stem cell differentiation: emergence of a new era in biology and medicine. Gen Dev, 2005, 19:1129

[24] Griffiths A, Gelbart W, Miller J, *et al.* Modern Genetic Analysis. New York: WH Freeman & Co,1999

[25] Hezel AF, Bardeesy N, Maser RS. Telomere induced senescence: end game signaling. Curr Mol Med, 2005, 5(2):145

[26] Hiroo F. Gignals that control plant vascular cell differentiation. Nat Rev Mol Cell Biol, 2004, 5: 379

[27] Isayeva T, Kumar S, Ponnazhagan S. Anti-angiogenic gene therapy for cancer. Int J Oncol, 2004, 25(2):335-343

[28] Jacks T. The expanding role of cell cycle regulators. Science, 1998, 280:1035-1036

[29] Darnell J, Lodish H, Berk A, *et al.* Molecular cell biology. Fifth Edition. 2004

[30] Johnson F B, David AS, Leonard G. Molecular biology of senescence. Cell, 1999, 96:291

[31] Johnston WK, *et al.* RNA-catalyzed RNA polymerization: accurate and general RNA-template primer extension. Science, 2001, 292 (5520): 1319-1325

[32] Levenson JM, Sweatt JD. Epigenetic mechanisms in memory formation. Nature, 2005, 6:108-118

[33] Katharine LA, Amanda GF. Epigenetic aspects of differentiation. Cell Sci, 2004, 117, 4355

[34] Katja N. Lineage-specific transcription factors and the evolution of gene regulatory networks. Brief Funct Genomic, 2010, 9 (1): 65

[35] Klug W, Cummings M. Essentials of Genetics. 4th ed. Beijing: Higher Education Press, 2002

[36] Korostelev AA. Structural aspects of translation termination on the ribosome. RNA, 2011, 17 (8): 1409-1421

[37] Lincoln TA, Joyce GF. Self-sustained replication of an RNA enzyme. Science, 2009, 323 (5918): 1229-1232

[38] Lodish H, Berk A, Zipursky SL, *et al.* Molecular Cell Biology. Fourth Edition. New York: W H Freeman Company, 2001

[39] Malumbres M, Barbacid M. To cycle or not to cycle: a critical decision in cancer. Nature Rev Cancer, 2001, 1: 222-231

[40] D' Angelo MA, Gomez-Cavazos JS, Mei A, *et al.* A change in nuclear pore complex composition regulates cell differentiation developmental. Cell, 2012, 22(2): 446

[41] Ming Guo, Bruce AH. Cell proliferation and apoptosis. Cur Opin Cell Biol, 1999, 11:745

[42] Ning Cai, Mo Li, Jing Qu, *et al.* Post-translational modulation of pluripotency. J Mol Cell Biol, published online, doi: 10.1093/jmcb/mjs031, 2012
[43] Nurse P. Universal control mechanism regulation onset of M-phase. Nature, 1990, 344:503-507
[44] Paolo SC. Unique chromatin remodeling and transcriptional regulation in spermatogenesis. Science, 2002. 296(5576): 2176
[45] Patrick F C, David CS, Joanna E, *et al.* Accumulation of mitochondrial DNA mutations in ageing, cancer, and mitochondrial disease: is there a common mechanism? Lancet, 2002, 360(9342):1323
[46] Phillip N, Yau-Huei Wei. Ageing and mammalian mitochondrial genetics. Trends Genetics, 1998, 14(12):513
[47] Purves WK, Sadava D and Orians GH, *et al.* Life: The Science of Biology. Seventh Edition. New York: Sinauer Associates, Inc & WH Freeman Company, 2003
[48] Reik W. Stability and flexibility of epigenetic gene regulation in mammalian development. Nature, 2007, 447, 425
[49] Robert SB, Shino N, Toren F. Mitochondria, oxidants, and senescence. Cell, 2005, 120(4):483
[50] Beck S, Olek A and Walter J. From genomics to epigenomics. Nature Biotechnol, 1999, 17 (12):1144
[51] Sandeep K, Donna LF, George CT. T cell rewiring in differentiation and disease. Immunol, 2003, 171: 3325
[52] Scott WL, Athena WL. Apoptosis in cancer. Carcinogenesis, 2000, 21(3) :485
[53] Solange D. Mitochondria as the central control point of apoptosis. Trends Cell Biol, 2000, 10:369
[54] Sunil KM, Angie R. Emerging roles of microRNAs in the control of embryonic stem cells and the generation of induced pluripotent stem cells. Dev Biol, 2010, 344(1): 16
[55] Takahashi K, Yamanaka S. Induction of pluripotent stem cells from mouse embryonic and adult fibroblast cultures by defined factors. Cell, 2006, 126: 663
[56] Tannishtha R, Sean JM, Michael FC, *et al.* Stem cells, cancer, and cancer stem cells. Nature, 2001, 414: 105
[57] Audesirk T, Audesirk G, EB B. Biology: Life on Earth. 5th Edition. 2002
[58] Thompson CB. Apoptosis in the pathogenesis and treatment of disease. Science, 1995, 267: 1456
[59] Tom KK. Polycomb group complexes-many combinations, many functions. Trends Cell Biol, 2009, 19(12): 692
[60] Wei Yau-Huei, Lu Ching-You, Lee Hsin-Chen, *et al.* Oxidative damage and mutation to mitochondrial DNA and age-dependent decline of mitochondrial respiratory function. Ann NY Acad Sci, 1998, 854: 155
[61] Yusupova G, *et al.* Structural basis for messenger RNA movement on the ribosome. Nature, 2006, 444 (7117): 391-394
[62] Zhai Zhonghe, Wang Xizhong and Ding Mingxiao. Cell Biology. Beijing: Higher Education Press, 2000
[63] Pfanner N, Warscheid B, Wiedemann N. Mitochondrial proteins: from biogenesis to functional networks. Nat Rev Mol Cell Biol, 2019, 20(5): 267-277
[64] Alberts B, Johnson A, Lewis J, *et al.* Molecular Biology of the Cell. 6th ed. New York: Garland Science, Taylor & Francis Group, 2015

[65] Alvarado-Kristensson M, Rossello CA. The biology of the nuclear envelope and its implications in cancer Biology. Int J Mol Sci, 2019, 20(10): 2586-2590
[66] Bersaglieri C, Santoro R. Genome organization in and around the nucleolus. Cells, 2019, 8(6); 579-599
[67] Canela-Pérez I, López-Villaseñor I, Mendoza L, *et al.* Nuclear localization signals in trypanosomal proteins. Molecular & Biochemical Parasitology, 2019, 229: 15-23
[68] Cautain B., Hill R, Pedro N, *et al.* Components and regulation of nuclear transport processes. FEBS Journal, 2015, 282: 445-462
[69] Cerqueira A, Lemos B. Ribosomal DNA and the nucleolus as keystones of nuclear architecture, organization, and function. Trends Genet, 2019, 35(10): 710-723
[70] Ekundayo B, BleichertI F. Origins of DNA replication. PLoS Genet. 2019, 15(9):e1008320
[71] Canier O, Prorok P, Akerman I, *el al.* Metazoan DNA replication origins. Current Opinion in Cell Biology, 2019, 58: 134-141
[72] Hampoelz B, Andres-Pons A, Kastritis P, *et al.* Structure and assembly of the nuclear pore complex. Annu. Rey. Biophys, 2019, 48(1): 515-536
[73] Iwasa J & Marshall W. Karp'S Cell and Molecular Biology: Concepts and Experiments. Fighth Edition. New York: John Wiley & Sons, 2016
[74] Janssen A, Colmenares SU, Karpen GH Heterochromatin: guardian of the genome. Annu. Rev. Cell Dev. Biol, 2018, 34(1); 265-88
[75] Koyama M, Kurumizaka H. Structural diversity of the nucleosome. J Biochem, 2018, 163(1): 85-95
[76] Lin D, Hoelz A. The Structure of the nuclear pore complex (An update). Annu. Rev. Biochem, 2019, 88(1); 725-83
[77] Liu J, Ali M, Zhou Q. Establishment and evolution of heterochromatin. Ann N Y Acad Sci, 2020, 1476(1): 59-77
[78] Lodish H, Berk A, Kaiser C, *et al.* Molecular Cell Biology. Eighth Edition, New York: W H Freeman and Company, 2016
[79] Luger K, Mäder AW, Richmond RK et al. Crystal structure of the nucleosome core particle at 2.8 Å resolution. Nature, 1997, 389: 251-260
[80] Machida S, Takizawa Y, Ishimaru M, *et al.* Structural basis of heterochromatin formation by human HP1. Molecular Cell, 2018, 69(3): 385-397
[81] Maeshima K, Ide S, Babokhov M. Dynamic chromatin organization without the 30-nm fiber. Current Opinion in Cell Biology, 2019,58(1): 95-104
[82] Marie-Noëlle Prioleaul, David M. MacAlpine. DNA replication origins — where do we begin? Genes & Development, 2016, 30(15): 1683-1697
[83] Meier I, Richards ED. Cell biology of the plant nucleus. Annu. Rev. Plant Biol, 2017, 68(1): 139-72
[84] Nemeth A, Grummt I. Dynamic regulation of nucleolar architecture. Current Opinion in Cell Biology, 2018, 52(1): 105-111
[85] Popov A, Smirnov E, Kováčik L, *et al.* Duration of the first steps of the human rRNA processing. Nucleus, 2013,4(2): 134-141
[86] Potapova T, Gerton J. Ribosomal DNA and the nucleolus in the context of genome organization. Chromosome Res, 2019, 27(1-2): 109-127
[87] Razin SV, Iarovaia OV, Vassetzky YS. A requiem to the nuclear matrix: from a controversial concept to 3D organization of the nucleus. Chromosoma, 2014, 123: 217-224

[88] Rosemary Wilson R, Coverley D. Relationship between DNA replication and the nuclear matrix. Genes to Cells, 2013, 18(1): 17-31
[89] Seongmin JS, Song JJ. The big picture of chromatin biology by cryo-EM. Current Opinion in Structural Biology, 2019, 58(1): 76-87
[90] Smith EM, Pendlebury DF, Nandakumar J. Structural biology of telomeres and telomerase. Cellular and Molecular Life Sciences, 2020, 77(1): 61-79
[91] Song Feng, Chen Ping, Sun Dapeng *et al.* Cryo-EM study of the chromatin fiber reveals a double helix twisted by tetranucleosomal units. Science, 2014, 344(6182): 376-380
[92] Tomita K. How long does telomerase extend telomeres? Regulation of telomerase release and telomere length homeostasis. Current Genetics. 2018, 64: 1177-1181
[93] Zetka M, Paouneskou D, Jantsch V. "The nuclear envelope, a meiotic jack-of-all-trades". Current Opinion in Cell Biology. 2020, 64(1): 34-42
[94] Zidovska A. The rich inner life of the cell nucleus: dynamic organization, active flows, and emergent rheology. Biophysical Reviews. 2020, 12: 1093-1106
[95] Atilgan Yilmaz, Nissim Benvenisty. Defining Human Pluripotency. Cell stem cell, 2019, 25(1): 9-22
[96] Nirupama Shevde. Stem Cells Flexible Friends, Nature, 2012, 483: 22-26
[97] F. Brad Johnson, David A. Sinclair, and Leonard Guarente. Molecular Biology of Senescence. Cell, 1999, 96(1): 291 -302
[98] R. Iyengar, M. Diverse-Pierluissi, D. Weinstein. Cell signaling systems: A course for graduate students. Sci. STKE, 2005, 2005(269): 1126-1129
[99] Rodier F, and Campisi J: Four faces of cellular senescence. J Cell Biol, 2011, 192:547-556
[100] Nicolás Herranz and Jesús Gil: Mechanisms and functions of cellular senescence. J Clin Invest, 2018; 128(4): 1238-1246

Glossary

图书在版编目(CIP)数据

细胞生物学 = Cell Biology：英文 / 李瑶主编. —3 版. —上海：复旦大学出版社，2022.1
ISBN 978-7-309-16062-8

Ⅰ. ①细… Ⅱ. ①李… Ⅲ. ①细胞生物学-英文 Ⅳ. ①Q2

中国版本图书馆 CIP 数据核字(2021)第 272319 号

Cell Biology 细胞生物学(第三版)
李 瑶 主编
责任编辑/张志军

复旦大学出版社有限公司出版发行
上海市国权路 579 号 邮编：200433
网址：fupnet@fudanpress.com http://www.fudanpress.com
门市零售：86-21-65102580 团体订购：86-21-65104505
出版部电话：86-21-65642845
上海丽佳制版印刷有限公司

开本 787 × 1092 1/16 印张 19.75 字数 550 千
2022 年 1 月第 3 版第 1 次印刷

ISBN 978-7-309-16062-8/Q · 113
定价：88.00 元